Chemistry, Biology and Pharmacology of Lichen

Chemistry, Biology and Pharmacology of Lichen

Edited by

Ashoke Kumar Das
Department of Botany
Abhayapuri College, Guwahati University
Bongaigaon, India

Ajay Sharma
Department of Chemistry
Chandigarh University, Gharuan, Punjab, India

Deepika Kathuria
Department of Chemistry
University Centre for Research and Development (UCRD)
Chandigarh University, Gharuan, Punjab, India

Mohammad Javed Ansari
Department of Botany
Hindu College Moradabad
Moradabad, India

Garima Bhardwaj
Department of Chemistry
Sant Longowal Institute
Sangrur, India

WILEY

This edition first published 2024

Registered Office(s)
John Wiley & Sons, Inc., 111 River Street, Hoboken, NJ 07030, USA
John Wiley & Sons Ltd, The Atrium, Southern Gate, Chichester, West Sussex, PO19 8SQ, UK

For details of our global editorial offices, customer services, and more information about Wiley products visit us at www.wiley.com.

Wiley also publishes its books in a variety of electronic formats and by print-on-demand. Some content that appears in standard print versions of this book may not be available in other formats.

Library of Congress Cataloging-in-Publication Data
Names: Das, Ashoke Kumar, 1977–editor. | Sharma, Ajay, 1989–editor. | Kathuria, Deepika, 1990–editor. | Ansari, Mohammad Javed, 1978–editor. | Bhardwaj, Garima, 1988–editor.
Title: Chemistry, biology and pharmacology of lichen / edited by Ashoke Kumar Das, Ajay Sharma, Deepika Kathuria, Mohammad Javed Ansari, Garima Bhardwaj.
Description: First edition. | Hoboken, NJ : Wiley, 2024. | Includes bibliographical references and index.
Identifiers: LCCN 2024016841 (print) | LCCN 2024016842 (ebook) | ISBN 9781394190676 (hardback) | ISBN 9781394190683 (adobe pdf) | ISBN 9781394190690 (epub)
Subjects: LCSH: Lichens. | Lichens–Ecology. | Lichens–Therapeutic use. | Lichen products.
Classification: LCC QK583 .S53 2024 (print) | LCC QK583 (ebook) | DDC 579.7–dc23/eng/20240510
LC record available at https://lccn.loc.gov/2024016841
LC ebook record available at https://lccn.loc.gov/2024016842

Cover Design: Wiley
Cover Image: © gorwol/Shutterstock

Set in 9.5/12.5pt STIXTwoText by Straive, Pondicherry, India
Printed and bound by CPI Group (UK) Ltd, Croydon, CR0 4YY

C9781394190676_300524

Dedication

This book is dedicated to my beloved family.
Ashoke Kumar Das

Contents

List of Contributors

Muhammad Afzaal
Department of Food Science, Government College University, Faisalabad, Pakistan

Fareed Afzal
Department of Food Science, Faculty of Life Sciences, Government College University, Faisalabad, Pakistan

Mateen Ahmad
Department of Food Sciences, Faculty of Agricultural Sciences, University of the Punjab, Lahore, Pakistan

Abdul Baquee Ahmed
Department of Pharmaceutics, Girijananda Chowdhury Institute of Pharmaceutical Science-Tezpur, Girijananda Chowdhury University, Sonitpur, Assam, India

Muhammad Zeeshan Ahmed
Department of Biochemistry, Bahauddin Zakariya University, Multan, Pakistan

Farak Ali
Girijananda Chowdhury Institute of Pharmaceutical Science, Tezpur campus, Girijananda Chowdhury University, Assam 784501, India

Department of Pharmaceutical Sciences Dibrugarh University, Dibrugarh Assam, India

Sheikh Rezzak Ali
Department of Pharmaceutical Sciences Dibrugarh University, Dibrugarh Assam, India

Shahnaz Alom
Girijananda Chowdhury Institute of Pharmaceutical Science, Tezpur campus, Girijananda Chowdhury University, Assam 784501, India

Department of Pharmaceutical Sciences, Dibrugarh University Dibrugarh, Assam, India

Mohammad Javed Ansari
Department of Botany, Hindu College Moradabad (Affiliated to MJP Rohilkhand University), Bareilly, India

Habiba Arooj
Department of Food Science Government College University Faisalabad, Pakistan

Muneeba Aslam
Department of Biochemistry, Abdul Wali Khan University, Mardan, Pakistan

Sourav Bhattacharjee
Department of Botany, Pandit Deendayal Upadhyaya Adarsha Mahavidyalaya, Tulungia, Bongaigaon, Assam, India

Munmi Borkataky
Department of Life Sciences
Dibrugarh University, Dibrugarh
Assam, India

Tridip Boruah
P.G. Department of Botany, Madhab Choudhury College Barpeta,
Assam, India

Sandipan Choudhury
Girijananda Chowdhury Institute of Pharmaceutical Science, Tezpur campus,
Girijananda Chowdhury University,
Assam 784501, India

Bhaben Chowardhara
Department of Botany, Arunachal University of Studies, Namsai, India

Ashoke Kumar Das
Department of Botany, Abhayapuri College, Abhayapuri, Assam, India

Puja Namo Das
Plant Ecology Laboratory, P.G. Department of Botany, Madhab Choudhury College
Barpeta, Assam, India

Barsha Devi
Department of Botany, Pandit Deendayal Upadhyaya Adarsha Mahavidyalaya
Tulungia, Assam, India

Himasri Devi
P.G. Department of Botany, Madhab Choudhury College, Barpeta
Assam, India

Lily Devi
Department of Botany, Bijni College, Bijni
Assam, India

Papori Devi
Department of Botany, Arya Vidyapeeth College, Guwahati, Assam, India

Sanjay J. Dhoble
Department of Physics, R.T.M. Nagpur University, Nagpur, India

Krity Dulal
Plant Ecology Laboratory, P.G. Department of Botany, Madhab Choudhury College
Barpeta, Assam, India

Lilla Nur Firli
Drug Utilisation and Discovery Research Group, Faculty of Pharmacy, Universitas Jember, Jember, Indonesia

Swati Gajbhiye
Department of Physics, R.T.M. Nagpur University, Nagpur, India

Ridip Kumar Gogoi
Bioinformatics Infrastructure
Facility Centre
Rajiv Gandhi University, Doimukh
Arunachal Pradesh, India

Apurba Gohain
Department of Chemistry, Assam University, Silchar, Assam, India

Nilayan Guha
Department of Pharmaceutical Sciences
Dibrugarh University, Dibrugarh
Assam, India

Shirin Gull
Department of Nutrition Science
Faculty of Life Sciences, Government College University, Faisalabad, Pakistan

Rimsha Gulzar
Department of Food Science, Government College University, Faisalabad, Pakistan

Chetana Hasnu
Department of Botany, Pandit Deendayal Upadhyaya Adarsha Mahavidyalaya,
Tulungia, Bongaigaon, Assam, India

Maryam Ilyas
Department of Human Nutrition & Dietetics
Faculty of Life Sciences, Minhaj University
Lahore, Pakistan

Mariam Islam
Department of Food Science
Government College University
Faisalabad, Pakistan

Rosni Jabin
Department of Botany, Nowgong College
(Autonomous), Nagaon, Assam

Miral Javed
College of Biosystem Engineering and
Food Science, Zhejiang University
Hangzhou P.R. China

Bibhuti Busan Kakoti
Department of Pharmaceutical Sciences
Dibrugarh University, Dibrugarh
Assam, India

Paul A. Keller
School of Chemistry and Molecular
Bioscience, Molecular Horizons University
of Wollongong
Wollongong, Australia

Waseem Khalid
University Institute of Food Science and
Technology, The University of Lahore
Lahore, Pakistan

Nihad Ashraf Khan
Department of Biosciences, Faculty of
Natural Sciences, Jamia Millia Islamia
University, New Delhi, India

Bhaskor Kolita
Department of Botany
Jorhat Kendriya Mahavidyalaya
Jorhat, Assam, India

Prashant Kumar
Department of Botany, Hindu College
Moradabad (Affiliated to MJP Rohilkhand
University), Bareilly, India

Beena Kumari
Department of Botany, Hindu College
Moradabad (Affiliated to MJP
Rohilkhand University)
Bareilly, India

Shuby Kumari
Department of Pharmaceutical
Sciences, Dibrugarh University
Dibrugarh, Assam, India

Thi Hai Yen Lam
Department of Bioorganic Chemistry
Leibniz Institute of Plant Biochemistry
Halle (Saale), Germany

Ratul Nath
Department of Life Sciences, Dibrugarh
University, Dibrugarh, Assam, India

Ari Satia Nugraha
Drug Utilisation and Discovery Research
Group, Faculty of Pharmacy, Universitas
Jember, Jember, Indonesia

School of Chemistry and Molecular
Bioscience, Molecular Horizons, University
of Wollongong, Wollongong, Australia

Yunita Sari Pane
Department of Pharmacology &
Therapeutics, Faculty of Medicine
Universitas Sumatera Utara, Medan
Indonesia

Alka Rajput
Department of Botany, School of Applied,
Basic and Biosciences, RIMT University
Mandi-Gobindgarh, India

Tazeen Rao
Department of Biochemistry, Bahauddin Zakariya University, Multan, Pakistan

Izza Faiz ul Rasool
Department of Food Science, Faculty of Life Sciences, Government College University, Faisalabad, Pakistan

Hadiqa Faiz ul Rasul
Center of Agricultural Biochemistry and Biotechnology (CABB), University of Agriculture, Faisalabad, Pakistan

Muhammad Ahtisham Raza
Department of Food Science, Government College University, Faisalabad, Pakistan

Farhan Saeed
Department of Food Science, Government College University, Faisalabad, Pakistan

Shilpa Sarkar
P.G. Department of Botany, Madhab Choudhury College, Barpeta
Assam, India

Subrata Sarkar
Department of Botany, Abhayapuri College, Abhayapuri, Assam, India

Girish Kumar Sharma
Department of Botany, Hindu College Moradabad (Affiliated to MJP Rohilkhand University), Bareilly, India

Varsha Sharma
Department of Zoology, School of Basic, Applied and Biosciences, RIMT University Mandi-Gobindgarh, India

Darshita Sinha
Department of Life Sciences
Dibrugarh University, Dibrugarh
Assam, India

Ludmilla Fitri Untari
Department of Tropical Biology
The Faculty of Biology, Gadjah Mada University, Yogyakarta, Indonesia

Amit Vaish
Department of Botany, Hindu College Moradabad (Affiliated to MJP Rohilkhand University), Bareilly, India

Hendris Wongso
Research Center for Radioisotope Radiopharmaceutical, and Biodosimetry Technology, Research Organization for Nuclear Energy, National Research and Innovation Agency, Banten
Indonesia

Research Collaboration Center for Theranostic Radiopharmaceuticals
National Research and Innovation Agency, Sumedang, Indonesia

Farishta Yasmin
Department of Botany, Nowgong College (Autonomous), Nagaon, Assam

Preface

Lichens are symbiotically associated living organisms composed of algae (phycobiont) and fungi (mycobiont). They are ubiquitous in distribution, including some of the most extreme environmental conditions on the earth. Lichens grow on tree trunk, rock surface, old concrete building wall, and even in iron and other metal surfaces. They comprise about 8–10% terrestrial land cover of the earth's surface. About 20,000 species of lichens have been discovered, where mycobiont predominates with 90% of the thallus volume. Lichens are a source of unique metabolites which possess vast therapeutic applications such as anti-inflammatory, antioxidant, anticancer, and antimicrobial. Lichens are also a good source of human food and food for other animals due to presence of different nutritionally potential compounds. Lichens contain natural pigments and chemical components because of which they are used potentially in dyes and in perfume and cosmetic industries. They have also been recognized as bioindicators of environment pollutants, climate changes, and ecological continuity. Owing to their wide applications, it is important to understand the chemistry and biology of this distinct entity.

This book presents detailed aspects of lichens from their biology, morphology, taxonomy, ecology, physiology, etc. Unique ecological niche of lichens provides a habitat for other organisms to grow, which have been discussed in detail in the book. The impact of climate change on the lichen flora and various adaptive strategies embraced by lichens owing to the harsh environment have also been described. Lichens are sensitive to hazardous chemicals and thus popular as natural bioindicators, and thus details have been summarized. Lichens have an inimitable position in the pharmaceutical industry because of their bioactive class of compounds such as depsides, depsidones, dibenzofurans, terpenes, and xanthones, which have been discussed in detail in this book. This book emphasizes on various therapeutic properties of lichens such as antioxidant, antifungal, antibacterial, and anticancer with key discussion on important secondary metabolites. Various other applications such as food, beverage, perfumery, and cosmetic of lichens have also been discussed in detail. Important species of lichens (*Parmelia*, *Cetraria*, *Cladonia*, and *Usnea*) with promising therapeutic properties have been discussed as well in a separate chapter. In recent years, lichen-based nanoparticles have gained significant interest owing to their distinctive attributes and potential uses across diverse fields, and hence all the related aspects have been discussed. The book serves as a valuable asset for researchers and graduate students of chemistry, botany, ecology, lichenology, biotechnology, microbiology, nanoscience, cosmetologist, and pharmaceutical sciences and for doctors (especially those engaged in Ayurveda and Traditional Chinese Medicine).

Ashoke Kumar Das
Ajay Sharma
Deepika Kathuria
Mohammad Javed Ansari
Garima Bhardwaj

1

Overview of Lichen

Ashoke Kumar Das[1], Subrata Sarkar[1], and Papori Devi[2]

[1] *Department of Botany, Abhayapuri College, Abhayapuri, Assam, India*
[2] *Department of Botany, Arya Vidyapeeth College, Guwahati, Assam, India*

1.1 Introduction

The term "lichen" was coined by Theophrastus (Father of Botany) more than two thousand years ago. Till the 19th century, lichens were thought to be individually recognized organisms. In 1869, only it was accepted that it was composed of two different organisms [1, 2]. Lichens are composed of different species of fungi (Mycobiont), algae, or cyanobacteria (Photobiont) [3–9], and some microorganisms like bacteria are also associated with them [10–14]. There is a long debate and study regarding the combination of association of different components of lichen symbiosis and their physiology [15–17]. Though lichen is composed of different components, they form morphologically constant forms that are designated as species [18]. They are frequently specified as the best example of mutualistic partnerships. Photobionts have the capacity to photosynthesize; they provide carbohydrates to the fungal partner, and the mycobiont creates a physical scaffold that encloses and supports the growth of photobionts [5]. According to Hawksworth, "a lichen is a stable self-supporting association of a mycobiont and a photobiont in which mycobiont is the exhabitant" [4]. Through a process called lichenization, a fungus and a photosynthetic partner transformed into a lichen thallus, from a free-living to a symbiotic state [19].

Lichens are distributed throughout the world and found to grow in almost all climatic conditions ("Lichens," [20]). They are dominating the earth's terrestrial ecosystem, particularly in the subarctic and arctic regions, covering about 8–10% of the total area [21]. About 13,500–20,000 species of lichens have been recognized globally so far [22–25]. According to the Botanical Survey of India, 19,500 species have been discovered so far ("Lichens," [20]). Lucking et al. made a prediction, which is 26,000 lichen taxa [26]. Among approximately 20,000 species of lichen-forming fungus known to exist worldwide, Ascomycetes make up 98% of all known lichenized fungi, followed by Deuteromycetes (1.6%), and Basidiomycetes (0.4%). On the other hand, there are only roughly 156 species of photobionts across 56 taxa. The majority of photobionts are either cyanobacteria (Cyanoprokaryota-35 species, 22.3%)

Chemistry, Biology and Pharmacology of Lichen, First Edition. Edited by Ashoke Kumar Das, Ajay Sharma, Deepika Kathuria, Mohammad Javed Ansari, and Garima Bhardwaj.

or green algae (Chlorophyta-116 species, 73.9% of total photobiont diversity). The three most common photobiont genera in lichens are *Trebouxia*, *Trentepohlia*, and *Nostoc* [5].

Lichens are separated into two groups based on their size: macrolichens and microlichens. They are classified as corticolous, ramicolous, lignicolous, saxicolous, musicolous, terricolous, and foliicolous based on the substrate on which they develop. They are also separated into three groups based on their physical characteristics: foliose, fruticose, and crustose [27]. Anatomically, the majority of lichen thallus is stratified into the upper cortex, photobiont layer, medulla, and lower cortex. The cortex of many foliose and fruticose lichens is made up of pseudoparenchymatous or prosoplectenchymatous tissues of fungi. The medullary layer consists of loosely interwoven long-celled hyphae having internal airspace. The photobiont layer is formed by the upper part of the medulla. Crystalline secondary products often encrusted the hyphal cell walls of the algal and medullary layers. The lower cortex is well-developed in some typical foliose lichen groups like Parmeliaceae. Reproduction of lichen is mainly expressed by its fungal partner with sexual and asexual methods, while it is reduced in case of the algal partner in the lichenized state [28].

The significant role played by lichens includes natural soil formation and nutrient cycling [29]. For monitoring anthropogenic disturbances over time, such as air pollution, acid rain, nitrogen deposition, and several other environmental variables, lichens have been utilized as a crucial biological indicator [7, 30]. Since long time, a wide range of lichen species have been used as traditional medicine and in different folk cultures in countries like North America, Europe, India, Nepal, and China [31–35]. They are also good sources of human food and are also taken by various wild and domesticated animals as feed, containing a range of nutrients and biologically active compounds [36–38]. Some lichens are also used as raw materials in various industries like cosmetics and perfume, minerals, brewing, distilling, and essential oil [32, 38].

The globally declining trends of abundance and diversity of lichens have been documented by various authors across the world. Pollution due to industrialization, large-scale modern agriculture, urbanization, deforestation, habitat loss, overexploitation, global warming, and climate change are the main causes of their decline [22, 39].

1.2 Distribution

Lichens have some surprising capacity of ecological resilience and adaptability. They can absorb and retain moisture from various sources, due to which they can easily grow in exposed substratum like leaves and barks of trees, rocks, etc., and can survive in extreme conditions like hot deserts, barren rocky cliffs, and frozen environments of the polar regions. They can also grow in the marble of old buildings and monuments [5, 25].

Lichens are an old group of fungi that may be traced back to 400 MA to the Early Devonian in the Rhynie chert deposits in Scotland and to 600 MA in marine phosphorite of the Doushantuo Formation at Weng'an in South China. The earliest ascomycetes may symbiotically associate with the already available algae and cyanobacteria, which formed the hypothetical *Protolichen* group. Eighteen (18) patterns of lichen distributions are described by Galloway, which are Cosmopolitan, Endemic, Austral, Bipolar, Paleotropical, Neotropical,

Pantropical, Australasian, Circum-pacific, Atlantic, Eastern North America–western European (amphi-Atlantic), Western North American–western European, Mediterranean, American–Asian, South American–African, Southern xeric and Boreal arctic–alpine taxa [40]. The cosmopolitan taxa are widespread in their occurrence and found in all land masses and various oceanic islands. The family Parmeliaceae has a worldwide distribution having species like *Parmelia sulcata*, *Flavoparmelia caperata*, *Hypotrachyna sinuosa*, *Parmotrema perlatum*, etc. The endemic lichen taxa are present in some particular geographical areas with limited distribution. Lichen flora of India comprises over 2900 species, among which 18% (540 species) are endemic. ("Lichens," [20]). In New Zealand, 23% of their lichen flora is endemic. Some of the endemic lichen species of New Zealand are *Austrella brunnea*, *Caloplaca erecta*, *Lobaria asperula*, *Umbilicaria murihikuana*, etc. A high percentage of endemic lichen flora is found in South Georgia (24%) to continental Antarctica (50%). The Austral taxa are represented by the southern hemisphere land masses, and the lichen flora is divided into Paleoaustral and Neoaustral lichens. The Paleoaustral lichens are represented by the primitive Gondwanan groups, which are poorly adapted for long-distance dispersal; examples are—*Bartlettiella fragilis*, *Brigantiaea phaeomma*, *Bryoria austromontana*, *Caloplaca cribrosa*, etc. The Neoaustral lichens are dispersed after Gondwanaland fragmentation, which takes place between the post-Oligocene and the present. Some examples of Neoaustral lichens are—*Caloplaca cirrochrooides*, *Leifidium tenerum*, *Parmelia cunninghamii*, etc. The lichens which are distributed in high latitudes of both the northern and southern hemispheres are the bipolar taxa. Some examples of bipolar lichen taxa are *Cladonia ecmocyna*, *Caloplaca tornoensis*, *Bellemera alpine*, etc. The lichen flora found in Africa, the Indian subcontinent, the Arabian peninsula, the Malesian Archipelago, and some islands of the Pacific Ocean are called Paleotropical taxa. *Bactrospora metabola*, *Cladia aggregate*, *Parmelinopsis swinscowii*, etc., are examples of Paleotropical taxa. The lichen species represented in some regions of South America and the Caribbean islands are known as Neotropical taxa. Some of the Neotropical lichen taxa are species of *Erioderma*, *Leptogium*, and *Peltigera*. Pantropical taxa are found in most of the tropical regions and show affinities with warm temperate characteristics, for example—species of *Parmotrema*, *Glyphis*, *Graphis*, etc. Australasian lichen taxa have similar characteristics to the lichen flora represented in Australia and New Zealand. Species of the genus *Nothofagus*, *Placopsis*, are included in the Australasian taxa. The Western Pacific lichen taxa are found in the extent northwards to Japan, westwards to India, and in some parts of Africa of Australia. *Calopadia subcoerulescens*, *Parmelia erumpens*, and *Rinodina reagens* are some examples of Western Pacific lichen taxa. The lichens found to grow around the Pacific Ocean are called Circum-Pacific taxa, for example—*Hypogymnia pulverata*, *Placopsis cribellans*, *Mastodia tessellate*, etc. The Atlantic lichen taxa are found in the islands of the Atlantic Ocean. Some Atlantic taxa are—*Byssoloma croceum*, *Pyrenula hibernicum*, *Porina atlantica*, etc. Lichen flora of present-day North Atlantic regions are known as Eastern North America–western European or amphi-Atlantic taxa. Comparatively, fewer lichen species are distributed in this region, some of them are *Cladonia strepsilis*, *Rhizocarpon timdalii*, *Lasallia pustulata*, etc. Various lichen taxa are restricted to the areas of western North America and Western Europe, for example—*Cliostomum leprosum*, *Lecidella laureri*, *Rinodina disjuncta*, etc. The Mediterranean lichen taxa are comprised of characteristics of different elements—northern, temperate, humid sub-tropical element, and arid. Some of the lichen species found in this region are *Diploschistes diacapsis*, *Placidium fringens*, *Toninia tristis*, etc. The lichens are represented in the regions of eastern North American–eastern Asian pattern called American–Asian taxa, which are—species under the genus *Cetrelia*, *Collema*, *Allocetraria*, etc. The South American–African taxa of lichen are isolated between South America and southern Africa. Some of the lichen flora of this type are—*Peltula clavata*, *Umbilicaria haplocarpus*, *Caloplaca isidiosa*, etc. The Southern xeric taxa are distributed in the regions of southern Africa, Western Australia, South Australia, and southern New Zealand with characteristic climatic conditions like winter rainfall and summer drought. *Diploschistes hensseniae*, *Xanthoparmelia subimitatrix*, *Digitothyrea rotundata*, etc. are some of the species recorded under this region. The boreal arctic-alpine taxa are confined to the areas of the northern hemisphere, such as North America, Europe, and some parts of Asia. The lichen species found in this region are mostly growing in woodlands, heathlands, and tundra. Some of the lichen species found in this region are under the genus—*Cladonia*, *Cetraria*, *Vulpicida*, *Brodoa*, etc. [40].

Although the distribution range sizes and patterns of many lichen species resemble those of vascular plants, lichens have far more continental disjunctions. Species of "enigmatic disjunctions" are primarily characterized by having rather large distributional areas in each of the continents where they occur. An even more enigmatic portion of this group of disjunct species exhibits a normal-sized distributional area in one or more continents but an extremely restricted, point-like distribution in another. For example, *Alectoria imshaugii* grows in large areas of the west coast of North America from south Alberta to northeast California but is only known from Gomera to Hierro in the Canary Islands in Africa. *Cetraria odontella*, commonly found in Finland and Sweden has a holarctic distribution that can also be found on Australia's Mount Kosciusko. A most extreme type of enigmatic disjunct distribution is shown by *Coleopogon abraxus*, which was only known from a mountain on the east coast of Cape Town, South Africa, but has recently been observed in a forest of central Chile. Another species, *Acroscyphus sphaerophoroides*, is found to grow in 13 different areas of Bhutan, China, Canada, Japan, Patagonia, Peru, South Africa, and the United States [23].

1.3 Morphology and Anatomy

In most of the lichen, the fungal partner mainly determines the morphological appearance. In only a few lichen thallus, it is determined by the algal partner. There are three morphological groups of lichens based on their habit-crustose, foliose, and fruticose [28, 41, 42]. The simple and undifferentiated thallus, with irregularly distributed algae, are known as homoiomerous, and more complex thallus, where algae are restricted to a particular layer in the thallus and medullas without algae are called heteromerous.

In crustose lichens, the thallus is tightly attached to the substratum and is difficult to detach from the surface. In most cases, the thallus contains a well-defined upper cortex, an algal layer, and a medulla. There are various subtypes of crustose lichens—powdery, endolithic, endophloeodic, squamulose, peltate, pulvinate, lobate, effigurate and suffruticose crusts. The powdery crusts or leprose type thallus are simple in structure, where fungal hyphae cover algal cells, and they have no definite algal or fungal layer, e.g. *Lepraria* genus.

The endolithic lichen (e.g. *Acrocordia conoidea* and *Verrucaria baldensis*) grows inside rock, while endophloeodic lichens grow underneath the cuticle of leaves and stems of higher plants. They are more organized in structure and form an upper cortex consisting of a densely conglutinated hyphal layer named "lithocortex." In the squamulose type of crustose lichen, the areolae are enlarged in the upper portion and become partly free from the substrate, often form overlapping scale-like squamules (e.g. genera like *Catapyrenium* and *Peltula*). In general like *Mobergia*, squamules are extremely inflated; they are called bullate type. The peltate type has more or less central attachment area on the lower surface of flat scales of squamulose thalli (e.g. *Peltula euploca* and *Anema nummularium*). In effigurate type of thallus, the marginal lobes are prolonged and are radially arranged, e.g. genera like *Caloplaca* and *Acarospora*. When the thallus becomes radially striate with marginal lobes, they are called lobate type of thallus [28].

The foliose lichens are flat and leaf-like, partially attached to the substratum. They have well-defined upper and lower surfaces with dorsiventral organization. The branching thallus bears several lobes. The foliose lichens may be of two types—laciniate and umbilicate. Laciniate lichens are lobate with various sizes. In *Parmelia* species, the lobes are radially arranged, and in *Peltigera*, lobes overlap, similar to tiles on a roof. In *Menegazzia*, thallus lobes are inflated with a hollow medullary center. Umbilicate lichen thallus is circular in look and consists of either one single unbranched lobe or a multilobate with a limited branching pattern. Umbilicate type of thallus has a central umbilicus, which arises from the lower surface and is attached to the substratum.

In Fruticose lichens, the thallus lobes are hair-like, strap-shaped, or shrubby; lobes are either flat or cylindrical. In some of the fruticose lichens, thallus is dorsiventrally arranged, e.g. *Evernia prunastri*. Some of them have radially symmetrical thalli, e.g. *Usnea* and *Ramalina* species [28].

The cortex is the outermost protective layer of stratified lichen thallus. In some foliose lichen, the cortex may also be present in the lower side of the thallus but is absent in the squamulose type. The photobiont layer of stratified lichen is formed just beneath the cortex. The medulla occupies most part of the thallus in stratified lichen composed of fungal hyphae.

1.4 Reproduction

In a lichenized state, the fungal partner usually expresses full sexual and to a certain extent, asexual mode of reproduction, whereas in the case of an algal partner, it is a reduced type [28]. Since lichen cannot exist without the symbiotic relationship between a mycobiont and a photobiont, either both partners must be dispersed at the same time, or specific adaptations must guarantee contact and relichenization following the independent dispersal of mycobionts and photobionts. The lichen's symbiotic relationship allows it to thrive on a broad range of substrates under a range of climatic circumstances, but its primary means of dispersal—sexual or non-sexual—is determined by its mode of reproduction. After being released, the fungal spores germinate on an appropriate substrate, take up algae that are compatible with them, and grow new vegetative thalli. Mycobionts and photobionts coexist and reproduce to produce vegetative propagules such hormocysts, isidia, and soredia.

Even the typical means of propagation that photobionts experience in their free-living stage may be absent or extremely limited in lichenized conditions.

In lichens, one of the most common forms of vegetative reproduction is fragmentation. There is the potentiality to act as a source of regeneration of any portion of lichen thallus containing both the symbionts. The soredia are formed in specialized organ sorelia; they are the most common diaspore of foliose and fruticose lichen. Another diaspore is formed by isidia, which are finger-like projections with well-developed fungal tissue enclosed by algal cells. They are very commonly found in crustose, foliose, and fruticose lichen species. When the photobiont is a cyanobacteria, hormocytes are formed, which consist of trichomes or individual cells with a gelatinous sheath [43]. The sexual reproduction of Ascomycetes and Basidiomycetes fungi as a lichen partner is analogous to that in free-leaving Ascomycetes and Basidiomycetes [44].

1.5 Lichen Phytochemicals

The secondary metabolites produced by lichen are known as lichen phytochemicals [45]. All these phytochemicals are of fungal origin [46]. Other than primary metabolites (proteins, amino acids, polyols, carotenoids, polysaccharides, and vitamins), lichen also produces over 700–1050 (including under culture) different secondary metabolites. The main categories of lichen phytochemicals are depsides, depsidones, dibenzofurans, anthraquinones, xanthones, chromones, pulvinic acid derivatives, terphenylquinones, terpenes and steroids. The medulla portion of lichen thallus mostly produces colorless depsides and depsidones. Usnic acid is also formed in the medulla portion [25, 47].

Lichen phytochemicals play some important ecological and medicinal roles, which determine the relative ecological success of individual lichen species. In lichen thallus, these phytochemicals are responsible for light-screening, chemical withering, biological defense, anti-herbivore defense, and allergenic. The yellow-colored cortical pigment, usnic acid is produced by thousands of lichen that are exposed to the sun, while the lichen containing a low concentration of depsides is grey-green in color and shade loving [48]. Many lichen substances exhibit multiple biological activities, such as the dibenzofuran usnic acid, which has characteristic antimicrobial, larvicidal, and anticancer properties, and is also known for its ultraviolet absorption. The phytochemicals are genetically regulated and, in certain cases are related to an individual's morphology and location within a species or genus. Since secondary metabolite distribution patterns are typically species-specific, they are frequently employed in lichen systematics and taxonomy. Secondary metabolites from lichens can act as allelopathic agents, which means they can have an impact on the growth and development of nearby lichens, mosses, and vascular plants as well as microbes. Some lichens can significantly inhibit the growth of higher plants. It has been demonstrated that two common species found in boreal forests, *Cladonia stellaris* and *Cladonia rangiferina*, exhibit allelopathic effects on white spruce and jack pine (*Pinus banksiana*) [25].

The most studied lichen substance is usnic acid, which was isolated in 1884 from Usnea and later from other lichen genera- *Cladonia*, *Hypotrachyna*, *Lecanora*, *Ramalina*, *Evernia*, *Parmelia* and *Alectoria*. Usnic acid is famous for its anti-proliferative activity and has anticancer potential. Cellular apoptosis (programmed cell death) of carcinogenic cell lines observed after treatment of usnic acid. It has the ability to affect cell lines of ovarian, hepatic, gastric, and breast cancer. Another compound atranorin extracted from some lichen species like *Everniastrum vexans* was tested for its anticancer properties. Regarding toxicity studies in the human body, there are some records in case of usnic acid [49].

1.6 Economic Importance

Lichens are good sources of food for humans and feed for other animals. *Cetraria islandica*, *Lecanora esculenta*, *Umbilicaria esculenta*, *Peltigera canina*, *Parmelia* sp., and *Ramalina sinensis* are consumed as food in different counties. *Cladonia rangiferind*, *C. rangiferina*, *Cetraria islandica*, species of *Parmelia*, *Evernia*, etc., are used as fodder in many countries. Some of the lichen species are used as flavoring agents, e.g. *Heterodermia tremulans*, species of *Pyxine*, *Physcia*, etc. Some lichen species are commercially used in the Litmus dye industry. Litmus mixture is obtained from *Roccella tinctoria*. Moreover, they are also used in the textile dye industry. Red-colored natural dye is obtained from *Rubia tinctorum* and *Rubia cordifolia*. *Evernia prunastri* has been used for making perfume and the cosmetic industry.

There are several lichen species that have medicinal properties, such as antimicrobial, antiviral, anti-inflammatory, anticancer, insecticidal, antipyretic, etc. *Usnea* sp. (anticancer), *U. esculenta* (anti-HIV), *Parmelia* sp. (wound healing), etc., are some medicinal lichens [37]. Scientific investigations have identified a large number of lichen species, indicating their biological activity and application in traditional medicine. About 60 lichen genera that have been traditionally employed by various cultures worldwide, primarily in North America, Europe, and Asia. Traditionally, the most commonly used lichen genera is *Usnea*, which is used around the world. The lichen genera *Cladonia*, *Ramalina*, *Lobaria*, *Pertigera*, *Evernia*, *Pseudevernia*, *Umbilicaria*, *Xanthoparmelia*, *Letharia*, *Cetraria*, *Parmotrema*, *Thamnolia* are used for various medical conditions like external injuries, skin infections, respiratory ailments, etc. *C. islandica* has been used in European pharmacopoeias to treat lung disease, cold symptoms, and gastroenteritis since 1500. *Pseudevernia furfuracea* was used by the Egyptians in the process of embalming mummies for aromatic purposes and is now effectively used in the perfume industry. In India, *Usnea longissima* has been used in traditional medicine for its analgesic, cardiotonic, digestive, and wound-healing properties [49].

Due to their sensitivity to various pollutants (nitrogen, sulfur, etc.) and heavy metals, lichens are widely recognized as bioindicators of environmental pollution. The health condition of lichen can indicate the accumulation of heavy metals showing its negative effect [25, 29].

1.7 Conservation

The rate of species extinction in the Anthropocene has been 100–1000 times higher than the background rate, or 0.1–1 million species per year. They have coincided with reductions in the useful biodiversity [50]. Experts in biodiversity calculated that since 1500, over

30% (uncertainty range: 16–50%) of species have faced worldwide threats or have been driven extinct. Habitat loss and climate change are predicted to worsen the extent of biodiversity loss. According to expert estimates, either 41% (range: 30–60%) or 80% (range: 63–95%) of species are threatened or driven to extinction when 50% or 90% of their habitat is lost. Additionally, the experts calculated that a 2 °C or 5 °C increase in global warming would drive approximately 25% (range: 15–40%) or 50% (range: 32–70%) of species to extinction [51]. Lichens are ubiquitous in terrestrial ecosystems across the globe and are ecologically significant symbioses that are well-known to non-scientists. They play an immense role in a variety of crucial ecological processes and ecosystem performance including as soil formation, nutrient cycling, rock weathering, and humidity regime regulation. The abundance and diversity of lichen across the globe have been declining. The main threats that apply to biodiversity in general are also true for lichens [39]. Lichens suffer from a loss of diversity and abundance on a variety of levels, ranging from entire communities to individual species and populations. The variety and abundance of lichens are adversely affected by numerous well-established human-mediated activities. Climate change and habitat loss are two of these challenges that affect almost all forms of biodiversity. Lichens are disproportionately affected by other hazards, such as air pollution, in comparison to other types of organisms. It has long been known that overbrowsing of the *Cladonia* by growing populations of reindeer in Scandinavia and Alaska is a major contributing element to the drastic loss in lichen cover, which could pose a significant threat to the husbandry of reindeer. The species richness and composition of lichen communities are significantly impacted by both deforestation and the deterioration of lichen habitats caused by the planting of plantation forests in place of natural forests [39].

Lichens are protected under the Endangered Species Act at the federal level in the USA. Currently, only two endemic species from the Southeast of the United States are listed as endangered species: *Cladonia perforata* and *Cetradonia linearis*. Since these species were added to the list more than 20 years ago, researchers have paid considerable attention to them, focusing mostly on mapping their ranges and learning about the genetics and demographics of the populations [22]. Effective lichen conservation plans typically integrate attempts to preserve or enhance the size, demographics, and genetic makeup of populations with the goal of protecting environments. The preservation of habitat size, connectivity, and quality should be the key goals of lichen conservation. In contrast to dynamic habitats like woods, grasslands, and gravel fields, permanent ecosystems like mountain ridges can best be preserved by static protection, which is also very simple to accomplish [39].

1.8 Conclusion

Lichens are the predominant flora on Earth exhibiting extensive biological, physiological, and chemical peculiarities. The fungal partner is the main component of the lichen thallus with regard to its morphology, but the algal partner is necessary from a nutritional standpoint. With the successful partnership of symbiosis, they are dominating a large terrestrial ecosystem on Earth. A wide range of secondary products has made them an essential natural treasure house for human food, medicine, and various industrial products.

They are useful organisms for ecosystem health monitoring. Various anthropogenic factors have made them threatened in natural habitats, especially some commercially used lichen species.

References

1 Armaleo, D., Müller, O., Lutzoni, F. et al. (2019). The lichen symbiosis re-viewed through the genomes of Cladonia grayi and its algal partner *Asterochloris glomerata*. *BMC Genomics* 20 (1): 605. https://doi.org/10.1186/s12864-019-5629-x.

2 Aprile, G.G., Catalano, I., Migliozzi, A., and Mingo, A. (2011). Monitoring epiphytic lichen biodiversity to detect environmental quality and air pollution: the case study of Roccamonfina park (Campania Region - Italy). In: *Air Pollution—New Developments* (ed. A. Moldoveanu). InTech. https://doi.org/10.5772/17907.

3 Asplund, J. and Wardle, D.A. (2017). How lichens impact on terrestrial community and ecosystem properties. *Biological Reviews* 92 (3): 1720–1738. https://doi.org/10.1111/brv.12305.

4 Hawksworth, D.L. (1988). The variety of fungal-algal symbioses, their evolutionary significance, and the nature of lichens. *Botanical Journal of the Linnean Society* 96 (1): 3–20. https://doi.org/10.1111/j.1095-8339.1988.tb00623.x.

5 Saini, K.C., Nayaka, S., and Bast, F. (2019). Diversity of lichen photobionts: their coevolution and bioprospecting potential. In: *Microbial Diversity in Ecosystem Sustainability and Biotechnological Applications* (ed. T. Satyanarayana, S.K. Das, and B.N. Johri), 307–323. Singapore: Springer. https://doi.org/10.1007/978-981-13-8487-5_13.

6 Sanders, W.B. and Masumoto, H. (2021). Lichen algae: the photosynthetic partners in lichen symbioses. *The Lichenologist* 53 (5): 347–393. https://doi.org/10.1017/S0024282921000335.

7 Bhagarathi, L.K., DaSilva, P.N.B., Subramanian, G. et al. (2023). An integrative review of the biology and chemistry of lichens and their ecological, ethnopharmacological, pharmaceutical and therapeutic potential. *GSC Biological and Pharmaceutical Sciences* 23 (3): 092–119. https://doi.org/10.30574/gscbps.2023.23.3.0223.

8 Lutzoni, F. and Miadlikowska, J. (2009). Lichens. *Current Biology* 19 (13): R502–R503.

9 Zhao, Y., Wang, M., and Xu, B. (2021). A comprehensive review on secondary metabolites and health-promoting effects of edible lichen. *Journal of Functional Foods* 80: 104283. https://doi.org/10.1016/j.jff.2020.104283.

10 Aschenbrenner, I.A., Cernava, T., Berg, G., and Grube, M. (2016). Understanding microbial multi-species symbioses. *Frontiers in Microbiology* 7: 180. https://doi.org/10.3389/fmicb.2016.00180.

11 Duran-Nebreda, S. and Valverde, S. (2023). Composition, structure and robustness of lichen guilds. *Scientific Reports* 13 (1): 3295. https://doi.org/10.1038/s41598-023-30357-w.

12 Grube, M. (2018). The lichen thallus as a microbial habitat. *Biosystems Ecological Series* 34: 528–545.

13 Morillas, L., Roales, J., Cruz, C., and Munzi, S. (2022). Lichen as multipartner symbiotic relationships. *Encyclopedia* 2 (3): 1421–1431. https://doi.org/10.3390/encyclopedia2030096.

14 Leavitt, S.D. and Lumbsch, H.T. (2016). Ecological biogeography of lichen-forming fungi. In: *Environmental and Microbial Relationships* (ed. I.S. Druzhinina and C.P. Kubicek), 15–37. Springer International Publishing. https://doi.org/10.1007/978-3-319-29532-9_2.

15 Grimm, M., Grube, M., Schiefelbein, U. et al. (2021). The lichens' microbiota, still a mystery? *Frontiers in Microbiology* 12: 623839. https://doi.org/10.3389/fmicb.2021.623839.

16 Mitchell, M.E. (2007). Signposts to symbiosis: a review of early attempts to establish the constitution of lichen. *Huntia* 13 (2): 101–120.

17 Stanton, D.E., Ormond, A., Koch, N.M., and Colesie, C. (2023). Lichen ecophysiology in a changing climate. *American Journal of Botany* 110 (2): e16131. https://doi.org/10.1002/ajb2.16131.

18 Alexopoulos, C.J., Mims, C.W., and Blackwell, M. (2007). *Introductory Mycology*, 4e. Wiley.

19 Pichler, G., Muggia, L., Carniel, F.C. et al. (2023). How to build a lichen: from metabolite release to symbiotic interplay. *New Phytologist* 238 (4): 1362–1378. https://doi.org/10.1111/nph.18780.

20 Lichens (n.d.). Lichens of India. National Information Centre, Govt. of India. https://bsi.gov.in/page/en/lichens.

21 Payette, S. and Delwaide, A. (2018). Tamm review: the North-American lichen woodland. *Forest Ecology and Management* 417: 167–183. https://doi.org/10.1016/j.foreco.2018.02.043.

22 Allen, J.L., McMullin, R.T., Tripp, E.A., and Lendemer, J.C. (2019). Lichen conservation in North America: a review of current practices and research in Canada and the United States. *Biodiversity and Conservation* 28 (12): 3103–3138. https://doi.org/10.1007/s10531-019-01827-3.

23 Feuerer, T. and Hawksworth, D.L. (2007). Biodiversity of lichens, including a world-wide analysis of checklist data based on Takhtajan's floristic regions. *Biodiversity and Conservation* 16 (1): 85–98. https://doi.org/10.1007/s10531-006-9142-6.

24 Lakatos, M. (2011). Lichens and bryophytes: habitats and species. In: *Plant Desiccation Tolerance*, vol. 215 (ed. U. Lüttge, E. Beck, and D. Bartels), 65–87. Berlin Heidelberg: Springer. https://doi.org/10.1007/978-3-642-19106-0_5.

25 Molnár, K. and Farkas, E. (2010). Current results on biological activities of lichen secondary metabolites: a review. *Zeitschrift Für Naturforschung C* 65 (3–4): 157–173. https://doi.org/10.1515/znc-2010-3-401.

26 Lucking, R., Plata, E.R., Chavej, J.L. et al. (2009). How many tropical lichens are there. . . Really? *Bibliotheca Lichenologica* 100(Diversity of Lichenology –Jubilee Volume.): 399–418.

27 Singh, S., Arya, M., and Vishwakarma, S.K. (2019). Advancements in methods used for identification of lichens. *International Journal of Current Microbiology and Applied Sciences* 8 (08): 1450–1460. https://doi.org/10.20546/ijcmas.2019.808.169.

28 Büdel, B. and Scheidegger, C. (2008). Thallus morphology and anatomy. In: *Lichen Biology*, 2e (ed. T.H. Nash), 40–68. Cambridge University Press. https://doi.org/10.1017/CBO9780511790478.005.

29 Yang, J., Oh, S.-O., and Hur, J.-S. (2023). Lichen as bioindicators: assessing their response to heavy metal pollution in their native ecosystem. *Mycobiology* 51 (5): 343–353. https://doi.org/10.1080/12298093.2023.2265144.

30 Oksanen, I. (2006). Ecological and biotechnological aspects of lichens. *Applied Microbiology and Biotechnology* 73 (4): 723–734. https://doi.org/10.1007/s00253-006-0611-3.

31 Crawford, S.D. (2015). Lichens used in traditional medicine. In: *Lichen Secondary Metabolites* (ed. B. Ranković), 27–80. Springer International Publishing. https://doi.org/10.1007/978-3-319-13374-4_2.

32 Upreti, D.K., Divakar, P.K., and Nayaka, S. (2005). Commercial and ethnic use of lichens in India. *Economic Botany* 59 (3): 269–273. https://doi.org/10.1663/0013-0001(2005)059[0269:CAEUOL]2.0.CO;2.

33 Devkota, S., Chaudhary, R.P., Werth, S., and Scheidegger, C. (2017). Indigenous knowledge and use of lichens by the lichenophilic communities of the Nepal Himalaya. *Journal of Ethnobiology and Ethnomedicine* 13 (1): 15. https://doi.org/10.1186/s13002-017-0142-2.

34 Pyakurel, D., Smith-Hall, C., Bhattarai-Sharma, I., and Ghimire, S.K. (2019). Trade and conservation of Nepalese medicinal plants, fungi, and lichen. *Economic Botany* 73 (4): 505–521. https://doi.org/10.1007/s12231-019-09473-0.

35 Yang, M.-X., Devkota, S., Wang, L.-S., and Scheidegger, C. (2021). Ethnolichenology—the use of lichens in the Himalayas and Southwestern Parts of China. *Diversity* 13 (7): 330. https://doi.org/10.3390/d13070330.

36 Llano, G.A. (1948). Economic uses of lichens. *Economic Botany* 2 (1): 15–45. https://doi.org/10.1007/BF02907917.

37 Sharma, M. and Mohammad, A. (2020). Lichens and lichenology: historical and economic prospects. In: *Lichen-Derived Products*, 1e (ed. M. Yusuf), 101–118. Wiley. https://doi.org/10.1002/9781119593249.ch4.

38 Wa, E., De, E.-G., and Gm, D. (2022). Lichens uses surprising uses of lichens that improve human life. *Journal of Biomedical Research & Environmental Sciences* 3 (2): 189–194. https://doi.org/10.37871/jbres1420.

39 Scheidegger, C. and Werth, S. (2009). Conservation strategies for lichens: insights from population biology. *Fungal Biology Reviews* 23 (3): 55–66. https://doi.org/10.1016/j.fbr.2009.10.003.

40 Galloway, D.J. (2008). Lichen biogeography. In: *Lichen Biology*, 2e (ed. T.H. Nash), 315–335. Cambridge University Press. https://doi.org/10.1017/CBO9780511790478.017.

41 Jahns, H.M. (1973). Anatomy, morphology, and development. *The Lichens* 3–58. https://doi.org/10.1016/B978-0-12-044950-7.50006-4.

42 Sanders, W.B. (2006). A feeling for the superorganism: expression of plant form in the lichen thallus. *Botanical Journal of the Linnean Society* 150 (1): 89–99. https://doi.org/10.1111/j.1095-8339.2006.00497.x.

43 Krishnamurthy, K.V. and Upreti, D.K. (2001). Reproductive biology of lichens. In: *Reproductive Biology of Plants* (ed. B.M. Johri and P.S. Srivastava), 127–147. Berlin Heidelberg: Springer. https://doi.org/10.1007/978-3-642-50133-3_7.

44 Bowler, P.A. and Rundel, P.W. (1975). Reproductive strategies in lichens. *Botanical Journal of the Linnean Society* 70 (4): 325–340. https://doi.org/10.1111/j.1095-8339.1975.tb01653.x.

45 Yamamoto, Y., Hara, K., Kawakami, H., and Komine, M. (2015). Lichen substances and their biological activities. In: *Recent Advances in Lichenology* (ed. D.K. Upreti, P.K. Divakar, V. Shukla, and R. Bajpai), 181–199. India: Springer. https://doi.org/10.1007/978-81-322-2235-4_10.

46 Culberson, C.F. and Elix, J.A. (1989). Lichen substances. In: *Methods in Plant Biochemistry*, vol. 1, 509–535. Elsevier. https://doi.org/10.1016/B978-0-12-461011-8.50021-4.

47 Podterob, A.P. (2008). Chemical composition of lichens and their medical applications. *Pharmaceutical Chemistry Journal* 42 (10): 582–588. https://doi.org/10.1007/s11094-009-0183-5.

48 Karunaratne, V. (1999). Lichen substances: biochemistry, ecological role and economic uses. *Ceylon Journal of Science: Physical Sciences* 6 (1): 13–28.

49 Poulsen-Silva, E., Gordillo-Fuenzalida, F., Atala, C. et al. (2023). Bioactive lichen secondary metabolites and their presence in species from Chile. *Metabolites* 13 (7): 805. https://doi.org/10.3390/metabo13070805.

50 Dasgupta, P. and Levin, S. (2023). Economic factors underlying biodiversity loss. *Philosophical Transactions of the Royal Society B: Biological Sciences* 378 (1881): 20220197. https://doi.org/10.1098/rstb.2022.0197.

51 Isbell, F., Balvanera, P., Mori, A.S. et al. (2023). Expert perspectives on global biodiversity loss and its drivers and impacts on people. *Frontiers in Ecology and the Environment* 21 (2): 94–103. https://doi.org/10.1002/fee.2536.

2

The Biology of Lichen

Ludmilla Fitri Untari

Department of Tropical Biology, The Faculty of Biology, Gadjah Mada University, Yogyakarta, Indonesia

2.1 Introduction

Lichens are mutualistic associations or symbioses between algae (photobiont) and fungi (mycobiont) that produce new and specific entities both morphologically and physiologically. However, if the two are separated and grown in a laboratory, they will not form lichens because lichens only live in natural environments. The algae function as a producer through the process of photosynthesis. On the other hand, fungi act as suppliers of minerals and water from the environment, providing mass, structure, water, and mineral availability; determining the morphological form of the lichens resulting from the symbiosis; and performing reproductive and protective functions. Fungi also provide a refuge for algae to live. This symbiosis tends to increase the ability of both to adapt to different environments due to the thallus structure, physiology, and specific chemical compounds [1, 2].

Lichens can absorb water from the atmosphere. However, unlike plants, they have no way of retaining water inside the lichen during dry periods. When humidity rises, the thallus absorbs water and the cortex becomes transparent. If there is sufficient light, the algae begin to photosynthesis within minutes. The carbohydrates produced by photosynthesis are then taken up by the fungus in the form of sugar alcohols or glucose. They are rapidly converted to mannitol for storage, making them inaccessible to photobionts.

Lichens are known to be able to produce secondary metabolites that are very specific to each lichen species. It is thought that these chemical compounds act as a mechanism for the lichen thallus to maintain its symbiosis. The lichen symbiosis is therefore successful because it can adapt to almost all tropical and even polar terrestrial habitats. Lichens can thrive in a wide range of moisture, temperature, and light conditions, often in places where few other macroscopic life forms can survive. The cosmopolitan distribution of lichens means that they use different habitats all over the world and dominate about 8% of terrestrial ecosystems. There are more than 20,000 species worldwide [3].

Lichens containing cyanobacteria also benefit from their ability to fix inorganic nitrogen from the air, a nutrient that is often limiting but essential for protein synthesis. The ability

Chemistry, Biology and Pharmacology of Lichen, First Edition. Edited by Ashoke Kumar Das, Ajay Sharma, Deepika Kathuria, Mohammad Javed Ansari, and Garima Bhardwaj.

to fix nitrogen enables lichens to live in nutrient-poor environments. *Nostoc* is the most common genus of cyanobacterial lichen symbionts that can photosynthesize and fix atmospheric nitrogen into ammonia, nitrates, or nitrites that can be taken up by the fungal partner [4].

2.2 Lichen Life Form

Lichen thalli grow in a wide range of thallus habit. The organization of the fungal filaments determines the life forms of the thallus, while the color of the thallus is influenced by the photobiont. In several lichen genera, e.g. the filamentous genera *Coenogonium*, *Ephebe, Cystocoleus*, and *Racodium*, the photobiont determined the thallus shape [5]. Based on thallus habit, lichens can be classified into five lichen's life forms, i.e. crustose, foliose, fruticose, leprose, and squamose (Figure 2.1).

2.2.1 Crustose

Crustose is a crust-like thallus, like a layer of skin, pressed against the substrate and adhere to the substrate of tree bark or rock. The colors of the thallus are generally green, greyish-white, and sometimes pale brown, or even vivid colors such as yellow, orange, and red. The organization of the thallus is either homoiomerous or heteromerous, meaning that the algal partner is evenly distributed throughout the thallus or scattered distributed throughout the thallus [2, 5]. Examples of crustose lichens are *Protoparmeliopsis muralis*, *Lepraria lobificans*, *Pertusaria hymenea*, and *Chrysothrix candelaris.*

Included in this crustose group is placodioids and areolate. Placodioid lichens are crustose lichen but with lobes toward the edge and have an upper cortex but lack a lower cortex. They differ from foliose lichens, which may look similar but foliose lichens have both upper and lower cortices. Examples of placodioid lichens are *Variospora flavescens*, a saxicolous crustose lichen, and *Lecanora valesiaca* [6].

Areolate lichens have a crusty, paint-like appearance. The thallus consists of numerous areolae. These are polygonal parts of the thallus that contain both symbionts and can look a little like cracked mud, peeling paint, or small islands. These areolae are covered by a thin layer of fungal hyphae called the hypothallus. The hypothallus is usually dark in color. In the dry season, the hypothallus forms many areolae, but in wet conditions, the areolae merge. The areolae can develop either from a primary thallus with a closed surface, which

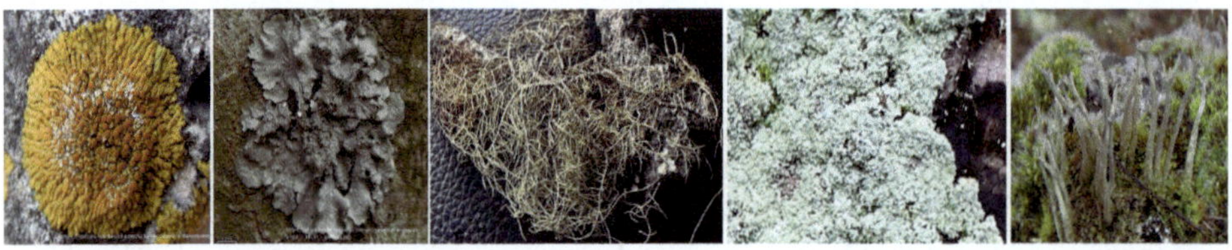

Figure 2.1 Thallus life form in lichens.

splits secondarily from the surface to the base into several areolae, or from a single thallus primordia developing on a prothallus. A thallus is called effigurate when the marginal lobes are extended and radially arranged, as in many species of the genera *Aspicilia*, *Caloplaca*, *Rhizocarpon*, and *Acarospora* [2, 5].

2.2.2 Foliose

Foliose lichens are leaf-like thalli that later develop into lobes. Foliose thalli are either homoiomerous or heteromerous with a dorsiventral organization of distinct upper and lower surfaces (cortex). This lichen life form is easy to separate from its substrate because it has rhizines as a thallus attachment system. The growth rate of the thallus into lobes is about 1–2 cm/year [2]. Examples of foliose lichens are *Phaeophyscia orbicularis*, *Parmelia sulcata*, and *Leptogium cyanescens*.

There are two types of foliose thallus, i.e. the umbilicate and the laciniate. Umbilicate lichens are those that attach to their substrate with a single point or central umbilicus from the lower surface, unbranched lobe, or multilobed thalli with unbranched patterns. The umbilicus usually consists of tightly packed, parallel, and conglutinated hyphae without photobiont cells and can be recognized by a slightly concave shape. Examples are *Lasallia papulosa*, *Umbilicaria esculenta*, and *Dermatocarpon muhlenber* [2, 5].

Laciniate lichens are lobed, lobes radiating (e.g. *Parmelia* spp., *Lobaria* spp.) or overlapping (e.g. *Lobaria* spp., *Peltigera* spp.); a wide range of sizes; lower cortex covered by rhizines, cilia or a tomentum for attachment to the substrate; may be gelatinous-homoiomerous (e.g. *Collema* spp., *Leptogium* spp., *Physma* spp.) or heteromerous. The anatomy of laciniate lichens is the most complex of all lichens, with some resembling large plants and others only a few millimeters in size [2, 5].

2.2.3 Fruticose

The thallus of fruticose lichens is shrub-like, strap-like, or hair-like with a holdfast. The lobes or strap-like may be flattened thalli (e.g. *Evernia* spp.) or terete thalli (e.g. *Usnea* spp. and *Ramalina* spp.). The branching of the lobes varies considerably between different systematic groups and even within a single genus. The thalli vary in size from several meters long to tiny species only 1 or 2 mm high. Fruticose lichens are found in a wide range of climates, from desert to humid rainforest, and on a variety of substrates [2, 5].

2.2.4 Leprose

A leprose lichen with a powdery or dust-like appearance is typically considered to be a form of crustose lichen. Morphologically, leprose lichen is the simplest growth form and lacks an organized thallus with a mixture of fungal hyphae and scattered photobiont cells. The surface of the thallus is covered by globose soredia-like granules (a cluster of photobiont cells surrounded by fungal hyphae called goniocysts) and lacking a cortex or any definable layers. Leprose thalli do not have a distinct fungal or algal layer. Species of the genus *Lepraria* and *Chrysothrix* are widespread leprose lichens [2, 5].

2.2.5 Squamulose

A squamulose lichen thallus is composed of small, scale-like plates called squamules, 1–15 mm in diameter. These scales may adhere to the substrate over their entire undersurface or along one edge only. A squamulose lichen typically lacks a lower cortex, but in a few species, the lower cortex is present. Included in this squamulose group is cladoniform. A cladoniform lichen is a dimorphic thallus in which the thallus is a combination of a double thallus, i.e. a vertical apothecia-bearing stalk (a branched, spike-like or cup-like structure is known as a podetium) and a horizontal crustose-squamulose to foliose-squamulose thallus composed of small and overlapping scales. Lichens with this growth form are found in the families of Cladoniaceae and Baeomycetaceae [5, 7].

2.3 The Internal Structure of Lichen

Based on the distribution of photobionts within the thallus, there are two organizations of thallus lichens, i.e. homoiomerous and heteromerous. In the homoiomerous thallus, the fungal hyphae and the algal cells are evenly distributed throughout the entire thallus. The homoiomerous lichen thallus is particularly common in crustose lichens where the photobiont belongs to the cyanobacteria (cyanobiont). The mycobiont and photobiont partners mix and form a protective outer cortex layer. Examples include *Psoroma*, *Placopsis*, *Collema*, and *Leptogium*.

In heteromerous lichen thallus, the organization of the lichen thallus consists of three distinct layers, i.e. the outer cortex, the medulla, and an algal layer. The outer cortex is a protective layer of densely packed fungal filaments. The cortex consists of densely woven, packed, and agglutinated fungal filaments. Crustose and squamulose lichens have only an upper cortex, with the medullary layer of the lichen in direct contact with the substrate, including the edges of the thallus of placodioid lichens, lacking a lower cortex. However, a layer of cortex envelops the strap-like or hair-like structure of fruticose lichens. Foliose lichens have an upper cortex and a lower cortex on the underside. Filamentous, byssoid, leprose, and gelatinous lichens lack a cortex, which is called ecorticate.

The medulla layer consists of interwoven fungal hyphae, with the algae usually scattered throughout the medulla. The fungal hyphae envelop the algal or cyanobacterial cells, and then the complex fungal tissues (haustorium) attach to the algal or cyanobacterial cells, typical of lichen associations. The photobiontic layer in crustose and foliose lichens is diffuse between the fungal filaments. In fruticose lichens, however, the photobiontic layer is sharply separated from the underlying layer [1, 2, 8] (Figure 2.2).

The internal structure of the fruticose lichens, such as *Usnea*, has a unique composition. The thalli are terete or flattened, with the outer cortex layer enveloping the thalli. Inside the outer cortex are the medulla (composed of algal cells and fungal hyphae) and the central chondroid axis (composed of compactly arranged fungal hyphae).

2.4 Reproduction of Lichen

Lichens can reproduce by vegetative, asexual, and sexual reproduction. Many lichens reproduce by vegetative dispersal, either by breaking off the thallus into fragments and each fragment being able to grow to form a new thallus, or by separating some lobes or

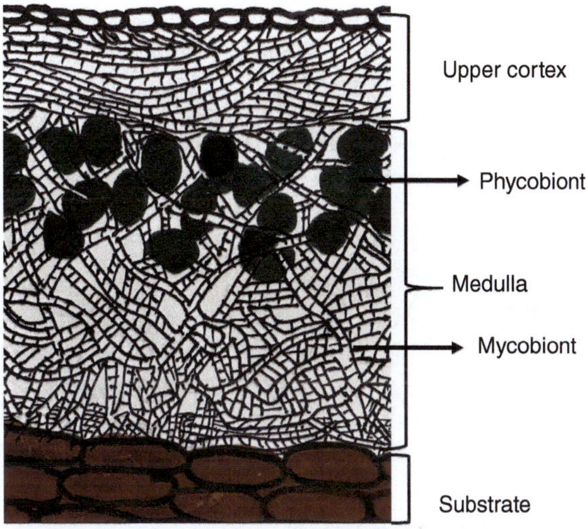

Figure 2.2 Thallus structure of crustose lichen.

branches and growing to form a new thallus. Asexual reproduction occurs when the specific form of symbiotic propagules (diaspores), i.e. soredia (singular: soredium), isidia (singular: isidium), and pycniospore, is dispersed [2, 9].

Soredium is composed of one or a few algal cells surrounded by fungal hyphae and usually develops in a particular form of a pustule-like organization called a soralia. Soredia can be dispersed by wind. Isidium consists of a group of algal cells surrounded by fungal hyphae. They form a protrusion that develops on the upper surface of the thallus. Isidia differ from soredia in that they have an outer cortical layer continuous from the thallus. Isidia vary in shape and may be scale-like, small-sphere stalked, simple coral-like, or branched coral-like. They are easily broken and dispersed. The isidium can germinate and grow into a new thallus under the right conditions.

Some lichens develop pycnospores or spermatids inside a small structure called a pycnidium (plural: pycnidia). The pycnidium is an asexual fruiting body produced by mitosporic fungi. The pycnidium varies in appearance, from small black dots to dark orange pimples on the surface of the thallus lichen. The pycnidiospores germinate under certain conditions, and then fungal hyphae develop. These fungal hyphae develop into a new lichen thallus when they come into contact with the appropriate algal partner.

Lichens can reproduce sexually by producing spores, which are formed by meiosis and the fusion of gametes. After dispersal, the fungal spores germinate and must meet a compatible algal partner for the formation of a functional lichen. However, only mycobiont partners of lichens can reproduce sexually, forming fruiting bodies on the surface or embedded in the thallus lichen tissue.

Most mycobionts belong to the ascomycetes (ascolichens), so the mode of sexual reproduction in ascolichens is similar to that of members of the ascomycetes. In ascolichens, ascospores are produced in spore-producing structures called ascomata. The most common structures of ascomata are the apothecium (plural: apothecia), the perithecium (plural: perithecia), and the podetium (plural: podetia). Apothecia (Figure 2.3) are usually cups or plate-like discs on the upper surface of the lichen thallus. They are called lirellae when the apothecia are shaped like squiggly line segments rather than discs. Perithecia

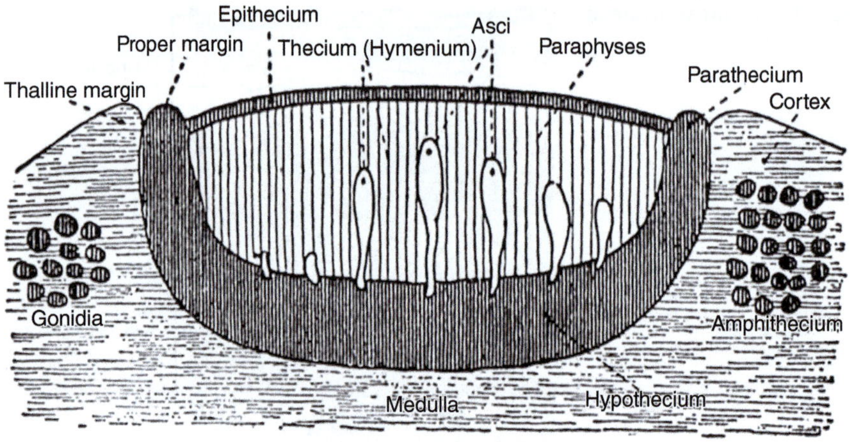

Figure 2.3 Schematic diagram of lichen apothecia (*Source:* Encyclopedia Britannica 11th Ed.: 16, p. 582 (1911)/Internet Archive).

are pot-shaped structures that dip into the lichen thallus tissue, which has an ostiole for spore release to the environment, and appear as black dots on the lichen surface.

Podetia are lichenized, hollow, pointed, stalk-like structures that emerge from the primary thallus of the lichen. This fruiting body is common in *Cladonia* spp. The podetium may be unbranched, branched, and sometimes cupped. In some species of *Cladonia*, the podetia bear an ascocarp at the tip of the stalk-like podetia.

2.5 Lichen Substrates

Lichens are cosmopolitan organisms that can be found in almost all parts of the terrestrial world, although a few aquatic lichens are known. Lichens grow in a wide range of ecosystems, from the ice-free polar regions to the tropics, from tropical rainforests to desert ecosystems. Although some species can grow in a wide range of habitats, others require specific microclimate conditions of water, air, humidity, shade, light, substrate, or nutrient enrichment to grow. Furthermore, the substrates on which lichens grow vary from natural elements (bark, wood, soil, rock, bone) to man-made materials (asphalt, walls, gravestones, roofs, metal). Lichen substrates are the foundation of lichen communities, allowing these remarkable organisms to thrive in different habitats. Understanding the different types of lichen substrates and their importance provides us with valuable insights into ecosystem health, climate change, and environmental monitoring. Based on their substrate, lichens can be classified as saxicolous (on rock), corticolous (on tree trunk), lignicolous (on dead wood), muscicolous (on Moss), terricolous (on soil), and foliicolous (on leaf).

Lichens that grow on the surface of rocks are called saxicolous lichens. The type of rock can be different from one species of lichen to another. Some can grow on siliceous (acid) rock, granite rock, limestone rock, or others. Examples of saxicolous lichens are *Aspicilia*, *Caloplecta*, *Cystocoleus*, *Lepraria*, and *Verrucaria*.

Corticolous is a lichen that grows on the surface of trees (bark). This lichen is restricted to tropical and subtropical areas. They grow on the surface of trees because the bark is moist and stable in terms of pH, texture, and water. They usually consist of crustose, fruticose, and foliose lichens. Examples of corticolous lichens are *Graphis* spp., *Evernia* spp., *Usnea* spp., and *Parmelia* spp. Whereas lignicolous lichen grows on wood from which the bark has been removed. An example of a lignicolous lichen is *Athallia holocarpa*.

Terricolous is a lichen that grows on the surface of the soil. They include *Cladonia ciliata*, *Cladonia squamosa*, *Cladonia uncialis*, *Leptogium britannicum*, *Peltigera canina*, and *Peltigera didactila*. There are some lichens that sometimes grow on moss. An example of muscicolous lichen is *Parmelia saxatilis*.

Foliicolous lichens grow on the surface of living angiosperm leaves, usually on the upper (epiphyllous) surface, but sometimes on the lower (hypophyllous) surface. These lichens are often found below the rainforest canopy in tropical areas with high humidity, higher temperatures, and lower light intensity. Many foliicolous species are strictly foliicolous, but some can be found on other parts of plants (twigs, branches, trunks) or even on non-vegetative substrates such as rocks. Examples of foliicolous lichens are *Porina*, *Byssoloma*, and *Strigula*.

References

1 Ahmadjian, V. (1967). *The Lichen Symbiosis*. United States of America: Blaisdell Publishing Company.

2 Ahmadjian, V. (1973). *The Lichens*. New York: Academic Press.

3 Paracer, S. and Vernon, A. (2000). *Symbiosis: An Introduction to Biological Association*, 2e. New York: Oxford University Press.

4 Kaasalainen, U., Fewer, D.P., Jokela, J. et al. (2012). Cyanobacteria produce a high variety of hepatotoxic peptides in lichen symbiosis. *Proceedings of the National Academy of Sciences* 109 (15): 5886–5891. https://doi.org/10.1073/pnas.1200279109.

5 Büdel, B. and Scheidegger, C. (2008). Thallus morphology and anatomy. In: *Lichen Biology*, 2e (ed. T.H. Nash III), 40–68. Cambridge: Cambridge University Press. https://doi.org/10.1017/CBO9780511790478.005.

6 Smith, C.W., Aptroot, A., Coppins, B.J. et al. (2009). *The Lichens of Great Britain and Ireland*, 1e, 1–1046. London, UK: British Lichen Society.

7 Brodo, I., Sharnoff, S., and Sharnoff, S. (2002). Lichens of North America. *Blue Jay* 60 (1). https://doi.org/10.29173/bluejay5827.

8 Nash, T.H. (2012). *Lichen Biology*. Cambridge University Press. https://doi.org/10.1017/CBO9780511790478.

9 Honegger, R. and Scherrer, S. (2008). Sexual reproduction in lichen-forming ascomycetes. In: *Lichen Biology*, 2e (ed. T.H. Nash III), 94–103. Cambridge: Cambridge University Press. https://doi.org/10.1017/CBO9780511790478.007.

3

Taxonomy of Lichen

Darshita Sinha[1], Munmi Borkataky[1], Bhaben Chowardhara[2], and Ratul Nath[1]

[1] *Department of Life Sciences, Dibrugarh University, Dibrugarh, Assam, India*
[2] *Department of Botany, Arunachal University of Studies, Namsai, India*

3.1 Introduction

Lichens are fascinating organisms that evolved in a manner distinct from that of any other plant group. Their evolutionary history is amazing, having no counterpart in the universe of living beings. They are organisms formed by the symbiotic relationship between a fungus (mycobiont) and an alga (photobiont) or cyanobacterium [1]. They are distinct in terms of structure, ecology, and categorization. Lichens are classified within the broader field of mycology (the study of fungi) and phycology (the study of algae) [2]. Therefore, they have a double affinity, being closely related to the fungi on the one hand and algae on the other [3]. Taxonomy is the study of categorizing and identifying organisms, and it aids in the knowledge of species variety and interactions [4]. The taxonomy of lichen is influenced mainly by its fungal companion [5, 6]. The taxonomy of lichens considers several characteristics for classification, including thallus (the body of the lichen), chemistry, ecological adaptations, reproductive structures, and various other factors. The thallus, which can vary in form, color, and texture, is crucial for lichen identification and classification. Overall, when it comes to the identification of lichen, the role of their reproductive strategies, morphological characteristics, and genetic relationships are to be considered [7–9]. Lichens were categorized according to their external characteristics into fruticose (shrubby or bushy) or squamulose (forming small scales), foliose (leafy or lobed), and crustaceous groups (forming crusts on the substrate) when they were recognized as distinct from algae or fungi [10]. Reproductive structures in lichens are also considered crucial for their classification as they are found to reproduce through specialized reproductive structures called isidia, soredia, and apothecia. Isidia are finger-like projections, whereas soredia are tiny collections of fungal cells that also contain algal cells. Apothecia are cup-shaped structures that produce pores. The presence and characteristics of these reproductive structures aid in identifying and classifying lichens [9]. Chemical composition plays a significant role in lichen taxonomy as well. This unique organism secretes a large number of chemical substances known

Chemistry, Biology and Pharmacology of Lichen, First Edition. Edited by Ashoke Kumar Das, Ajay Sharma, Deepika Kathuria, Mohammad Javed Ansari, and Garima Bhardwaj.

as secondary metabolites or lichen acids, the majority of which are found in the cortex and form unique crystals on the surface of the fungal hyphae [11]. These chemical compounds serve ecological roles, such as protecting against environmental stresses, and can be employed as diagnostic characteristics for identification and categorization [12]. In recent years, advances in molecular techniques have revolutionized lichen taxonomy. DNA sequencing and phylogenetic analysis have highly contributed to understanding the evolutionary relationships and genetic diversity of lichens. These molecular technologies have aided in the refinement of lichen taxonomy, notably in the resolution of complicated species complexes and the identification of cryptic species [13]. Overall, the taxonomy of lichens is a dynamic and evolving field, constantly expanding our knowledge of this remarkable symbiotic organism. Through careful observation of morphological characteristics, reproductive structures, chemical composition, and genetic analysis, scientists continue to unravel the diversity and relationship within the lichen kingdom, shedding light on their ecological roles and evolutionary history.

3.2 Identification of Lichen

Lichens possess tremendous properties due to the presence of a variety of natural components, including primary and secondary metabolites, which are considered unique [14]. They have been used for a wide range of purposes throughout the world. Due to their complex structure, they are quite difficult to identify and consume a lot of time. Therefore, in order to benefit from these properties, proper identification of lichen is necessary. Lichens can only be explored if they are accurately, appropriately, simply, and promptly identified. Lichen identification involves a variety of techniques, and these techniques are always being improved to produce better results. In order to identify lichen, taxonomists used various methods, which have been used for so many years and have been modified to get more information. Primarily, three distinct methods are employed to identify lichen, including microscopic methods, chemical methods, and molecular methods [9, 10]. The best and quickest methods for accurately and quickly identifying lichens are those that are done using molecular approaches [9]. Moreover, all the useful information gathered during identification is combined and saved in a computer program in the form of a digital library, where it may be quickly accessed for use in the future [15]. This process is made possible by information technology and computer programs. This aids taxonomists in the fast identification, description, and categorization of lichens.

3.2.1 Microscopic Approaches

3.2.1.1 Morphology

The different morphological characteristics of lichens are identified under a dissection microscope or stereomicroscope These morphological characters can be divided into two categories – thallus characters and fruiting body characters. Under a dissection microscope, the upper and bottom surfaces of the lichen thallus are examined thoroughly while performing morphology. The thalli can initially be categorized based on their respective forms of growth (crustose, squamulose, foliose, leprose, fruticose, and dimorphic), shapes

Figure 3.1 Different types of lichen are identified based on the appearance of thallus. (a) https://britishlichen society.org.uk/resources/species-accounts/ricasolia-virens, Foliose lichen. *(Copywrite P. Whelan)*, (b) *Usnea articulata* | The British Lichen Society. Fruticose lichen. *(Copywrite N.A. Sanderson)*, (c) *Cladonia subcervicornis* | The British Lichen Society Squamulose thallus lichen. *(Copywrite N.A. Sanderson)*.

of thalli that are irregular or circular, the texture that is found to be rough, smooth, and warty, size, color, and the presence of finger-like projections (Figure 3.1). It's critical to pay attention to the marginal lobes' length, breadth, and branching pattern as well as any ciliated hairs [10].

It is necessary, however, to observe and study the fruiting bodies morphology separately. The shape that is sometimes rounded or elongated, its attachment sometimes stalked; size, texture, and color of the disc's edge; shape of the disc that may be concave or sometimes convex; whether the disc consists of powder are all noted in case of apothecia. Things that are to be observed in case of perithecia are its shape, color, size, and opening location (ostiole, apical, or lateral): solitary or clustered. Only foliose lichens have a bottom surface visible; crustose lichens do not. At the same time, dorsiventral differentiation is not shown by fruticose and dimorphic lichens. It is important to take notice of the color of the bottom surface, the existence of any pores, such as those seen in cyphellae and pseudocyphellae, the presence or absence of rhizines, which are a root-like structure, and their distribution, branching, and abundance. Some lichens exhibit fluorescence because of a chemical called lichexanthone, and to view this, lichens are studied within a closed UV chamber at 254 and 365 nm [10].

3.2.1.2 Anatomy

Under a compound microscope with a magnification range of 4–40×, the anatomy of the thallus of the lichen and its fruiting bodies is studied. The anatomy of the thallus is examined in order to identify the type of algae present and where it is distributed (stratified – heteromerous or uniform-homeomerous). In addition to the thickness of different layers that include the upper cortex, medulla, algal layer, and the lower cortex, the orientation of fungal hyphae that might be vertical or horizontal within the thallus is noted. Although microtome sections are helpful, the procedure is time consuming. It is not necessary to make a section of the thallus in order to identify the species of algae that are present in the thallus. Instead, the type of algae present within the thallus may be determined by its color. Grayish, yellowish, brownish, and greenish thallus lichens often contain green alga, whereas blackish, blue, and slate grey thallus lichens typically include blue-green algae. However, scraping the top cortex of the thallus with a blade may be carried out to determine the type of algae present. Particularly in the case of crustose lichens, the anatomical characteristics of fruiting bodies (ascocarp) are highly significant. Simple and septate spores, their color (hyaline or brown), size, form, number of spores in an ascus, the color of the exciple (ascocarp wall), the presence of algal cells and crystals in the wall, and the height and color of the various layers (hymenium, epihymenium, and subhymenium) within the ascocarp are to be studied for identification. Important characteristics include the paraphyce arrangement and branching pattern, as well as the apical cell's form and color. A stereo binocular microscope is used to examine the morphology. Using a compound microscope, the thallus anatomy and fruiting bodies can be examined in detail [9, 10].

3.2.2 Chemical Approaches

Around 600 secondary metabolites are produced by lichens that are referred to as lichen substances. Out of these, 550 substances are solely found in lichens and are not seen in any groups of plants. These substances are considered important for lichen identification (chemotaxonomy). By using a color spot test, high-performance liquid chromatography (HPLC), or thin layer chromatography (TLC), the lichen components can be detected [16].

3.2.2.1 Color Spot Test

Aqueous solutions of Potassium hydroxide (K), paraphenylenediamine (Pd), and bleaching powder or calcium hypochlorite (C) are the most common chemicals used for the color spot test. The K-test is carried out on the cortex of the thallus, by scrapping the surface with a blade. A drop of K solution is poured on the surface of the cortex, and the reaction of the color is noted. Similar methods are used for C and Pd tests. KC test can be carried out by first applying solution K, followed immediately by applying solution C over the drop of K solution. The presence of certain lichen compounds in the thallus like anthraquinones, parietin, melanin, etc., causes the color of the cortex or medulla to alter [9, 10, 17].

3.2.2.2 Thin Layer Chromatography (TLC)

Color spot tests can fail to identify numerous lichen compounds or provide inaccurate results. TLC must be carried out in these circumstances. In TLC, different lichen materials found in lichen thallus are separated and shown as spots on a plate. With the use of TLC guides, the patches are classified as various lichen compounds by noting their hue and determining the distance moved by them [10, 18, 19].

3.2.2.3 High Performance Liquid Chromatography (HPLC)

The isolation and identification of lichen-related compounds can be accomplished successfully using high-performance liquid chromatography (HPLC). This approach has the added benefit of providing quantitative data on the compounds in the whole lichen extracts. However, the procedure is to some extant expensive [10, 20].

3.2.3 Molecular Approaches

To obtain more valuable, dependable, and precise information, molecular techniques are the best ways to identify lichens since genes responsible for all physical characteristics and chemical compositions of organisms are researched in depth in these methods. Biological systematics has been increasingly reliant on DNA sequence analysis with the advent of DNA sequencing tools and the discovery of the polymerase chain reaction (PCR) for DNA amplification. Lichen samples are collected from the field. This can involve carefully removing a small portion of the lichen thallus or taking a small piece of the lichen's reproductive structures. The genetic material is then extracted from the collected samples using specialized laboratory techniques. This process aims to isolate the DNA of both the fungal and photosynthetic partners. The next step that includes is the PCR. In lichen studies, specific genetic markers, such as the Internal Transcribes Spacer (ITS), are commonly targeted for amplications. The amplified DNA fragments are then subjected to DNA sequencing, such as Sanger sequencing or Next Generation Sequencing (NGS) technologies like Illumina sequencing. The DNA sequence data obtained from the lichen samples are compared to existing databases, such as the International Nucleotide Sequence Database Collaboration (INSDC) or specialized lichen database. By comparing the obtained data with the known, researcher can identify the lichen species or determine its closest relatives. These data can finally be used to construct phylogenetic trees depicting the evolutionary relationships among different lichen species [9].

3.3 Nomenclature of Lichen

Nomenclature of lichen means the naming of lichen. Normally, lichens are named on the basis of fungal partners rather than the type of algal partner associated with them [21]. In a case study on lichen name practices among the Sherpa, Scots, Sami, and Okanagan ethnobotanics, there is enough evidence for comparing varied biological nomenclature and examining strategies to preserve alternative naming practices [22]. A grey nomenclature is one that uses names that are either intentionally or accidentally non-Linnaean. According to the "International Code of Nomenclature for Algae, Fungi, and Plants" (ICN), botanists have classified lichen thallus outside the Linnaean System [22]. Comparing the knowledge of academically formalized "grey names" and indigenous nomenclature in a way that retains their sources of particular informative content while exploring the history of naming of lichen and early ethnolichenological study, the British Lichen Society database is one of the most widely used databases for naming lichen [23].

A taxon name that is currently in use may not always be the most accurate in terms of taxonomy or adherence to the "code" (ICN for Algae, Fungi, and Plants, formerly International Code of Botanical Nomenclature) [24]. Recent research-based recommendations for adjustments are frequently not implemented right away. For instance, many of the evolutionary divides in the big genus *Verrucaria* have been adopted, but molecular methods still need to be used to study many of the species [25]. This does not mean that species remain to be studied molecular technique splits are not taxonomically meaningful. However, the changed names are given as synonyms. A few unpublished names are included and denoted by the suffixes "ad int" or "ined" for taxa recognized by recorders.

Nomenclature Synonyms that are sometimes referred to as homotypic synonyms are names that are two or more based on the same types, e.g., *Punctelia reddenda* (Stirt) Krog and *Parmelia reddenda* Stirt. Here, a species is transferred from one genus to another, and its species epithet is unchanged [26].

3.4 Classification of Lichen

When lichens were identified as different creatures from algae or fungi, they were categorized into three classes based on their exterior characteristics: fruticose, foliose, and crustaceous [27]. Some lichenologists classified some lichens as gelatinous lichens because their thallus became gelatinous when they were damp. This technique was helpful in the era before the development of the compound microscope and allowed for the differentiation of common lichens [28]. External and vegetative characters were once thought to be the most important in phanerogamic taxonomy, but complete emancipation from these views has now been achieved that internal and reproductive characters have been shown to be the most consistent, and their use is now held to be a leading principle in a natural classification [27]. However, in lichen taxonomy, complete emancipation from such ideas is far from achieved, though it has been seen to be partially accomplished. The International Code of Botanical Nomenclature (ICBN) stipulates that the algal component of lichen has no taxonomic significance. The fungi that they are classified with are the only factors. Mifller-Argau's classification suggested in 1862, as well as Reinke's significant contributions in 1894, made considerable headway towards the construction of a natural system of lichen categorization, but Zahlbruckner's system has been widely adopted by all British lichenologists [27]. It was based on all of the plant's dual characteristics, yet the spores and the thecia contained inside them have a dominant impact. Rather than the traditional classifications based on the thallus's fruticose, foliose, or crustaceous nature, a more accurate classification has progressively emerged, and Zahlbruckner included the majority of the

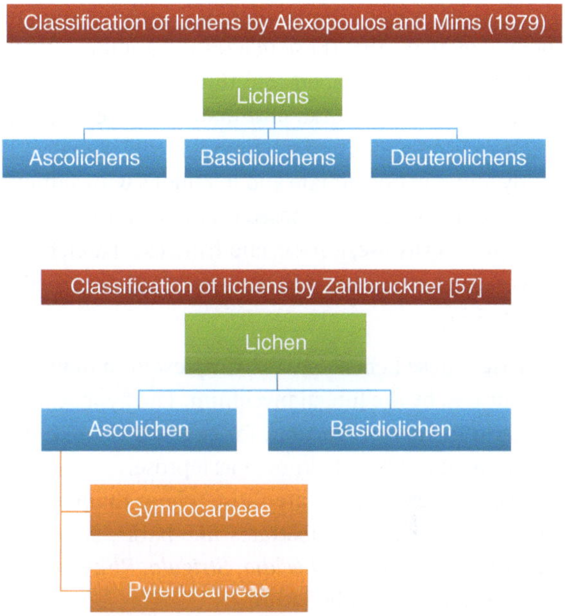

Figure 3.2 Classification of lichen by Alexopoulos and Mims [29] and Zahlbruckner.

advancements made in his scheme of taxonomy. Lichens were classified based on the kind and nature of fruiting bodies of the fungal partner. Zahlbruckner divided lichens into two classes based on the anatomy of their fruit bodies: Ascolichens and Basidiolichens (Figure 3.2). Ascolichens are lichens whose fungus members are members of the Ascomycotina. It includes the vast majority of lichens whose spores are produced in asci. He then classified Ascolichens into two groups based on the structure of their fruit bodies: Gymnocarpeae and Pyrenocarpeae. Gymnocarpeae is when the fruit body is an apothecium that is exposed at maturity. It is also known as Discolichen (for example, *Permelia*). In pyrenocarpeae, the fruit body is flask-shaped, indicating that it is perithecial, with a perithecium in which the hymenium is neither exposed nor communicates with the exterior by a small apical pore only. Pyrenolichen (e.g. *Dermatocharport*) is another name for this form of lichen. Basidiolichen's fungal members are Basidiomycotina, such as Dictyonema and Corella. Basidiolichens (Hymenolichens) are a subclass of lichens in which the spores are carried by basidia. The first category is vast, and because it comprises lichens with three different forms of apothecia, these three groups Graphidineae, Coniocarpineae, and Cyclocarpineae are recognized. The apothecia in the Ascolichens are frequently stalked, and when mature, the spores leave the asci and form a sporal mass (mazedium) on the surface. In the second, the apothecia are elongated or have a slit-like disc. At maturity, the apothecium in Cyclocarpineae has an open orbicular disc [27].

Later, lichens are classified and categorized into three groups by Alexopoulos and Mims [29] (Figure 3.2). They are Ascolichens, Basidiolichens, and Deuterolichens. Ascolichens, where the fungus partner is of the Ascomycetes class and the majority of the lichen is discovered to belong within this class, Basidiolichens are an additional group and the fungus in this instance belongs to the class Basidiomycetes and produce basidia and basidiospores during reproduction. There are very few lichens that fit to this description.

His final grouping, Deuterolichens, includes the fungus companion Deuteromycetes. There is evidence that these lichens cannot reproduce sexually.

The lichens are grouped into three types depending on the growing region: Corticoles, which grow on tree barks and mainly in the tropical and subtropical regions; Saxicoles, which grow on a rock and are found in cold climates; Terricoles are those that grow on soil, mostly found in hot climate and in regions with sufficient rain and dry summer. Lichens are classified into three types based on their exterior appearance, general growth, and mode of attachment: crustose, foliose, and fruticose. Later, based on detailed structures lichens were categorized by Hawksworth and Hill [30] classified them into five main types or forms (Figure 3.3):

Leprose: These lichens are the simplest kind in which a single or tiny cluster of algal cells is encased by the fungal mycelium. The hyphae of a fungus do not completely round the algal cell. *Lepraria incana*, for example, is a kind of lichen that appears as a powdery mass on the substrate known as leprose.

Crustose: These lichens are those whose thallus is discrete and flat and appear as a thin coating or crust on substrates like barks, stones, boulders, etc. For example, *Graphis, Lecanora, Verrucaria, Lecidia, Strigula, Rhizocarpon,* etc., are either entirely or partially embedded in the substratum.

Foliose: Lichens of this type are leaf-shaped, have a flat thallus, are horizontally spread, and have lobes. As a result of hyphal outgrowth, some parts of the thallus are attached to the substratum, and rhizines are often developed from the lower surface, e.g., *Parmelia, Anaptychia, Physcia, Gyrophora, Hypogymnia, Collema, Peltigera*, etc.

Fruticose (Frutex, Shrub): These lichens are shrubby, where the thalli are found to be well developed, cylindrically branched, either hang from the substratum (Usnea) or grow erect (Cladonia). They are attached to the substratum through a basal disc, e.g. *Cladonia, Usnea*, etc.

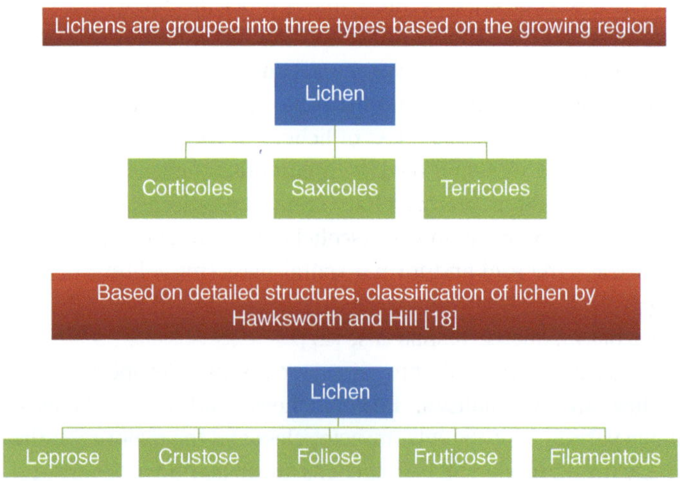

Figure 3.3 Classification of lichens based on growing regions, detailed structure by Hawksworth and Hill [30].

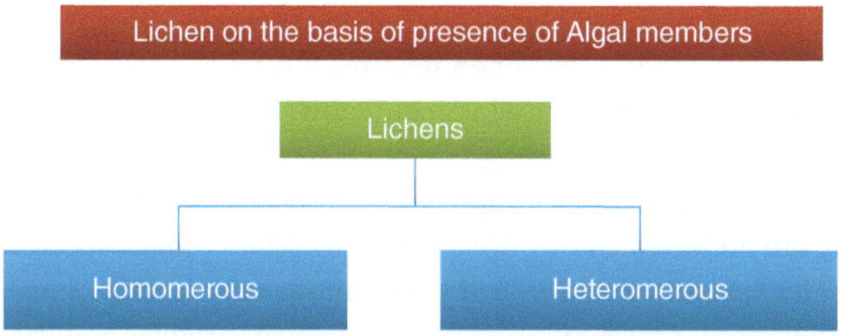

Figure 3.4 Classification of lichens based on the presence of Algal members.

Filamentous: Algal members of this kind are filamentous and well-developed. Only a few fungal hyphae remain to sheath or cover the algal filaments. Here, the dominant partner is still an algal member of the filamentous type, such as *Ephebe, Cystocoleus, Racodium,* etc.

Lichens can be divided into two types based on how the algal members are distributed inside the thallus: Homoisomerous and the other is Heteromerous (Figure 3.4).

Homoisomerous: The thallus is evenly distributed with both fungal hyphae and algal cells. Algal members are categorized as cyanophyta. Crustose lichens exhibit orientation like this. A thin layer of protection is formed by both partners intermingling. E.g.: *Leptogium, Collema* etc.

Heteromerous: There are four distinct layers in the thallus: upper cortex, algal zone, medulla, and lower cortex. Members of the algal zone are the only members of the algal zone. This type of orientation is found to be seen in foliose and fruticose. E.g. *Physcia, Parmelia* etc.

3.5 Phylogeny of Lichen

Phylogeny is the study of the connections and evolutionary history of species [31]. Understanding the evolutionary origins, diversity, and connections of lichen-forming fungus and their related photosynthetic partners (algae or cyanobacteria) is the goal of the study of lichen phylogeny [32]. Our knowledge of the evolutionary connections among lichens has improved greatly over time due to developments in molecular methods like DNA sequencing. The following are important aspects of the phylogeny of lichens:

Dual Nature of Lichens: Lichens are unique organisms formed through a symbiotic relationship between a fungus (the mycobiont) and a photosynthetic partner (the photobiont). The photobiont can be a green alga or a cyanobacterium. Lichenization represents an evolutionary adaptation in which these two partners establish a mutualistic association [33].

Fungal Diversity: Lichen-forming fungi belong to different fungal groups, including Ascomycota, Basidiomycota, and occasionally Zygomycota. The fungal phylum Ascomycetes is most frequently connected to lichens. According to molecular research, lichen-forming fungus frequently belongs to certain lineages within these fungal families, suggesting a common evolutionary past [34].

Phylogenetic Patterns: Studies of molecular phylogeny have shed light on the connections between various lichen species, genera, and families. The evolutionary links within and between lichen-forming fungus taxa have been clarified by these investigations. Instances of parallel evolution and convergence in lichen features have also been found, indicating that certain ecological adaptations may have independently developed in several lichen lineages [34].

Co-evolution and Selectivity: Lichens exhibit a high level of specialization in their fungal-algal relationships. There are preferred connections between certain fungus species and particular algae partners. The specificity and selectivity of these relationships have likely developed through co-evolutionary processes, driven by ecological variables and the compatibility of the partners, as shown by molecular research [35].

Cryptic Diversity: Molecular methods have revealed lichen diversity that was concealed or cryptic. Previously thought to be a single species, morphologically related lichens have been found to be made up of several genetically different lineages. These findings underline the significance of molecular methods in revealing hitherto unknown variety and improving our comprehension of lichen taxonomy and phylogeny [36].

Accurate species identification, comprehension of their ecological functions, and research into their evolutionary adaptations all depend on knowledge of the phylogeny of lichens. As a result of ongoing studies using molecular techniques, classifications of lichens are being revised, and our knowledge of their evolutionary past is being expanded. These studies also reveal new insights into the phylogenetic connections among lichens.

3.6 Molecular Taxonomy of Lichens

A subfield of taxonomy called molecular taxonomy uses molecular information to categorize and identify different kinds of organisms [37]. The symbiotic relationship between a fungus and one or more photosynthetic partners, such as algae or cyanobacteria, results in the formation of lichens, which are distinct species [38]. Lichen diversity, evolutionary relationships, and classification have been greatly enriched by molecular taxonomy.

Traditionally, lichens have been categorized according to their morphological traits, such as growth shape, color, and structure [10]. However, by exposing hidden diversity and addressing challenging taxonomic problems, genetic approaches have fundamentally changed our understanding of lichen taxonomy. DNA sequencing and phylogenetic analysis are the two primary molecular methods utilized in lichen classification.

Finding the nucleotide sequence of particular parts of the lichen's genome, such as the internal transcribed spacer (ITS) region of the ribosomal DNA, is known as DNA sequencing [39]. Due to its great variability, this area has been extensively employed in lichens as a DNA barcode for species identification and phylogenetic research. Researchers may identify species, ascertain their genetic links, and find obscure or novel species by comparing the ITS sequences of various lichen specimens.

An evolutionary tree, or phylogeny, is constructed based on molecular data in phylogenetic analysis. These trees show the evolutionary connections between various lichen species and shed light on their diversification and evolutionary history [40]. A variety of techniques, including maximum likelihood and Bayesian inference, as well as several molecular markers, including ITS, rpb1, and mitochondrial small subunit (mtSSU), among others, can be used to undertake phylogenetic research [41].

Through molecular taxonomy, several important findings have been made in the study of lichens. Firstly, molecular data have revealed that lichens are not a natural group but rather a convergence of multiple fungal lineages with different photosynthetic partners. This discovery has led to the reevaluation and reclassification of many lichen taxa based on their molecular phylogenies.

Furthermore, molecular studies have unveiled the presence of cryptic species within traditionally recognized lichen species. Cryptic species are morphologically similar but genetically distinct species that can only be differentiated using molecular techniques. This hidden diversity has significantly increased the known species richness of lichens and highlighted the importance of molecular approaches in their taxonomy [36]. Molecular taxonomy has also facilitated the identification of lichen-forming fungi that lack a known algal or cyanobacterial partner. These fungi, known as "lichenicolous" fungi, live on or within lichens, utilizing them as a substrate [42]. Molecular techniques have helped in elucidating the phylogenetic relationships of these fungi and their associations with specific lichen hosts. In addition to species identification and phylogenetic analysis, molecular taxonomy has been instrumental in studying lichen symbiosis. By comparing the genetic diversity and evolutionary patterns of the fungal and algal partners within lichens, researchers can gain insights into the coevolutionary processes and ecological interactions between these organisms.

In conclusion, molecular taxonomy has revolutionized the field of lichenology by providing powerful tools for species identification, resolving taxonomic complexities, and unraveling the evolutionary relationships of lichens. Through DNA sequencing and phylogenetic analysis, researchers have gained a deeper understanding of lichen diversity, including the discovery of cryptic species and the identification of lichenicolous fungi. Molecular approaches continue to contribute significantly to our knowledge of lichen taxonomy and are likely to uncover further insights into the intricate world of lichen symbiosis and evolution.

3.7 Therapeutic and Commercial Values of Lichen

Lichens have been utilized for their therapeutical and commercial value in various cultures throughout history [43]. Here is a list of some lichens that are known for their medicinal and economic value (Table 3.1). It's important to note that while lichens have been used traditionally for their medicinal properties, their efficacy and safety should be evaluated through rigorous scientific research and consultation with healthcare professionals. Additionally, the economic importance of lichens extends beyond their medicinal uses, as some species are collected and traded for ornamental, dyeing, and food purposes [58].

3.8 Conclusion

Lichens are captivating and complex organisms that have long intrigued scientists and nature enthusiasts alike. Combining the unique characteristics of both fungi and algae or cyanobacteria, they form a mutually beneficial symbiotic relationship, resulting in their distinct and resilient structures. Through their taxonomic classification, scientists have

Table 3.1 List of some common lichens and their importance.

Serial number	Name of lichens	Importance
1.	*Usnea* spp.	Perfumes and traditional crafts, antimicrobial properties, respiratory health benefits, immune system support, and anti-inflammatory effects [44]
2.	*Cetraria islandica*	Source of food, dietary supplements, and natural dyes [45] used for respiratory conditions, digestive issues, and as a demulcent and antimicrobial agent [46, 47]
3.	*Cladonia rangiferina*	Forage resource for reindeer and other grazing animals [48]; used in respiratory ailments, gastrointestinal disorders, and as a wound dressing agent [46]
4.	*Evernia prunastri*	Natural fixative and scent enhancer [49]; antimicrobial and anti-inflammatory properties
5.	*Pseudotulostoma volvata*	*Pseudotulostoma volvata*, a species of truffle, has limited economic importance. It is not commonly cultivated for culinary purposes due to its small size and rarity [50]
6.	*Lobaria pulmonaria*	Medicinally, it has been used in traditional herbal medicine for respiratory ailments, but scientific evidence is lacking [46]
7.	*Parmelia perlata*	Medicinally, it has been used in traditional herbal medicine for its potential antimicrobial and anti-inflammatory properties [51]
8.	*Ramalina* spp.	Medicinally, some species have been used in traditional herbal medicine for respiratory conditions, digestive issues, and wound healing [43]
9.	*Cetraria cucullata*	Medicinally, it has been used in traditional herbal medicine for respiratory ailments and as a demulcent [43]
10.	*Cladonia* spp.	Some species, like *Cladonia rangiferina* (reindeer lichen), serve as a forage resource for grazing animals [52] Medicinally, they have been used in traditional medicine for respiratory and digestive issues [46]
11.	*Peltigera* spp.	Medicinally, certain species have been used in traditional herbal medicine for their potential antimicrobial and anti-inflammatory properties [53]
12.	*Umbilicaria* spp.	In some cultures, they have been used as a food source during times of scarcity [54]. Medicinally, certain *Umbilicaria* species have been used in traditional herbal medicine for their potential antibacterial and antifungal properties [55], but further research is needed to validate their medicinal benefits
13.	*Sticta* spp.	Medicinally, certain *Sticta* species have been used in traditional herbal medicine for respiratory conditions, but scientific evidence is limited, necessitating further research to validate their medicinal properties [43]
14.	*Hypogymnia* spp.	Medicinally, some species have been used in traditional herbal medicine for their potential antimicrobial and anti-inflammatory properties, but further research is needed to validate their medicinal benefits [56]

Table 3.1 (Continued)

Serial number	Name of lichens	Importance
15.	*Physcia* spp.	Medicinally, some *Physcia* species have been used in traditional herbal medicine for their potential antimicrobial and anti-inflammatory properties, but scientific evidence is limited, necessitating further research to validate their medicinal properties [55]
16.	*Xanthoparmelia* spp.	Medicinally, certain *Xanthoparmelia* species (commonly known as **green rock shields** or **rock-shield lichens**) have been used in traditional herbal medicine for their potential aphrodisiac and antimicrobial properties, but scientific evidence is limited, requiring further research to validate their medicinal benefits [46]
17.	*Letharia vulpina*	It is toxic and not suitable for commercial use. Medicinally, it has been historically used by indigenous cultures for its potential antimicrobial properties, but due to its toxicity, it is not commonly used in modern herbal medicine [57]

gained valuable insights into their diversity, ecological roles, and potential applications. Furthermore, ongoing research on lichens promises exciting discoveries in various fields, from medicine to environmental monitoring and beyond. Taxonomically, lichens belong to the Kingdom of Fungi, classifying them as a distinct group within this diverse realm of organisms. The taxonomy of lichens is challenging due to their dual nature, as they are composed of two or more different entities living together in a complex relationship. Traditionally, lichens were classified primarily based on their morphological characteristics, such as growth form, color, and reproductive structures. However, with advancements in molecular techniques, researchers have been able to delve deeper into the genetic makeup of lichens, unraveling their evolutionary relationships and providing a more refined taxonomic framework.

The taxonomic studies have revealed a remarkable diversity within the lichenized fungi suggesting that there may be between 13,500 and 17,000 known species. The wide range of growth forms, from crustose and foliose to fruticose lichens, reflects their adaptability to different ecological niches. Lichens can be found thriving in diverse environments, ranging from deserts and arctic tundra to rainforests and rocky coastal regions. The ability of lichens to survive in such extreme conditions and their role as pioneer species in ecological succession make them invaluable indicators of environmental health and climate change.

Further research on lichens holds great promise in various fields. In medicine, lichens have demonstrated their potential as a source of bioactive compounds with antimicrobial, antioxidant, and anticancer properties. For instance, lichen compounds such as usnic acid have shown promising results in inhibiting the growth of pathogenic bacteria, including drug-resistant strains. Exploring the untapped potential of lichen metabolites could lead to the development of novel therapeutic agents or natural alternatives to conventional antibiotics. Lichens also offer unique opportunities for environmental monitoring and biomonitoring. Due to their sensitivity to air pollution and other environmental stressors, lichens have been widely used as indicators of air quality and ecosystem health. By analyzing the presence or absence of certain lichen species in a given area, researchers can assess the impact of pollution and monitor the effectiveness of conservation efforts. Ongoing research focuses on developing lichen-based biomonitoring programs that can provide real-time data on air quality and ecosystem changes, aiding in the development of more effective environmental policies. Moreover, lichens have caught the attention of researchers in the field of astrobiology. Their ability to survive in extreme environments and their reliance on simple nutrients and water sources make them potential candidates for supporting life in harsh extraterrestrial environments. Studying the adaptations and mechanisms that allow lichens to thrive in such conditions could provide valuable insights for future space exploration missions and the search for life beyond Earth.

In conclusion, further research on lichens is crucial for advancing our knowledge of their taxonomy, diversity, ecological roles, and potential applications. Through taxonomic studies, exploration of their chemical properties, and investigations into their ecological interactions, researchers can unlock the vast potential that lichens hold.

Authors' Conflict

No conflict of authors.

Authors' Contribution

DS prepared the manuscript, MB, BC, and RN have prepared the basic structure and reviewed the manuscript. Communication of the manuscript done by RN.

References

1. Wang, Y.Y., Liu, B., Zhang, X.Y. et al. (2014). Genome characteristics reveal the impact of lichenization on lichen-forming fungus *Endocarpon pusillum* Hedwig (Verrucariales Ascomycota). *BMC Genomics* 15: 1–18.
2. Hawksworth, D.L. (1988). The variety of fungal-algal symbioses their evolutionary significance and the nature of lichens. *Botanical Journal of the Linnean Society* 96 (1): 3–20.
3. Smith, H.B.D., Grande, F., Muggia, L. et al. (2020). Metagenomic data reveal diverse fungal and algal communities associated with the lichen symbiosis. *Symbiosis* 82: 133–147.
4. Fuentes, D. and Fiore, N. (2013). A taxonomic service for species identification. *International Journal of Computers & Technology* 12 (5): 3496–3499.
5. Aschenbrenner, I.A., Cernava, T., Berg, G., and Grube, M. (2016). Understanding microbial multi-species symbioses. *Frontiers in Microbiology* 7: 180.
6. Nazem-Bokaee, H., Hom, E.F.Y., Warden, A.C. et al. (2021). Towards a systems biology approach to understanding the lichen symbiosis: opportunities and challenges of implementing network modelling. *Frontiers in Microbiology* 12: 1–12.

7 Crespo, A., Kauff, F., Divakar, P.K. et al. (2010). Phylogenetic generic classification of parmelioid lichens (Parmeliaceae, Ascomycota) based on molecular morphological and chemical evidence. *Taxon* 59 (6): 1735–1753.

8 Matos, P., Pinho, P., Aragon, G. et al. (2015). Lichen traits responding to aridity. *Journal of Ecology* 103 (2): 451–458.

9 Singh, S.K., Arya, M., and Vishwakarma, S.K. (2019). Advancements in methods used for identification of lichens. *International Journal of Current Microbiology and Applied Sciences* 8 (8): 1450–1460.

10 Nayaka S. 2005. Studying Lichens. Sahyadri E-News Western Ghats Biodiversity Information System -Issue XV. http://wgbis.ces.iisc.ernet.in/biodiversity/bio_iden/lichens.htm

11 Goga, M., Elečko, J., Marcinčinová, M. et al. (2020). Lichen metabolites: an overview of some secondary metabolites and their biological potential. In: *Co-Evolution of Secondary Metabolites. Reference Series in Phytochemistry* (ed. J.M. Mérillon and K. Ramawat), 175–209. Cham: Springer.

12 Shukla, V., Upreti, D. K., and Bajpai, R. (2014). *Lichens to Biomonitor the Environment*. Springer eBooks.

13 Divakar, P.K. and Crespo, A. (2015). Molecular phylogenetic and phylogenomic approaches in studies of Lichen systematics and evolution. In: *Springer eBooks*, 45–60.

14 Zhao, Y., Wang, M., and Xu, B. (2021). A comprehensive review on secondary metabolites and health-promoting effects of edible lichen. *Journal of Functional Foods* 80: 104283.

15 Mietzsch, E., Lumbsch, H.T., and Elix, J.A. (1993). A new computer program for the identification on lichen substances. *Mycotaxon (USA)* 47: 475–479.

16 Elečko, J., Vilková, M., Frenák, R. et al. (2022). A comparative study of isolated secondary metabolites from lichens and their antioxidative properties. *Plants* 11 (8): 1077.

17 Culberson, C.F. and Elix, J.A. (1989). *Lichen SubstancesMethods in Plant Biochemistry Plant Phenolics* (1): 509–535.

18 Arup, U., Ekman, S., Lindblom, L., and Mattsson, J.E. (1993). High performance thin layer chromatography (HPTLC) an improved technique for screening lichen substances. *The Lichenologist* 25 (1): 61–71.

19 Culberson, C.F. (1972). Improved conditions and new data for identification of lichen products by standardized thin-layer chromatographic method. *Journal of Chromatography A* 72 (1): 113–125.

20 Patnana, D.P., Biswal, R.P., Dandamudi, R.B., and Pandey, M. (2021). Simple HPLCDAD-based method for determination of ergosterol content in lichens and mushrooms. *Journal of Liquid Chromatography and Related Technologies* 44 (3–4): 229–234.

21 Honegger, R. (2006). Water relations in lichens. In: *Fungi in the Environment* (ed. G.M. Gadd, S.C. Watkinson, and P. Dyer), 185–200. Cambridge University Press.

22 Kendig, C. (2020). Ontology and values anchor indigenous and grey nomenclatures: a case study in lichen naming practices among the Samí Sherpa Scots and Okanagan. *Studies in History and Philosophy of Science Part C: Studies in History and Philosophy of Biological and Biomedical Sciences* 84: 101340.

23 Helms, G., Friedl, T., Rambold, G., and Mayrhofer, H. (2001). Identification of photobionts from the lichen family Physciaceae using algal-specific ITS rDNA sequencing. *The Lichenologist* 33 (1): 73–86.

24 Wright, S.D. and Gillman, L.N. (2022). Replacing current nomenclature with pre-existing indigenous names in algae fungi and plants. *Taxon* 71 (1): 6–10.
25 John, D.M., Whitton, B.A., and Brook, A.J. (ed.) (2002). *The Freshwater Algal Flora of the British Isles: An Identification Guide to Freshwater and Terrestrial Algae*. Cambridge University Press.
26 Hawksworth, D.L. (ed.) (2010). *Terms Used in Bionomenclature: The Naming of Organisms and Plant Communities: Including Terms Used in Botanical Cultivated Plant Phylogenetic Phytosociological Prokaryote (Bacteriological) Virus and Zoological Nomenclature*. GBIF.
27 Watson, W.H. (1929). The classification of lichens. *New Phytologist* 28 (1): 1–36.
28 Ertz, D., Sanderson, N., and Lebouvier, M. (2021). Thelopsis challenges the generic circumscription in the Gyalectaceae and brings new insights to the taxonomy of *Ramonia*. *The Lichenologist* 53 (1): 45–61.
29 Alexopoulos, C.J. and Mims, C.W. (1979). *Introductory Mycology*, 3e. New Delhi, India: Wiley Eastern Limited.
30 Hawksworth, D.L. and Hill, D.J. (1984). *The Lichen-forming Fungi*. Glasgow: Blackie.
31 Qian, H. and Jin, Y. (2021). Are phylogenies resolved at the genus level appropriate for studies on phylogenetic structure of species assemblages? *Plant Diversity* 43 (4): 255–263.
32 Lücking, R., Leavitt, S.D., and Hawksworth, D.L. (2021). Species in lichen-forming fungi: balancing between conceptual and practical considerations and between phenotype and phylogenomics. *Fungal Diversity* 109 (1): 99–154.
33 Song, H., Kim, K.T., Park, S.Y. et al. (2022). A comparative genomic analysis of lichen-forming fungi reveals new insights into fungal lifestyles. *Scientific Reports* 12 (1): 10724.
34 Spribille, T., Resl, P., Stanton, D.E., and Tagirdzhanova, G. (2022). Evolutionary biology of lichen symbioses. *New Phytologist* 234 (5): 1566–1582.
35 Steinová, J., Škaloud, P., Yahr, R. et al. (2019). Reproductive and dispersal strategies shape the diversity of mycobiont-photobiont association in *Cladonia* lichens. *Molecular Phylogenetics and Evolution* 134: 226–237.
36 Fjelde M. O. (2021). Cryptic diversity in a seemingly paraphyletic lichen genus: a molecular phylogenetic study of *Calvitimela*. Master thesis. University of Oslo.
37 Singh, A.K. (2012). Molecular taxonomy: use of modern methods in the identification of a species. *Indian Journal of Life Sciences* 2 (1): 128–133.
38 Lutzoni, F. and Miadlikowska, J. (2009). Lichens. *Current Biology* 19 (13): R502–R503.
39 Coughlan, P., Carolan, J.C., Hook, I.L. et al. (2020). Phylogenetics of Taxus using the internal transcribed spacers of nuclear ribosomal DNA and plastid trnL-F regions. *Horticulturae* 6 (1): 19.
40 Veas-Mattheos, K., Almendras, K., Pezoa, M. et al. (2023). High Andean steppes of Southern Chile contain little-explored *Peltigera* lichen symbionts. *Journal of Fungi* 9 (3): 372.
41 Zhang, Y., Clancy, J., Jensen, J. et al. (2022). Providing scale to a known taxonomic unknown—at least a 70-fold increase in species diversity in a cosmopolitan nominal taxon of lichen-forming fungi. *Journal of Fungi* 8 (5): 490.
42 Moya, P., Molins, A., Chiva, S. et al. (2020). Symbiotic microalgal diversity within lichenicolous lichens and crustose hosts on Iberian Peninsula gypsum biocrusts. *Scientific Reports* 10 (1): 1–14.

43 Crawford, S.D. (2015). Lichens used in traditional medicine. In: *Lichen Secondary Metabolites* (ed. B. Ranković), 27–80. Cham: Springer.

44 Hoffmann, D. (2017). *Easy Breathing: Natural Treatments for Asthma Colds Flu Coughs Allergies and Sinusitis*. UK: Hachette.

45 Svanberg, I. and Ægisson, S. (2012). Edible wild plant use in the Faroe Islands and Iceland. *Acta Societatis Botanicorum Poloniae* 81 (4): 233–238.

46 Malhotra, S., Subban, R.A.V.I., and Singh, A. (2008). Lichens-role in traditional medicine and drug discovery. *The Internet Journal of Alternative Medicine* 5 (2): 1–5.

47 El-Darier, S.M. and Nasser, A.W. (2021). Cetraria Islandica as a pulmonary cytoprotective and supportive herbal remedy for lung complications related to COVID-19. *Sch Int J Tradit Complement Med* 4 (10): 168–174.

48 Uboni, A., Åhman, B., and Moen, J. (2020). Can management buffer pasture loss and fragmentation for Sami reindeer herding in Sweden? *Pastoralism* 10: 1–13.

49 Romagni, J.G. and Dayan, F.E. (2002). Structural diversity of lichen metabolites and their potential use. In: *Advances in Microbial Toxin Research and Its Biotechnological Exploitation* (ed. R.K. Upadhyay), 151–169. Boston, MA: Springer.

50 Spatafora, J.W., Aime, M.C., Grigoriev, I.V. et al. (2017). The fungal tree of life: from molecular systematics to genome-scale phylogenies. In: *The Fungal Kingdom*, 1–34. Wiley Online.

51 Akbar, S. (2020). *Dolichousnea longissima* (Ach) Articus/*Parmelia perlata* Esch. (Parmeliaceae). In: *Handbook of 200 Medicinal Plants*, 873–878. Cham: Springer.

52 Kumpula, J., Colpaert, A., and Nieminen, M. (2000). Condition potential recovery rate and productivity of lichen (*Cladonia* spp.) ranges in the Finnish reindeer management area. *Arctic* 53 (2): 152–160.

53 Studzińska-Sroka, E. and Dubino, A. (2018). Lichens as a source of chemical compounds with anti-inflammatory activity. *Herba Polonica* 64 (1): 56–64.

54 Adenubi, O.T., Famuyide, I.M., McGaw, L.J., and Eloff, J.N. (2022). Lichens: an update on their ethnopharmacological uses and potential as sources of drug leads. *Journal of Ethnopharmacology* 298: 115657.

55 Ranković, B., Mišić, M., and Sukdolak, S. (2007). Evaluation of antimicrobial activity of the lichens *Lasallia pustulata, Parmelia sulcata, Umbilicaria crustulosa*, and *Umbilicaria cylindrica*. *Microbiology* 76: 723–727.

56 Mitrović, T., Stamenković, S., Cvetković, V. et al. (2011). Antioxidant antimicrobial and antiproliferative activities of five lichen species. *International Journal of Molecular Sciences* 12 (8): 5428–5448.

57 Shrestha, G., Thompson, A., Robison, R., and St. Clair, L.L. (2016). *Letharia vulpine*, a vulpinic acid containing lichen targets cell membrane and cell division processes in methicillin-resistant *Staphylococcus aureus*. *Pharmaceutical Biology* 54 (3): 413–418.

58 Llano, G.A. (1948). Economic uses of lichens. *Economic Botany* 2 (1): 15–45.

4

Lichen as Habitats for Other Organisms

Ludmilla Fitri Untari

Department of Tropical Biology, The Faculty of Biology, Gadjah Mada University, Yogyakarta, Indonesia

4.1 Introduction

Lichens are one of the unsung heroes of the natural world. They are often mistaken for mosses or a type of plant. In fact, lichens are a symbiotic relationship between fungi and algae. They are found in almost every environment on Earth, from temperate forests to harsh deserts. Lichens are a complex symbiotic relationship often overlooked as a unique ecological niche that provides a habitat for other organisms to grow. The ability of lichens to adapt to almost all tropical and even polar terrestrial habitats makes it a successful symbiosis. This is the reason for its cosmopolitan distribution in many habitats.

Before a mutualistic symbiosis is formed, the fungus and the alga or cyanobacteria, which are the symbiotic constituents, can live freely. Thus, the process of lichen-forming involves a mechanism characterized by the photobiont growing in a relatively high-light environment [1]. The lichen association, or symbiosis, begins when the fungus grows between the algae, which then become parasitic on the substrate on which they are growing. This was a response to nutrient deprivation due to high competition at the time. As a result of the low nutrient uptake, the fungus has a tendency to surround the algae and is almost round in shape. This is followed by the intrusion of algal cells into the fungus. Almost all the algal cells that managed to get inside were killed (not accepted by the fungal cells), so initially, the algae–fungus relationship was indeed parasitic. Then, as the evolutionary process progressed, the relationship became mutualistic because of the mutual benefits [2].

4.2 Lichens as Habitat

4.2.1 Microorganism Living in Lichen

The fungus component of lichen forms the body and provides a protective shelter for the photosynthetic partner. The photosynthetic partner of lichen may be an alga, a cyanobacterium, or both working in unison. Together, the fungus and its photosynthetic partner

Chemistry, Biology and Pharmacology of Lichen, First Edition. Edited by Ashoke Kumar Das, Ajay Sharma, Deepika Kathuria, Mohammad Javed Ansari, and Garima Bhardwaj.

form an entirely new organism with unique characteristics and properties. Most lichens have layers comprising fungus cells, photosynthetic cells, and an intermediate gel-like layer that holds everything together.

Bacteria, yeast, and other microorganisms can also be found in lichens [3]. Some of these microbes produce antibiotics that protect the lichen from harmful pathogens. Living in a lichen requires unique adaptations. For example, some organisms have evolved specialized mouthparts that allow them to feed on lichen tissues without damaging the fungal partner.

4.2.1.1 Fungi

4.2.1.1.1 Lichen Mycobiont Mycobionts in lichens provide three important benefits. The first is that the mycobiont fungi control the lichen to take up and store nutrients from the environment, such as minerals and nitrogen compounds, which would otherwise be unavailable to the photobiont. This allows the mycobiont to produce secondary metabolites such as lichen acids and pigments. On the other hand, the mycobiont can provide a symbiotic niche for the photobiont to protect it from the harsh environment. The mycobiont fungi provide a protective layer against UV light, desiccation, and extreme temperatures through fungal filaments called hyphae. Thirdly, the survival and reproduction of the lichen depends on the persistence of the mycobiont. This makes the symbiotic relationship between the mycobiont fungi and the photobiont a very important part of the ecosystem as a whole.

The fungi have a unique way of acquiring their algal partners. However, the mycobiont's ability to lichenise a variety of photobionts allows it to colonize a wider range of ecological habitats [4]. Lichen fungi reproduce asexually and require a compatible photobiont for colonization. At this stage, a spore lands on a suitable substrate, germinates, and develops into a fungal thallus. The fungal thallus continues to grow mature. The mycobiont spreads out through the substrate and forms a network of hyphae around the photobiont. The fungal network creates a symbiotic niche for both organisms. In the lichen thallus aging process, the lichen thallus grows and thrives until it becomes overgrown and senescent. This phase marks the end of the lichen cycle.

There are three classes of mycobiont in lichen e.g. Ascomycetes, Basidiomycetes, and Lecanoromycetes. Ascomycetes are found in most lichens and are the largest group of mycobionts. They are known for their diverse morphological structures. This group plays an important role in nutrient uptake and storage in lichens. However, unlike Ascomycotina, Basidiomycotina need special conditions to form. They are more common in lichens that are found in temperate and tropical regions of the world. This group of fungi makes an important contribution to the structure and coloring of lichens. Lecanoromycetes are also remarkable for their ability to form a smooth, crusty surface on rocks and trees. Lecanoromycetes have a unique cell morphology. This allows lichens to survive in extreme environments. Several mycobionts have been identified in lichens, such as *Cyphellostereum jamesianum* Dal-Forno and Kaminsky, *Dictyonema darwinianum* Dal-Forno, Bungartz and Lücking, and *Dictyonema moorei* (Nyl.) Henssen [5].

4.2.1.1.2 Lichenicolous Fungi The lichenicolous fungi represent an important ecological group of species that form obligate associations with lichens. In nature, fungi are often associated with lichen thalli. However, these associations are often rather loose and unspecific.

By definition, these are fungi that grow exclusively on the thalli or ascomata of lichen that are obligate associated with lichen, either as saprotrophic colonists of dead lichen thalli, or as parasitic consumers of fixed carbon from living lichen hosts. Lichenicolous fungi are a highly specialized and successful group of organisms that develop on lichens and form three- or sometimes four- to-five-membered associations with them. However, lichenicolous fungi were poorly known until recently [6, 7].

Lichenicolous fungi are a diverse group of fungal species that coexist on lichens and play an important role in the formation of lichen communities. They produce specialized structures that attach to the surface of the lichen thallus and penetrate the fungal layer to provide additional nutrients. Lichenicolous fungi have unique morphological and physical characteristics that distinguish them from other fungi. They vary in color, shape, size, and texture. They can be cylindrical, flattened, or branched. They reproduce either sexually or asexually by the production of spores that can be dispersed by wind, rain, or animals.

Lichenicolous fungi are worldwide in distribution and have a strong affinity for lichens. They can be either host-specific, meaning that they grow on a particular lichen species, or generalists, able to grow on a variety of lichen species. They have a significant impact on lichen communities as they can be either symbiotic, providing benefits to lichens; parasitic, causing harm to their hosts; or somewhere in between. Some species help the lichen to grow faster and provide nutrients, while others consume the lichen's photobiont or invade its thallus and cause damage. There are also lichenicolous lichens that colonize a separate lichen host and maintain a separate photobiont. The lichen thallus structure and biodiversity of lichen communities can be significantly affected by these interactions. The relationship between lichens and lichenicolous fungi is fascinating as they coexist in complex ways [8] that are still poorly understood.

Lichenicolous fungi are not able to kill lichen thalli completely and are generally not restricted to a specific host species. Lichen destruction can result in the death of both, the host and the invading fungus. An example of a pathogenic lichenicolous fungus is *Athelia arachnoidea* (Berk.) Jülich, which causes massive brown lesions on colonies of protococcoid algae, *Lecanora conizaeoides* Nyl., *Physcia* (Schreb.) Michx., and *Xanthoria parietina* (L.) Th. Fr.

The classification of lichenicolous fungi is complex and is still evolving. They are currently divided into dozens of genera and over 700 known species belonging to several different taxonomic groups, including Ascomycota, Basidiomycota, and Zygomycota. Major genera include *Arthonia*, *Llimoniella*, and *Stigmidium*, which have unique characteristics that distinguish them from other lichenicolous fungi [6, 7]. More than 1800 species of lichenicolous fungus of the Ascomycota and Basidiomycota have been described, and it is estimated that 3000 species will later on be described [7]. Hawksworth (1983) published keys to 218 lichenicolous species known from the British Isles. These keys stimulated many lichenologists to study this fascinating group of fungi [9].

The first description of lichenicolous fungi is *Biatoropsis usnearum* Räsänen, a heterobasidiomycete (belong to the class of Tremellomycetes) that forms large gall-like structures on the thalli of *Usnea* spp. [10], as the gall-like structures were described by Acharius (1795) (Figure 4.1) [12]. *Biatoropsis usnearum* is very common and widely distributed on the thalli of *Usnea* spp. and *Protousnea* spp. Later, two other lichenicolous fungi were identified to occur on the thalli of *Usnea*, i.e. *Tremella santessonii* Diederi and *Tremella stevensiana* Diederic [11].

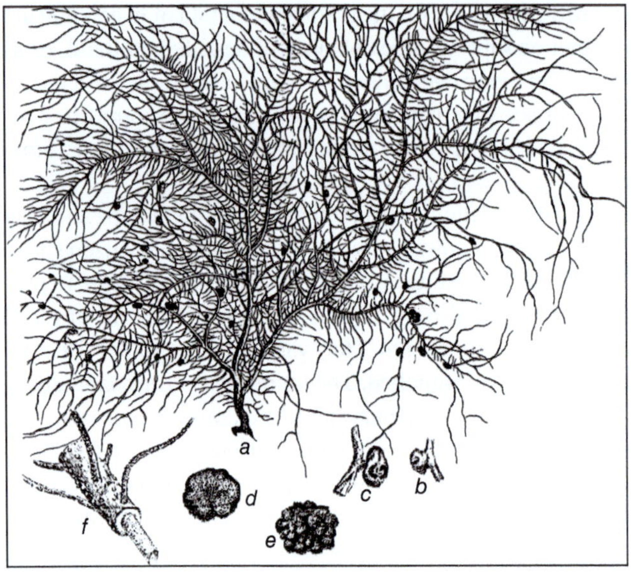

Figure 4.1 *Biatoropsis usnearum*, the lichenicolous fungi, a heterobasidiomycete forming large gall-like structures on thalli of *Usnea* spp (*Source:* Adapted from Diederich and Christiansen [11]).

Figure 4.2 *Abrothallus parmeliarum* (*Source:* Ave Suija and Hawksworth[13]/ with permission from John Wiley & Sons).

Another example of the lichenicolous fungus is *Abrothallus parmeliarum* (Sommerf.) Arnold on the thallus and apothecia of *Parmelia saxatilis* (L.) Ach. (Figure 4.2) [13]. The first scientific description of *A. parmeliarum* was in 1826 by the Norwegian botanist Søren Christian Sommerfelt under the scientific name *Lecidea parmeliarum* Sommerf. [14]. Later, in 2018, a proposal was submitted to conserve *L. parmeliarum* against *Endocarpon parasiticum* in order to maintain the use of the binomial *A. parmeliarum* [13].

As lichenicolous fungi are an important part of lichen communities, they play a variety of roles in shaping the ecology, morphology, and function of lichen communities. A better understanding of their taxonomy, biology, and ecology can therefore provide us with useful knowledge and insights into the complex relationships that exist in natural systems and their potential applications in various fields.

4.2.1.2 Lichen Photobiont

Lichen photobionts are the photosynthetic partners of lichens that form a symbiotic relationship with a fungus. Lichen photobionts play a vital role in the lichen symbiosis. They enable the lichen to survive in different habitats and contribute to its unique characteristics. The photobiont component of the lichen symbiosis is responsible for the synthesis of organic carbon and other substances essential for lichen growth, providing carbohydrates for the mycobiont and the primary color for the lichen. The role of the photobiont in the lichen symbiosis depends on the species of algae that make up the photobiont; the photobiont provides photosynthate to the fungal partner, which in turn provides a structured environment and nutrients essential for their continued coexistence. The photobiont partners will influence the growth and physiology of the lichen symbiotic structure. The growth and arrangement of the photobiont influence the physical structure of the lichen, affecting its morphology and overall appearance. In addition, the presence of photobionts provides the lichen with increased resistance to desiccation and UV radiation, allowing it to colonize harsh environments.

The most common photosynthetic partners of lichens are yellow-green algae (xanthophytes), green algae (phycobionts), and blue-green algae (cyanobionts). Green algae make up the vast majority of eukaryotic photobionts in lichens; they are best known for living in symbiosis with fungi in foliose lichens and best known for their ability to adapt to high levels of light intensity or UV radiation. Some cyanobacteria live in lichens as cyanobionts. Their contribution to the lichen symbiosis is mostly colonization of arid habitats. These blue-green algae can also contribute to fixing nitrogen from the air, improving soil fertility.

About 40 genera of green and blue-green algae have been reported as lichen photobionts. The most common photobionts are *Trebouxia*, *Trentepohlia*, and *Nostoc*. Lichen phycobionts that are eukaryotic algae and belong to the green algae (chlorophyta) or xanthophytes include *Dictyochloropsis*, *Stichococcus*, *Trebouxia*, *Trentepohlia*, *Coccomyxa*, *Pseudochlorella*, *Chlorella*, *Cephaleuros*, *Heterococcus*, *Asterochloris*, *Pleurococcus*, and *Botrydiopsis*. Whereas lichen cyanobionts are oxygenic photosynthetic blue-green algae (cyanobacteria), which include *Rhizonema*, *Nostoc*, *Anacystis*, *Scytonema*, *Stigonema*, *Calothrix*, *Gloeocapsa*, and *Chroococcum* [4, 15–18].

The mycobiont and photobiont interact in several ways to form a lichen thallus. The mycobiont helps to provide the structural basis for the thallus, regulates water and nutrient exchange, produces secondary metabolites such as lichen acids and pigments, and protects and nurtures the photobiont from predators and pests. The photobiont, on the other hand, will benefit by engaging in photosynthesis to produce glucose and other nutrients, fix nitrogen for use by the mycobiont, provide additional protection from UV light, and provide a carbon source for the mycobiont.

4.2.2 Insects and Other Invertebrates Living on Lichen

Insects, including beetles, moths, termites, barklice, and beetles, whose larvae feed on epiphytic lichen [19], including several species of termites of the genus *Hospitalitermes* [20]. They feed on lichen tissues and contribute to their decomposition. They collect lichen by scraping it into small balls and carrying it back to the nest. The lichens presumably provide a source of nitrogen, as analysis of the lichen balls has shown that they contain 10–60 times more nitrogen than wood [21].

Many small invertebrates, including species of bristletails (*Thysanura*), springtails (*Collembola*), termites (*Isoptera*), psocids or barklice (*Psocoptera*), grasshoppers (*Orthoptera*), snails and slugs (*Mollusca*), web-spinners (*Embioptera*), butterflies and moths (*Lepidoptera*), and mites (*Acari*) feed on lichens. They burrow into the lichen thallus, feed on fungi or algae, and reproduce in the protected microclimates of lichen habitats [22]. These include lichen grazing snails, i.e., *Balea perversa*, *Chondria clienta*, and *Helicigona lapicida*, which feed on crustose lichens such as *Aspicilia calcarea*, *Caloplaca flavovirescens*, *Lecanora muralis*, *Physcia adscendens*, *Tephromela atra*, and *X. parietina* [23]. In addition, a crab arthropod *Gecarcoidea natalis* has been recorded feeding on fruticose lichen *Ramalina dumeticola* [24]. In the intertidal area, lichens such as *Verrucaria* spp. are consumed by many marine species, including limpets (*Patella* spp.), *periwinkles*, barnacle snails (*Littorina*), and mussels [25].

In addition to the nutritional value of lichens, several invertebrate species benefit from being able to hide from predators by mimicking the thallus of a lichen as a camouflage, e.g. peppered moth (*Biston betularia*). Several invertebrates carry fragments of lichen as camouflage. In addition, to camouflage themselves from predators, some lacewing larvae cover themselves with moss or lichen fragments. Many other invertebrates modify their appearance to be inconspicuous when resting on lichen colonies. These include many invertebrates with coloration and patterns that enable them to blend seamlessly with their lichen habitats, protecting them from predators [26, 27].

Invertebrates can be picky eaters, with some showing a preference for certain species or parts. For example, the intertidal isopod *Campecopea hirsuta* only eats the blue-green algae cells of *Lichina pygmaea*. In addition, lichen-feeding psocids feed on *L. conizaeoides*, with one species grazing indiscriminately on one species and the other grazing only on the apothecia [28].

4.2.3 Vertebrates Using Lichen as a Shelter or Food

Lichens provide food, shelter, or nesting materials for wildlife. Some species of mammalian, such as reindeer (*Rangifer tarandus*) [29], Rocky Mountain elk (*Cervus elaphus*), white-tailed deer (*Odocoileus virginianus*), mule deer (*Odocoileus hemionus*) [30], moose (*Alces alces*) [31], caribou (*R. tarandus caribou*) [32], and mountain goats (*Oreamnos americanus*) [33], mule deer (*O. hemionus*), and white-tailed deer (*O. virginianus*) [30] feed on lichen for the food value or free water during winter season.

During the winter, in the absence of green vegetation, reindeer (*R. tarandus* L.) feed on lichens as part of their daily diet. These include *Alectoria sarmentosa, Bryoria* spp., *Cetraria islandica*, *Chlamydomonas nivalis*, *Carex ericetorum*, *Cladonia arbuscula*, *Cladonia stellaris*, *Centruroides gracilis*, *Lobaria pulmonaria*, *Peltigera aphthosa*, *Parmelia sulcate*, *Umbilicaria* spp., and *Stereocaulon paschale* [29].

Rocky Mountain elk (*C. elaphus nelsoni*), white-tailed deer (*O. virginianus*), and mule deer (*O. hemionus*) feed on *Bryoria* spp. and *Nodobryoria* spp. due to the availability of food during the winter season [30]. Aside of the lichen's food value, lichens are important during the winter season for the source of free water.

Lichens provide insulation or nesting material for some bird species. Lichens are light yet excellent insulators, keeping eggs and chicks warm even in harsh weather conditions. Lichens are also excellent nesting material, providing soft, comfortable bedding that birds can mold into the perfect shape for their nests. Lichens also have antibacterial properties, which help to keep the nest clean and prevent infections, and are a reliable source of food source for some bird species, especially during the winter months. Since many bird species are highly vulnerable during the nesting season, a reliable source of insulation and nesting material is essential for survival. Birds and lichens have a delicate relationship that has evolved over millions of years. By using lichens for insulation and nesting, many bird species have been able to thrive and survive in their environment.

Many bird species nest in the overlapping fronds of the lichens, including the American dipper (*Cinclus mexicanus*), hummingbirds (*Trochilidae*), and gnatcatchers (*Polioptila caerulea*) [34–36]. They collect foliose lichens such as *P. sulcate*, *Punctelia spp.*, or *Flavoparmelia caperata* as nesting material with their beaks. They therefore prefer epiphytic foliose lichens, whose fat lobes are loosely attached to the bark and easily picked up by beaks. However, the lichens can still grow for at least three years after being incorporated into a nest, suggesting that the birds may have a role in the spread of the lichens and the formation of a new colony.

4.3 Benefits of Living in Lichens

Lichens provide a sheltered environment that protects organisms from desiccation, UV radiation, and other environmental stresses. Lichens offer a rich source of nutrients, including amino acids and carbohydrates, that support a wide variety of organisms. Some lichen species have high concentrations of antioxidants, making them a valuable resource with medical potential. Some organisms that live in lichens can camouflage themselves to avoid predation. For example, some spiders blend in with the lichen thallus to avoid being detected by their prey.

However, lichen communities also benefit from lichen grazers. Firstly, the invertebrates contribute to the decomposition of the lichen, which helps to release and recycle nutrients within the ecosystem. In addition, some invertebrates, such as mites, assist in the dispersal of lichen propagules, allowing new areas to be colonized. Thirdly, certain insects can influence the composition and diversity of lichen communities by grazing on specific lichen species or by competing with other invertebrates.

4.4 Importance of Lichen for Biodiversity

Lichen photobionts contribute to nutrient cycling, carbon sequestration, and soil stabilization, making them essential players in ecosystem functioning. The diversity and distribution of lichen photobionts serve as indicators of environmental quality, allowing researchers to assess pollution levels and habitat suitability. Different lichen photobionts have specific habitat preferences and adaptations that allow them to thrive in specific ecological niches. In addition, cyanobionts play an important role in atmospheric nitrogen fixation, providing lichens and other organisms with usable nitrogen for their growth. Lichens control the humidity, moisture, and temperature of their habitat. Lichens also capture and slowly release nutrients from dew, fog, rain, and gases. Lichens are useful as indicators of climate and air quality. Finally, lichens provide food, cover, and shelter for insects, invertebrates, and mammals.

4.5 Threats to Lichen Habitats

Maintaining a diversity of lichens allows the ecological system to thrive and retain its biodiversity. Air pollution is not only harmful to human beings; it can also have a devastating effect on lichens. Reducing air pollution will help to maintain a healthy environment for lichens to continue to grow.

The invertebrates that live in lichen habitats are exposed to a variety of threats, including habitat loss, pollution, and climate change, which disrupt their delicate ecological balance. The conservation of lichen ecosystems is therefore essential not only for the conservation of invertebrates but also for the overall biodiversity and ecological stability of their habitats. Actions such as habitat restoration, monitoring, and public education are essential to protect these unique ecosystems and their inhabitants.

Exposure to heavy metals, acid rain, and other contaminants may disturb lichen communities; changes in temperature and precipitation patterns may alter the composition of lichen communities; logging, mining, and other forms of habitat destruction may remove lichen habitat and the organisms that depend on it.

4.6 Conclusion

Lichen habitats have a wide range of ecological benefits and are essential for the maintenance of biodiversity. Further research will provide a better understanding of the complex relationship between lichens and the organisms that depend on them. This will allow better conservation efforts and protection of these unique habitats.

References

1 Nash, T.H. (2012). *Lichen Biology*. Cambridge University Press. https://doi.org/10.1017/CBO9780511790478.

2 Ahmadjian, V. (1967). *The Lichen Symbiosis*. United States of America: Blaisdell Publishing Company.

3 Bates, S.T., Cropsey, G.W.G., Caporaso, J.G. et al. (2011). Bacterial communities associated with the lichen symbiosis. *Applied and Environmental Microbiology* 77 (4): 1309–1314. https://doi.org/10.1128/AEM.02257-10.

4 Tibell, L. (2001). Photobiont association and molecular phylogeny of the lichen genus chaenotheca. *The Bryologist* 104 (2): 191–198.

5 Masumoto, H. and Sanders, W.B. (2022). The lichen photobiont genus *Rhizonema* (Cyanobacteria) exhibits diverse modes of branching, both false and true. *Journal of Phycology* 58 (4): 612–625. https://doi.org/10.1111/jpy.13256.

6 Hawksworth, D.L. (1982). Secondary fungi in lichen symbioses: parasites, saprophytes and parasymbionts. *Journal of the Hattori Botanical Laboratory* 52: 357–366.

7 Lawrey, J.D. and Diederich, P. (2003). Lichenicolous fungi: interactions, evolution, and biodiversity. *Bryologist* 106 (1): 80–120.

8 Hawksworth, D.L. (1982). Co-evolution and the detection of ancestry in lichens. *Journal of the Hattori Botanical Laboratory* 52: 323–329.

9 Hawksworth, D.L. (1983). A key to the lichen-forming, parasitic, parasymbiotic and saprophytic fungi occurring on lichens in The British Isles. *The Lichenologist* 15 (1): 1–44.

10 Rasanen, V. (1934). Einige neue Flechtenarten und Flechtenparasiten. 4. *Biatoropsis usnearum* gen. et sp. n. *Annales Botanici Societatis Zoologicae-Botanicae Fennicae Vanamo* 5 (9): 8–10.

11 Diederich, P. and Christiansen, M.S. (1994). *Biatoropsis usnearum* rasanen, and other heterobasidiomycetes on usnea. *The Lichenologist* 26 (1): 47–66.

12 Acharius, E. (1795). Nya och mindre kända Svenska Laf-arter, beskrifne. *Kongl. Vetenskaps Academiens nya Handlingar, Ser. 2* 16 (1): 66–71.

13 Ave Suija, S.P.-O. and Hawksworth, D.L. (2018). The generic name *Abrothallus* (Abrothallales, Dothideomycetes), and names proposed in the genus by Giuseppe De Notaris, Søren Christian Sommerfelt, and Ignaz Kotte. *Taxon* 67 (6): 1169–1179. https://doi.org/10.12705/676.13.

14 Arnold, F. (1874). Flora, oder Allgemeine Botanische Zeitung herausgegeben von der Königl. Bayer. Botanischen Gesellschaft zu Regensburg. *Regensburg* 57 (7): 102. https://www.biodiversitylibrary.org/item/973.

15 Friedl, T. and Büdel, B. (2008). Photobionts. In: *Lichen Biology*, 2e (ed. T.H. Nash III), 9–26. Cambridge: Cambridge University Press. https://doi.org/10.1017/CBO97805 11790478.003.

16 De Carolis, R., Cometto, A., Moya, P. et al. (2022). Photobiont diversity in lichen symbioses from extreme environments. *Frontiers in Microbiology* 13: https://doi.org/10.3389/fmicb.2022.809804.

17 Ahmadjian, V. (1993). The lichen photobiont: what can it tell us about lichen systematics? *Bryologist* 96 (3): 310–313. https://doi.org/10.2307/3243858.

18 Kim, J.I., Kim, Y.J., Nam, S.W. et al. (2020). Taxonomic study of three new Antarctic *Asterochloris* (Trebouxio-phyceae) based on morphological and molecular data. *Algae* 35 (1): 17–32. https://doi.org/10.4490/algae.2020.35.2.23.

19 Gerson, U. (1973). Lichen-arthropod associations. *Lichenologist* 5 (5–6): 434–443. https://doi.org/10.1017/S0024282973000484.

20 Collins, N.M. (1979). Observations on the foraging activity of Hospitalitermes umbrinus (Haviland), (Isoptera: Termitidae) in the Gunong Mulu National Park, Sarawak. *Ecological Entomology* 4 (3): 231–238. https://doi.org/10.1111/j.1365-2311.1979.tb00580.x.

21 Jones, D.T. and Gathorne-Hardy, F. (1995). Foraging activity of the processional termiteHospitalitermes hospitalis (Termitidae: Nasutitermitinae) in the rain forest of Brunei, north-west Borneo. *Insectes Sociaux* 42 (4): 359–369. https://doi.org/10.1007/BF01242164.

22 Bokhorst, S., Asplund, J., Kardol, P., and Wardle, D.A. (2015). Lichen physiological traits and growth forms affect communities of associated invertebrates. *Ecology* 96 (9): 2394–2407. http://www.jstor.org/stable/24702345.

23 Hesbacher, S., Baur, B., Baur, A., and Proksch, P. (1995). Sequestration of lichen compounds by three species of terrestrial snails. *Journal of Chemical Ecology* 21 (2): 233–246. https://doi.org/10.1007/BF02036654.

24 McCarthy, P. (2001). Land crabs of Christmas Island eat lichen. *Australasian Lichenology* 48: 5. https://www.anbg.gov.au/abrs/lichenlist/AL_48.pdf.

25 Colman, J. (1940). On the faunas inhabiting intertidal seaweeds. *Journal of Marine Biological Association of the United Kingdom* 24 (1): 129–183. https://doi.org/10.1017/S0025315400054503.

26 Fang, H., Labandeira, C.C., Ma, Y. et al. (2020). Lichen mimesis in mid-Mesozoic lacewings. *Elife* 9: https://doi.org/10.7554/eLife.59007.

27 Gressitt, J.L., Sedlacek, J., and Szent-Ivany, J.J.H. (1965). Flora and fauna on backs of large papuan moss-forest weevils. *Science* 150 (3705): 1833–1835. https://doi.org/10.1126/science.150.3705.1833.

28 Wieser, W. (1963). Adaptations of two intertidal isopods II. Comparison between *Campecopea Hirsuta* and NAESA BIDENTATA (Sphaeromatidae). *Journal of the Marine Biological Association of the United Kingdom* 43 (1): 97–112. https://doi.org/10.1017/S0025315400005282.

29 Storeheier, P.V., Mathiesen, S.D., Tyler, N.J.C., and Olsen, M.A. (2002). Nutritive value of terricolous lichens for reindeer in winter. *Lichenologist* 34 (3): 247–257. https://doi.org/10.1006/lich.2002.0394.

30 Ward, R.L. and Les Marcum, C. (2005). Lichen litterfall consumption by wintering deer and elk in Western Montana. *Journal of Wildlife Management* 69 (3): 1081–1089.

31 Thomas, D.C. (1990). Moose diet and use of successional forests in the Canadian Taiga. *Alces* 26: 24–29.

32 Joly, K., Wasser, S.K., and Booth, R. (2015). Non-invasive assessment of the interrelationships of diet, pregnancy rate, group composition, and physiological and nutritional stress of barren-ground caribou in late winter. *PLoS One* 10 (6): e0127586. https://doi.org/10.1371/journal.pone.0127586.

33 Fox, J.L. and Smith, C.A. (1988). Winter mountain goat diets in Southeast Alaska. *Journal of Wildlife Management* 52 (2): 362–365. https://doi.org/10.2307/3801249.

34 Camacho-Alpízar, A., Eckersley, T., Lambert, C.T. et al. (2021). If it ain't broke don't fix it: breeding success affects nest-building decisions. *Behavioural Processes* 184: 104336. https://doi.org/10.1016/j.beproc.2021.104336.

35 Furuichi, S. and Kasuya, E. (2014). Costs, benefits, and plasticity of construction of nest defensive structures in paper wasps. *Behavioral Ecology Sociobiology* 68 (2): 215–221. https://doi.org/10.1007/s00265-013-1636-0.

36 Mulder, T., Campbell, C., and Ruxton, G. (2020). Evaluation of disruptive camouflage of avian cup-nests. *Ibis (Lond. 1859).* 163: https://doi.org/10.1111/ibi.12848.

5

Ecology of Lichen

Tridip Boruah, Krity Dulal, and Puja Namo Das

Plant Ecology Laboratory, P.G. Department of Botany, Madhab Choudhury College, Barpeta, Assam, India

5.1 Introduction

The diversity and distribution of lichen are based on climate and habitat conditions. Lichen responds according to changes in climatic conditions. Approximately the world's lichen flora comprises about 20,000 taxa, of which only 10% are found in India [1]. A wide distribution of lichen has been found in India stating about 80 families, which are further subcategorized into 234 genera after studying classified into 1850 species out of which macro lichen (Squamulose- foliose to fruticose forms) comprises 72 genera which are further categorized into 700 species while micro lichen (Crustose–Squamulose format) includes 163 genera and 1150 species whereas South Indian region exhibits the abundancy [2]. The South Indian region is leading in terms of lichen diversity by having a maximum percentage of 33.9%, which is preceded by the Himalayan region with 20% along with Central India and Andaman Island. If we consider the altitude only, the world lichen flora with special reference to India for our better understanding can be divided into tropical temperate alpine based on the variation in altitude [3]. In North Sikkim, alpine lichen growing in the Himalayas is located at an altitude of 3650 m and extends to a height of 5900 m. In the alpine zone, there is a lack of large trees, whereas species like *Rosa*, *Cotoneaster*, and many more species of *Rhododendron* impart favorable substratum to the species of corticolous lichen containing genera like *Ramalina*, *Usnea*, *Lecanora*, *Pertusaria*, *Ochrolechia*, *Heterodermia*, *Buellia*, etc. Terricolous or saxicolous lichen is more common in alpine regions. A few species of *Thamnolia*, *Umbilicaria*, *Cladonia*, and *Stereocaulon* are found growing on deteriorating moss. Those lichens that grow in temperate regions are restricted to the Western Ghats, The Himalayas, and parts of Assam and Manipur, between 1500 and 3500 m in elevation. The lichens of the temperate zone prefer the shrubs as their substratum for their favorable growth; this type of lichen includes genera like *Usnea*, *Ramalina*, *Parmelia*, etc. Approximately it involves the entire of South India (excluding the top of Nilgiri and Palni Hills); apart from South India, it also involves some parts of Northeast

Chemistry, Biology and Pharmacology of Lichen, First Edition. Edited by Ashoke Kumar Das, Ajay Sharma, Deepika Kathuria, Mohammad Javed Ansari, and Garima Bhardwaj.

India state including Arunachal Pradesh, Manipur, Assam, and mainly the foothills of Himalayas extending to an altitude of 1500 m.

The North Eastern Indian zone is mainly the portion of Assam and Bengal accompanied by the above-mentioned region. Species of lichen such as *Collema*, *Peltigera*, and *Leptogium* are all moisture-possing and develop abundantly in this sector. Many pyrenocarpous, graphidaceous, and corticolous grow conveniently in this forest [4]. Shola is the indigenous name of this forest patch, which provides luxuriant growth for lichen flora. Due to the even distribution of wind current, rain, and sunlight, it becomes comparatively easy for the fruticose and foliose lichen to grow. The pyrenocarpous and graphidaceous lichen prefer the suitable habitat of palm trees for their growth in coastal regions accompanied by the members of *Dirinaria*, *Pyxine*, and *Physcia* [5]. The trees of the deciduous forest containing broad leaves are the hope for the modest number of corticolous lichen. The average lichen growth in mid-Eastern India and the peninsular plateau include the suppression factors like flaky bark, dry bark, etc. In the Indo-Gangetic Plain, several prohibiting factors like high temperature, scarce rainfall less anthropogenic activities, etc., promote the reduced growth of lichen in some of the genera found growing on the trees like Citrus, *Mangifera indica* favored lichen like *Rinodina*, *Dirinaria*. A limited number of Corticolous genera mainly pyrenocarpous and graphidaceous possess the dominant lichen growing in this region. Tree species like *Ficus*, *Delonix*, etc., provide the luxuriant growth of foliicolous and corticolous lichen. In India, about 69 microlichens are endemic to various Phytogeographical regions. The majority of existing in South India, which ensures 17 microlichens succeeded by Andaman along with Central Himalaya possesses 13 micro lichens individually, whereas North and Northwest Himalaya include 11 micro lichens. Finally, South India and Eastern Himalayas possess the former 6, and the latter includes 3 micro lichens. Around 33 macro lichen are endemic to India, out of which 17 macro lichen possess the maximum number in Central Himalayan regions although North Western Himalaya encompasses 7 macro lichens preceded by South India, which includes 3 macro lichens, and eventually, Eastern Himalaya retains 3 macro lichens [6]. Although there are many research has been conducted to understand the ecological impact and importance of lichen by various researchers across the globe, only a few have been found to be consistently working in this area due to the amount of effort one needs to put into the area to understand the complex ecology of lichen. This chapter is an attempt to accumulate all the authentic data regarding ecology and adaptation of lichen under a single heading to encourage the young researchers to help in their initial literature survey to carry out their research in this field.

5.2 Ecological Habitats

Lichens can be found on every continent and in several aquatic and marine habitats as well as hot and dry deserts. Substrates include things like, trees, rocks, soil, buildings, gravestones, cars, old farm machinery, etc., which act as the most suitable substrates for lichens [7]. Trees, rocks, and soil are the three most prevalent natural substrates. Lichens can be classified based on their habitat-Lignicolous term is used for those lichens, which are typically found in the forests, the lichens that thrive on tree bark are referred to as corticolous lichens. When the lichens grow on rocks or stones, they are categorized under saxicolous lichens. Oceanic lichens can be seen growing on siliceous rocks close to the sea. Again, terricolous lichens are grown on soil or ground. Although dead wood is a crucial component of the habitat for lichens present in forest ecosystems, there is very little knowledge about the number and types of lichens that depend on it. Macrolichens are species are aerial thalli having three-dimensional orientations, including all squamulose species, fruticose and foliose lichen, microlichens are species with thalli so closely attached that their removal requires lifting of the substrate [8]. *Corticolous crustose* green algal lichens are widely distributed and plentiful in tropical low-laying forest ecosystems. The largest and least well-known group of lichens is the *Corticolous microlichen*. The necessity for in-depth research on this group makes it challenging for academics to compile taxonomic keys and also prevents them from identifying and using this material in their surveys or environmental studies. The phorophytes were located on the edge of the forest patches, inside the forest areas, or occasionally on isolated trees. The lichen samples were taken from their tree trunks. There were forty-one corticolous microlichen species in total, spread across 17 families and 25 genera. *Psoroglaena stigonemoides, Cryptothelium cecidiogenum, Herpothallon echinatum, Lecanora thysanophora, C. cecidiogenum, Distopyrenis composita* and *Graphis pseudocinerea* are six species that represent new entries for South American flora [9]. *Rinodina conradii* and *Pyrenula dissimulans* are two new species records for Brazil, and the following four species are new entries for the state of Rio Grande do Sul-*Graphis elongata*, *Graphis longula*, *Haematomma africanum*, and *Graphis furcata* [10]. A total of 31 new entries for the Southern area of Brazil were reported. Gran Piedra's montane rainforest is renowned for having a wide variety of plants. In boreal, Mediterranean, and tropical woods, corticolous lichen diversity and substrate preferences have been studied. However, certain physiognomic and atmospheric features of the woods in the insular Caribbean would argue against projecting lichen preferences seen around these forests. With their total surface connected to the substrate, crust-forming epilithic lichens are able to burrow rapidly and comparatively depth into the weathering surface of rocks. The amount of hyphal layer created by the symbiotic thallus, which is viewable at the base of the stone, can be 10 times as bulky as the lichen fungus that lives inside the rock [11]. An important aspect of the ecosystem of lichen that lives on crustose rocks is its endolithic portion. Lichens are the first noticeable colonizer of exposed rock due to their eye-catching colors and distinctive morphologies. According to estimates, lichens make up to 6% of the surface of the planet on the land, and they play a noteworthy role in the initial phases of soil development and rock weathering [12]. Marine lichens have different habitats than their terrestrial counterparts and are far less well understood, because they are frequently submerged in seawater. Within the *Cladonia pyxidata* group, *Cladonia monomorpha* is classified as a separate species. The thallus segments have slightly curled borders, disc-like shaped, bullate layers are present on the podetial face, and lengthy, occasionally branched proliferations of the scyphus borders holding the apothecium plates are characteristics of the species [13]. It is revealed from the Netherlands, which has the largest terrestrial lichen diversity, where it can be found in areas of acidic inland sandy beaches. It looks to be common on siliceous rock and acid sand in Europe. There is a description of the lichen flora connected to 20 salt lakes in England. Lichens were found on shingles, various rocks, partially submerged brick buildings, concrete structures, saline marsh vegetation, plywood sheets, and strand-line items;

however, they were frequently rare. Richer assemblages were frequently found in places that had been altered by human influence than in those that were in a more natural state. A mix of common species and maritime experts make up the lichen flora. The Fleet, a sizable lagoon near Chesil Beach, was the most significant location examined. Numerous marine lichens found here are typical of extremely protected beaches [14]. Scientists consider the two sites where *Caloplaca suaedae* was discovered to be novel. In vast parts of the Namib Desert, where fog is the only form of wetness and other flora is rare, terrifying lichens are the predominant flora. Although lichens are crucial to the primary producers of the Namib Desert ecosystem and to stabilizing the soil crust, little is known about their variety and spread in the northern, unknown areas of the Namib. The geographic distribution of terricolous lichens throughout 7 different types of landscapes and 60,000 square km in the western part of New South Wales. Additionally, information is provided about the abundance of lichens on red earth soil in Cobra within run-on zones and geomorphic series of flow. In arid and semi-arid parts of eastern Australia, 48 species from 23 genera were gathered from 282 locations. Of them, 74% were reported as either squamulose or crustose, and the remaining 26% accounting for foliose. A maximum of 57% of the species belonged to six genera: *Endocarpon*, *Diploschistes*, *Acarospora*, *Peltula*, *Catapyrenium*, and *Xanthoparmelia*. The lichen floristics or crust layers were poorly predicted by the type of substrate. However, all types of landforms contained a small group of species, including *Endocarpon simplicatum* var. *bisporum*, *Endocarpon rogersii*, *Endocarpon pallidum*, *Collema coccophorum*, *Heppia despreauxii*, *Peltula patellata*, *Psora decipiens*, *Synalissa symphorea*, and *Catapyrenium squamulosum* [15]. Terricolous lichens, especially those of the squamulose and crustose forms, build close relationships with other biota that live in the soil, including cyanobacteria, fungi, algae, and bryophytes, to create morphological soil crusts. *Verrucaria margacea*, a common species on siliceous rocks in the Alps, is an example of how the freshwater lichen flora appears to be very uniform throughout the world. This species is found in Europe, North America, India, and Australia. Freshwater lichens selected the surfaces of clay blocks that had been baked at a lower temperature, while exposed sections of local rocks were less frequently colonized. Since most freshwater lichens only spend part of the year underwater, they are actually amphibious organisms.

5.3 Ecological Factors for Lichen Development

Numerous environmental factors, such as the prolonged impact of the climatic condition, confined changes in the microclimate, and an array of substratum-related factors, such as the nature of soil, bark, rock, surface chemistry, and the intensity of salinity or pollution have an impact on lichens [16]. It is crucial to conduct research on how environmental conditions affect lichen growth for a number of reasons. The first step is to comprehend the variables affecting the allotment of uncommon or intriguing species and the makeup of significant communities of lichen. Therefore, if the environment is altered to a less-than-ideal state, lichen growth may be severely inhibited, causing thallus to quickly break down and vanish from a substrate. Additionally, the environment can influence species in a community more subtly by changing the balance of competition between them. In researching lichen growth, there have been two significant issues [17]. First, there are not many facilities for growing complete lichen thallus in the lab for extended periods of time, which has hindered research. In order to study the synergistic impacts of several variables, a small number of "factorial-type" experiments have been conducted. There are many different aspects of how the ecological factors govern the growth and development of lichens.

5.3.1 Effect of Climate on Growth of Lichen

One of the first to show the significant impact of climate on lichen growth was reported by researchers through lichenometric investigations in the late 1950s. Reports have emerged as the identical species of the genus crustose termed as *Rhizocarpon* grew much more gradually in Greenland as compared to the Swiss Alps, while re-assessment at those sites, 12 years subsequent to the initial study revealed that the locations in the Greenland farthest from the sea had *Rhizocarpon ographicum* (L.) [18]. As a result, it was discovered that in Scotland, UK, the rate of growth of the thalli of the *Rhizocarpon* generally decreased from western to eastern side. Additionally, a study has discovered that the growth rate gradients in southern Norway and Iceland are likely reflected in decreased moisture availability brought on by a declining marine effect [19]. However, not all of these investigations have discovered meaningful growth effects. For instance, when the development of transplanted *Dolichousnea longissima* specimens was examined in four diverse settings with a gain of dry weight and length being calculated after one year, all of the sites under study showed similar growth rates, even those where the habitat was not expected to be good for the species' expansion [20]. In a similar vein, *Cladonia portentosa* (Dufour) Coem dry weight increase was examined at five locations of heathland that varied in yearly rainfall, the acidity of precipitation, and nitrogen deposition, but no regional variations in growth were found.

5.3.2 Rainfall

In North Wales, *Melanelia fuliginosa* ssp. *fuliginosa* (Fr. ex Duby) Essl monthly relative rate of growth was assessed over the course of a year [21]. Major growth peaked in June, with lower peaks of growth in November and March, which coincided with the wettest months. According to general trends, monthly growth fluctuations are best correlated with average or total rainfall; however, monthly measures of radial growth through linear regressions against total rainfall only account for a tiny portion of the total variance in growth (typically 40%). The same researcher also studied the growth of *Xanthoparmelia conspersa* Hale in northern parts of Wales over the course of 22 consecutive months regarding total monthly rainfall. Reverse transplants have regularly been used to study how the climate affects the growth of tree bark. Therefore, using transplants of *Lobaria pulmonaria* (L.) Hoffm, the link between growth and thallus size was investigated in three successional forest patches located in three different climatic zones of boreal type [22]. The most significant factor causing an increase in thallus area was stand-specific water availability. Less investigation has been done on the seasonal growth of crustose species because radial growth rates are substantially lower. Nevertheless, *Lecidea tumida* Massal and *Rhizocarpon reductum Th. Fr.* budding on rocks in northern parts of Wales have their seasonal growth over a three-month time span, which was examined. Both had their slowest growth during August–October, moderate growth in November–January, and strongest growth between May and July, demonstrating that a combination of lower temperatures and higher rainfall was not favorable for the growth of the species in question [23]. The infamous Karenlampi was one of the first lichenologists to investigate how fruticose lichen development and climate interact. The mean rainfall in the summer on a daily basis was connected with the comparative growth rate of many fruticose species at sites in Finland, while the temperature was negatively correlated to growth. The relative growth rate was quantified as an increase in weight or length. Additionally, the growth of *Ramalina menziesii* (Taylor) in inland and coastal sites in California was, with rainfall seeming to be the main factor influencing growth [24].

5.3.3 Temperature

Other climatic conditions have been found to have an impact on lichen growth, even though rainfall is often the main variable in many places. Hence, *Lecanora caesiorubella* Ach. and *Lecanora chlarotera* Nyl., exhibit yearly fluctuations in radial growth, and it was observed significantly larger growth during 1959–1961 compared with 1961–1962 was mostly attributed to variations in mean temperature [18]. There were noteworthy nonlinear associations between maximum temperature and growth, the total number of air and ground frosts, mean temperature minimum, and amount of rain days, with an upper limit of temperature being the most significant. However, no significant linear relationships between climatic variables and growth were reported. Therefore, *Rhizocarpon geographicum* can grow to its maximum size in north Wales at any time of year, while July–September seems to be the most favorable. *Rhizocarpon superficiale* (Schaer.) seasonal variations in photosynthetic capability were examined by many researchers [25]. Although there were no appreciable periodic changes in photosynthetic capability, this species had a wide sensitivity to temperature and an elevated resilience to high-temperature stress, characteristics of species found in "boundary-layer" environments.

5.3.4 Snowfall

The severity of frost damage caused by winter and snowfall are significant issues at higher altitudes. Studies have discovered that *Rhizocarpon* thalli were smaller near snow patches as a result of a shorter growing season and substrate instability [26]. However, in some locations, the positive impacts of rising moisture offset the unfavorable impact of snow. Frost damage, together with "spalling" of the upper cortical and phycobionts, was noticeable in the Colorado Front Range, particularly in foliose species. Many of the crustose species like *Rhizocarpon riparium* Räsänen were not severely impacted, but some saplings did happen and alter the areolae of the surface. Studies further showed that although photosynthesis was inhibited under the snow because of reduced penetration of light, thalli did maintain to breathe [27]. Two of the semi-permanent snow portions in the Colorado Front Range were outfitted with lines of lichen-covered boulders to exhibit this effect.

5.3.5 Light Intensity

The field environment of woodlands has been the most frequently researched for its effect on lichen growth. In a transplant test where the impacts of canopy structure on *L. pulmonaria* were investigated, the growth was strongly correlated with canopy light transmission with dry weight gain over two years being less than 5% under canopies and rising to 20% on branches connected with canopy gaps [28]. When thalli were moist, a substantial correlation between growth and light intensity was seen in a Swedish study of five macro-lichens. Thus, clear-cutting, which frequently results in a rapid edge to strands of the forests, may have an impact on the formation of foliose lichen. In British Columbia, the impact of "hard edges" and a less rushed forest border expansion was investigated by employing transplants of the plant *Lobaria retigera* [29]. A considerable reduction in loss when a lesser amount of abrupt border was there, indicating that remaining plants at the edge could lessen the effect of reduction in forest cover on lichen variety was observed. Over the course of three years, severe mortality of lichen thallus and loss of fresh biomass at "hard" edge sites was noticed.

5.3.6 Interaction Between Factors

There has not been a lot of research on the correlative effects of climate and associated factors on the growth of lichen using factorial experimental designs. However, researchers used a three-factor experimental approach to examine the growth of *X. conspersa*. This design included four layers of radiation exposure, two contrasting climates, and two conditions for the input of nutrients. Radial growth was considerably influenced by all variables, but the climate had the biggest impact [30]. Significant connections also existed between the climate, radiation exposure, and nutritional levels.

5.3.7 Rock Chemistry

There are significant differences in the lichen composition of lime-rich, lime-poor, siliceous rocks, and only a small number of species can grow successfully on all types of substrata. A variety of substrata, including slate, granite, cement, and asbestos, were used to attach fragments of foliose lichens to test the idea that the chemistry of the rock significantly affects growth. Numerous elements of the diversity of *Xanthoria parietina* in the United Kingdom may be explained by a calcium-dependent pathway [31]. As a result, this species can be found on tree bark, limestone rocks, asbestos, walls that have calcium-rich streamflow, and rocks of siliceous forms close to the marine habitat. *Parmelia saxatilis* (L.) in contrast, is almost exclusively found on acidic substrates. This thallus takes on a "crescent-shaped" morphology after being treated with calcium carbonate.

5.3.8 Bird Dropping

Numerous studies indicate that lichen growth and community composition may be significantly impacted by the intensity of nutrient enhancement of a substrate, particularly by bird droppings [25]. Bird droppings may have an impact on growth by suffocating thalli, changing pH, or introducing inhibitory and stimulatory chemicals. A common species on well-lit, nutrient-rich rocks, *X. conspersa*, had its radial development stimulated by treatment with bird droppings. On bird perching stones, the paste and, to a smaller extent, the suspension prevented the expansion of the uncommon species *P. saxatilis* (L.) was also reported [32].

5.3.9 Nitrogen and Phosphorus

Few young researchers investigated the reaction of *X. parietina* to various concentrations of N_2 [33]. Without any discernible effects from additional nitrogen, all thalli's specific weights were similar, indicating that this plant may be more sensitive to changes in pH than to nitrogen, as suggested by earlier research. The association between growth and nitrogen levels was also investigated in the plants *Nephroma arcticum* (L.) Torss and *Peltigera aphthosa* (L.) Willd. [34]. Worse chlorophyll a and nitrogen concentrations, thus significantly lowering the ineffectiveness of light-energy exchange effectiveness, were shown to be the causes of *N. arcticum's* slower growth. In a study of *Cladonia stellaris* discovered that phosphate, more so than nitrogen, had a higher impact on fruticose lichen growth [11]. Additionally, nitrogen and phosphorus enrichment can affect the algal and fungal partners differently.

5.3.10 Salinity

Growth in oceanic sites may be influenced by additional chemical variables. Salinity had little impact in these habitats, where calcium was experimentally demonstrated to be a crucial ion, with the species reacting more strongly to periods of immersion [35]. In contrast, later it was discovered that the zonation on marine rocks and the responses of four distinct lichen fungi to salinity were highly associated.

5.3.11 Pollution

Even though lichens have returned to many urban sites during the past 20 years, the diversity of lichens in polluted areas is much lower than in rural areas, demonstrating the strong growth-restraining effects of air pollution [36]. In a study, the specimens of foliose and crustose species were transplanted on pieces of rock collected from north Wales to a city center site in Birmingham; no discernible development was observed, and the cultures showed signs of disintegration within seven days of transplantation before disappearing completely within one month [18]. The study also found that environmental conditions of urban sites have a stronger negative impact on development during the winter months, implying that toxicity from air pollution, rather than drought, was the most significant issue. Studies were conducted to examine the direct correlation between air pollution and *P. saxatilis* development [37]. Thallus was shown to exhibit a crescent shape when sulfur dioxide levels rose, and they persisted for several years before flaking off the substrate. On the outside margin, however, active growth was observed at this time, and radial growth did not seem to be impacted. Additionally, the medulla of *X. parietina* and *P. saxatilis* both produced vivid green or orange colors, respectively, signifying the breakdown of lichen acids [38].

5.4 Adaptations of Lichen

Lichen has unique characteristics that enable it to withstand extreme temperatures and humidity. When there is a suitable amount of water content, sunshine, and a confined humidity region, the majority of lichen species may develop successfully. Lichens obtain the nutrients they require to thrive straight from the atmosphere, unlike flowers and trees, which obtain their nourishment from the rain and soil. Lichen can germinate in practically any place since it obtains all of its nutritional elements from the atmosphere. Some of the important adaptations pertaining to lichen adaptations have been discussed in detail below.

5.4.1 Adaptation of Lichen to Anthropogenic Activities

Anthropogenic means environmental change altered by people both directly or indirectly. To study the impact of human-driven performance on the physiological functioning of epiphytic lichens in Uludag National Parks, Kirazlyayla, and Saralan picnic places were selected as the research locations in these ecological planning and applications because of their ecological fragility [39]. Lichens in sun-exposed areas often contain more chlorophyll than those in shaded settings. Lichen samples were collected at Kirazlyayla and Saralan picnic places in Uludag National Parks; these lichen samples were collected in two phases for investigation of photosynthetic pigment. The sampling for the first stage took place in May at the start of summer, when anthropogenic activities were low, and the second stage was completed in September toward the conclusion of the summer when anthropogenic activity was high; foliose lichen species such as *Hypogymnia physodes* (L.) Nyl. and *Parmelia sulcata* Taylor, as well as fruticose lichens like *Pseudevernia furfuracea* (L.) Zopf and *Evernia prunastri* (L.) Ach was collected from Kirazlyayla and Saralan to assess the photosynthesis pigment concentrations of epiphytic lichens. In comparison to the control locations, it was discovered that all of the species gathered from the picnic spots at Kirazlyayla and Saralan had more chlorophyll a. The specimens taken from the picnic spots towards the summer end have greater chlorophyll content than the specimens obtained at the start of summer. At both the beginning and the end of the summer, picnic sites had higher chlorophyll content than control areas in *Evernia prunastri* obtained from Kirazlyayla *Parmelia sulcata*, *Parmelia furfuraceous*, and *Hypogymnia physodes* had similar outcomes. A noteworthy difference was recorded in the amounts of Chlorophyll b, carotenes, and the Optical Density (OD) 435/415 ratio of *E. prunastri* between the picnic area and the control. Picnic sites were found to have greater OD 435/415 values than the control area. This is a crucial indicator in identifying the impact of human-driven activities on the photosynthetic pigment content of lichens of epiphytic types and the conversion rate of chlorophyll to pheophytin. Both at the start of the summer and at the conclusion of it, the picnic area had higher OD 435/415 values than the control area for *E. prunastri* and *P. furfuracea* was taken from Kirazlyayla. In comparison to the control regions, the epiphytic lichen samples from Saralan picnic places that were taken at the beginning or end of the summer had greater chlorophyll a level. While there were no significant changes between the chlorophyll concentration of *H. physodes* in the control and picnic areas. Significant concentrations of chlorophyll a and b were found in *P. sulcata* collected from Saralan. In comparison to the control regions *P. furfuracea*, a species of fruticose lichen had a steadily decreasing chlorophyll a/b ratio near the end of the summer. Since only cars can access the Kirazlyayla picnic spot, there are fewer people during the summer months than at Saralan. Even though the Saralan picnic location is conveniently reachable by automobile, great anthropogenic activity is seen there year-round because it is also reachable by cable car.

5.4.2 Adaptations in Freshwater

Freshwater lichens in particular are one of the least researched types of organisms. Rarely are the data on the species habit preferences and distribution accessible, despite the fact that the occurrence of the species is frequently documented [40]. The great majority of lichens inhibit terrestrial ecosystems and are unable to endure in environments with high levels of moisture. However, certain species that live in freshwater and marine habitats those are aquatic or semiaquatic. Worldwide, there are only a few lichen species that have direct connections to water. Many lichen species growing on water are thought to be rare or endangered in many parts of Europe as a result of elevated demands on water purity and susceptibility to pollution. They make up around 6.5% of all lichen species worldwide, although only a small percentage of them (about 1.5%) is found in freshwater [41]. It is widely accepted that in increment to the fluctuations of stream currents, a number of other habitat factors, the kind, and stability of the substratum, physicochemical properties of the water, light availability may all have a role in determining freshwater variety. The amount of time lichens spend submerged, the amount of light available, the lithology and stability of the substrate, and the chemistry all have an impact on lichen growth in freshwater settings. Lichens depend on size of the rocks and the kind of bedrock for their existence in streams, where the substrate is especially crucial. Even though they are naturally saxicolous, certain aquatic lichens may live on submerged roots. Freshwater lichens are distributed in the vertical zone. The majority of species only exist in the splash zone or have a poor tolerance for total and prolonged immersion in water. Conglutinate pseudo parenchyma, which is hydrophilic and passively absorbs water and dissolved nutrients, gives lichen thalli their mechanical resilience. Plectenchyma is made up of gas-filled zones with hydrophobic surfaces that are developed by aerial hyphae, *Verrucaria hydrophila*, *Verrucaria praetermissa*, and *Thelidium minutulum* were the only three species to be found in all streams and habitat types [42]. According to several reports of these lichens in the Eastern and Western Carpathians might be thought of as common and dependable residents of the region's clear mountain streams.

5.4.3 Terricolous Lichen

A lichen that develops on the soil as a substrate is referred to as terricolous. A few Peltigera species serve as an illustration. Reindeer and caribou (*Rangifer tarandus L.*) eat a variety of terricolous and fruticose lichens, which are a major part of their diet, particularly in the winter. Consequently, reindeer survival can be severely impacted by the overgrazing of heathlands with a lichen-dominated ecosystem. The circumpolar region's winter grazing pastures are presently in a precarious situation due to overgrazing, as well as increasing human activity and climate change [43]. The lichen community in Las Lomitas, a fog oasis in the Pan de Azucar National Park in the Atacama Desert, demonstrated morphological and ecophysiological traits that allowed them to persist under the current climatic and environmental conditions [44]. It is alleged that the terricolous lichens' morphological characteristics, including fungal and algal stacks or the ability of the epidermal layer to swell, are adjustments that allow them to flourish in this region of the Atacama Desert, while the epiphytic lichens' ecological characteristics having physiological backgrounds, such as a quick reactivation by fog and high air humidity, were a factor in their success.

5.4.4 Adaptation to Marine Lichens

Seashore lichens are most noticeable on rocks creating a continuous coverage coating. Abiotic elements such as variations in waves, solar radiation, daily cycles of drying, wind, and humidification, opposing temperatures, and salt have a major impact on lichens that they must survive [45]. The tolerance of lichens growing in sub-littoral and littoral zones to prolonged exposure to salt water varies. As a result, it is easy to distinguish between the littoral and supralittoral zone types. There are *Wahlenbergiella* species in the Northern Hemisphere's littoral zone. Several *Hydropunctaria* species live largely within the supralittoral zone above the periodic flooding zone, while *Verrucaria serpuloides* grow continuously immersed in seawater in the Southern Hemisphere while *V. serpuloides* prosper continually submerged in seawater in the Southern Hemisphere. *Lecanora polytropa* and *Lecanora intricata*, two often found supralittoral types of lichen species on the shores of the Barents and white Sea, showed morphological and anatomical variety, and morphotypes were observed [46]. These species' capacity to adapt in diverse morphological, physiological, and biochemical ways in response to various climatic circumstances is what accounts for their widespread distribution. They may readily fix ascospores and/or conidiospores to the rocky substrate along the beaches, and they have crustose thalli. *L. intricata*, on the other hand, generates thinner thallus and unimmersed apothecia. *L. polytropa*, which is found along the coast of the White Sea, has a discrete thallus and sparsely spaced areoles. The overall width of the thalli *L. intricata* was much larger than the width of the thalli *L. polytropa*, according to the analysis of their anatomical structure. In actuality, the mycobiont, which develops the lichen's thallus and provides close interaction with the environment, is primarily responsible for morphological modifications [47]. The authors described two different adaptation forms: structural, owing to the mycobiont's flexibility; functional, due to the photobiont's variable quantity of photosynthetic pigments. On rocky shores, two species of marine lichen—*Lichina confinis* and *Lichina pygmaea* can be found coexisting but in distinct coastal zones. They are connected to *Renularia*, a cyanobiont. The researchers discovered proof of the cyanobiont strains' strong species specificity for both lichen species. While patterns were significantly associated between the two lichen groups, selectivity within each species of fungi varied over the study region, affected by surrounding circumstances (both marine and air). The discrepancy connection of phycobionts with every littoral sector is thought to have caused ecological diversification in marine genus Lichina. Both the mycobiont and the phycobiont, which control the synthesis of a variety of compounds, such as osmolytes, which participate in the processes of lichen colonization in coastal areas, have helped lichens develop their structural and functional adaptations [48].

5.4.5 Corticolous Lichen

Lichen that develops on the bark is referred to as corticolous lichen. *C. crustose* green algal lichens are widely distributed and plentiful in tropical lowland forests. This might be a result of adjustment to the current environmental circumstances, such as heavy precipitation and low intensities of light. French Guiana researchers led by Lakatos et al. [49] considered the morphology of crustose lichens having green algal phycobiont in the understory of a tropical rainforest of lowland area and assessed chlorophyll fluorescence and gas exchange and discovered that the understory's low light levels allowed photosynthesis to adjust in the following ways: (i) The presence of water-repelling surface structures on mycobiont hyphae at the surface of the thallus and the medulla decreased the amount of time that the thallus was supersaturated with water, (ii) photosynthesis was rapidly activated in fluctuating light conditions, and (iii) With varying light, photosynthesis was quickly triggered (iv) Corticolous lichens can employ special morphological and physiological strategies thanks to the combination of these three processes, which may help them develop more quickly in the constrained understory environment of tropical rainforests. The findings offer numerous proofs that tropical lichens may thrive in the understory of tropical rainforests of lowlands despite the unfavorable microclimates.

Few hardwood species, namely *Sterculia foetida, Sterculia cordata*, and *Vitex parviflora*, have been able to flourish on the limestone soil of many islands throughout time. These barks of the trees have been colonized by corticolous lichens, mostly in crustose and narrow-lobed foliose forms. At the Islands, salty water splashes and high radiation subject lichens to desiccation and osmotic stress when the trees and their epiphytes are exposed to alternating high temperatures (during the day) and cold temperatures because they are in a coastal setting (during the night). This suggests that the rate of desiccation is higher during the day and lower at night, including osmolytes, which participate in lichen colonization processes in coastal areas, which have helped lichens adapt both structurally and functionally [50].

5.5 Lichens in Extreme Habitat

Lichen can withstand even in extremely dry and cold regions of the whole world (through dormancy) and can regain rapidly. They can even survive continuous rainfall, but usually, they survive moderate rainfall. Lichens can naturally go dormant when the surroundings are too severe; the fungi with the help of the filament protect the lichen when the events are improved. Although lichens have the typical characteristics of organisms that can withstand stress, including slow growth rates, long lifespan, a low food need, and the occurrence of particular adjustments to living in the most hostile surroundings. Lichen survives in extreme environments and then characterized primarily based on specific environments.

5.5.1 Antarctica

In the coastal Antarctic, yearly rainfall is not a limiting factor, but presence of a noticeable gradient of moisture from the coastal regions to interior sites with other factors like ice cover, oxygenation, availability of nutrients, and salinity [51]. As the availability of H_2O

and the duration of the emergent season are the most probable deciding variables, there is a consequent drop in the varieties of species, growth rate, and coverage from the continental dry valleys to the marine Antarctic region. Early growth of Antarctic lichens is extremely sluggish, and reproduction frequently occurs through vegetative diaspores like soredia rather than lichenization. Unlike the maritime Antarctic, places covered in snow frequently hold onto the bitter wintery environment for an extended duration of time, reducing the chance of any signs of untimely warming. Nonetheless, there is a marked fall in growth rates from the drier, colder valleys to the drier, warmer peninsula. These valleys exhibit a faster rate of photosynthesis and a broader range of temperature responses compared to their continental counterparts. Endolithic species in the arid valleys are especially characterized by an exceptionally low nPS rate with the species *Caloplaca* displaying a design to improve the hydration of thallus [52]. Since radiation has such powerful negative effects on nPS, the algal partner is also necessary for lichens to successfully adapt to the circumstances of the continental Antarctic. Since melanin, a photoprotective pigment was present in *Umbilicaria decussate* (Vill.) Zahlbrwhole's thalli, photoinhibition was not seen. It was also shown that *Umbilicaria antarctica* E. Frey was more sensitive to intense light exposure, which resulted in higher amounts of glutathione in oxidized form and the transformation of violaxanthin into zeaxanthin, which are crucial for resistance to antioxidants.

5.5.2 Chemically Rich Environments

Chemically rich environments are extreme for several species including lichens because these environments include heavy metals, and extremely eutrophicated and contaminated sites are also affected by salinity due to all these, many species are incapable to flourish in these surroundings [17]. Despite the resurgence of lichens in many metropolitan areas over the past 20 years, air pollution caused by Sulfur dioxide pollution-affected areas has a much lesser diversity of lichen than unaffected areas, demonstrating the strong growth-restraining effects of air pollution. *Lecanora conizaeoides* Nyl. Ex Cromb., an SO_2-tolerant species, decreased in the UK during the 1960s due to the reduction in pollution during this time. Lichens showed a variety of adaptations to these circumstances. First, *L. conizaeoides* seems to be "superhydrophobic" due to the occurrence of hydrophobic chemicals and a thallus surface with rough characteristics that sheds H_2O, which are characteristics of lichens resistant to SO_2 [53]. Furthermore, gas passages are present for photosynthesis, creating what is basically a surface that is "breathable." Second, when exposed to Sulfur at low pH, several lichens, notably *H. physodes* (L.) Nyl. showed oscillations in stress ethylene concentration, a reaction correlated with their tolerance level. Finally, resistance to acidic air pollution may be linked to secondary compounds such as fumarprotocetraric acid originating from lichens. High concentrations of heavy metals in lichens may lead to the establishment of metal tolerance through intrinsic cytoplasmic tolerance, cytoplasmic immobilization, or the transport of ions to regions outside of the plasmalemma and cell wall [54]. *Physcia adscendens* (Th. Fr.) Oliv. Em. Bitt. expressed more glutathione S-transferase (GST) and heat shock proteins than ATP-binding cassette transporters (ABC transporters), which were only significantly activated after prolonged exposure to rising cadmium concentrations. Hence, this species' ability to tolerate heavy metals may depend on the differential expression of many protein systems. Also, the build-up of heavy metals in *Diploschistes muscorum* (Scop.) R. Sant. cause membrane deterioration, oxidative stress, and an increase in glutathione (GSH) and superoxide dismutase (SOD) indicating that resistance to antioxidants might be engaged.

5.5.3 Extra-terrestrial Environments

There has been a lot of debate about whether lichens may thrive in interplanetary regions like space and Mars because of their capacity to adapt to a variety of harsh situations. A number of studies conducted at the European BIOPAN Facility (EBF) and the International Space Station (ISS) have now looked into this aspect of lichen adaptability [55]. UV radiation, insolation, vacuum, freezing temperature and cosmic radiation may have an impact on lichens in space. As a result, after being exposed for 559 days, *Xanthoria elegans* (Link) Th. Fr. demonstrated significant confrontation, with after-exposure activity 50–80% higher than baseline digits in the alga and up to 90% higher than baseline figures in the fungus [56]. However, desiccation-induced cell integrity collapse was also noted, which was extra pronounced in space than in a "simulated" Martian environment. Furthermore, commotion was seen post-exposure with elevated algal cell viability and minimal rates of damage when photosynthesis was particularly investigated. However, when isolated alga was exposed to radiation, a significant mutilation of photosynthesis and mechanisms responsible for photoprotection was seen highlighting the importance of the defense offered by the lichen symbiosis because of morphological adaptability, the existence of lichen chemicals, and the capacity of lichens to withstand desiccation. *Aspicilia fruticulosa* (Eversm.) Flagey, on the other hand, experienced a decrease in chlorophyll "a" output over the course of a 10-day space flight as a result of exposure to electromagnetic radiation between 100 and 200 nm [57]. This decrease, however, was reversed after the lichen was reactivated for 72 hours. Moreover, carbon dioxide, with just trace quantities of oxygen, argon, nitrogen, and carbon monoxide, predominates in the Martian atmosphere. Endolithic lichens found in deserts and arid valleys in Antarctica are frequently thought of as potential models for how lichens can survive on Mars. As a result, when colonies of black fungi that live on rocks and dehydrated Antarctic cryptoendolithic communities were exposed to conditions that mimicked Martian life for 18 months on the ISS, less than 10% of the colonies multiplied while 60% of the rock communities and cells were unaltered [58]. Therefore, endolithic lichens on Mars could be capable of withstanding the cold surface temperatures, and similar to Antarctica, the stones beneath is probably warmer and more stable than the surface of the rocks.

5.5.4 Alpine Regions

Alpine regions are based on varying high humidity and frozen temperatures. Researchers have found that there is great variation in the length along higher angles of the sun along with long-wave and worldwide emission. The benefit of alpine surroundings is the capability to examine the consequence of environmental inclinations with rising elevations. It was investigated the lichen flora of Nepal between 200 and 7400 m in height and discovered that the pinnacle of species richness occurred at higher elevations of 4100–4200 m [59].

The maximum species richness was reported between 3100–3400 and 4000–4100. In contrast, it was reported that medium elevations in Northeast India had the largest taxonomic diversity. Altitude could restrict the range of potential phycobionts for the fungal partner, which might have an impact on a wide range of distinct lichens. Lichens may show a variety of adjustments to the harsher environmental conditions present at elevated altitudes. Hence, at higher elevations, *Cetraria nivalis* (L.) Ach. accumulates the ultraviolet-absorbing phenol usnic acid in the upper cortex region [60]. Pools of polyols are expected to help maintain cellular components and preserve intracellular structures during desiccation since lichens at high elevations face discontinuous periods of hydration and desiccation.

5.5.5 The Arctic

The Arctic regions are the ones that are featured with quite low to intermediate inputs of high temperature. The temperature usually remains lower than −40 °C, and the growing season is restricted only up to 60 days throughout the year. As a result, the soil remains ice-covered except for that short summer span, where the top layer just remains unfrozen. The deep layers of the frozen soil are regarded as Permafrost. Throughout the area, the vegetation is a complication of dwarf shrubs, sedges of grasses, and lichens clubbed in the frame of wet grassland. A gradient also shows the transition from the tundra to the Arctic desert to areas covered in enduring ice. There are two zones of vegetation. First, the low arctic region is made up of a thick, squashy layer of alive, undying plants that are overflowing with water and have thawed frozen ponds. Secondly, elevated arctic regions may be completely stripped in contrast to the patches compiled with liches and grasses [61]. Climate change, which can result in severe but transient winter warming spells, is a relatively new stressor for Arctic lichens. The Arctic can experience intense summer heating, although evaporation rates are slower. As a result, widely dispersed species like *R. geographicum*, which is typical of well-bright southern-facing surfaces in climates of the temperate but rocks facing the north in the Arctic may alter their biological behavior [62]. The arctic lichens manifest a diversity of physiological adaptions to take the upper hand over minuscule positive environmental actions after being in dormant conditions for an extended period. In particular, trehalose, a non-reducing disaccharide that builds up in response to heat, cold, or osmotic stress, this might contribute to partial dehydration and acclimatization to cold temperatures is a soluble carbohydrate that is frequently highest in summer and spring in cryptograms. Trehalose is therefore more concentrated in *R. geographicum*'s periphery prothallus, which is more susceptible than the center areolae. The accessibility of nitrogen is also a significant restricted factor for the lichens of Artic regions clubbed with other low-temperature regions.

5.5.6 Desert

Deficiency of H_2O is distantly the major difficulty for lichens found in the deserts; lichens might be missing in some sandy deserts. Moreover, in the Negev, a difference in taxa found at the sides or top of the stones has been noted in the distribution of tiny pebbles. The average daily dew on the top of the pebbles is double that on their edges, but after rain, the margins are wet due to capillary reaction in the soil next to the stone. Weakly welded rhyolitic ignimbrite in Chile's Atacama Desert is regularly inhabited by lichens that germinate inside the rock, or "endolithic lichens," which not only promote hydration but also shield the thalli from harmful Ultra violet rays and high amounts of visible light. Moreover, the Sonoran Desert's "caliche plates," which may reach temperatures of over 60 °C in the summer and 0 °C in the winter and are home to the endolithic species *Verrucaria rubrocincta* Breuss, are underground precipitates dominated by calcite that are visible on the surface [63].

5.5.7 Lichen on Lava

Epiphytic lichen that grows on the branches and trunks of plants is identified as the main colonizer in the biodegradation of rocks. Organisms including cyanobacteria and lichen live even in stressful atmospheres named extremophiles. Some of the latest research on pioneer lichen analysis is based on the Hawaiian lava flows [64]. In Hawaii, there were primarily two different types of lava flows; one of them, called "Pahoehoe," has a smooth, billowy, or ropy surface, while "aa," lava has a surface that is rough, jagged, prickly, and typically clinkery. The predominant species of lichen *Stereocaulon vulcani* particularly scrubby adults along with some adolescents follow by less abundance of *Cladonia sp.*, *Parmelia subramigera*, and *Campylopus densifolius*. While on the pahoehoe lava, the abundant lichen species are *Parmelia citrate*, and *P. subramigera* followed by some crustose lichen. However, *S. vulcani* is found abundant in both "pahoehoe" and "aa" lava [65].

5.6 Conclusion and Future Prospects

Excessive levels of air pollution and nutrient discharges from agriculture and wastewater continue to harm ecosystems, reduce agricultural production, and cause eutrophication and acidification. Global levels of pollution are anticipated to rise sharply in the ensuing decades. Some sensitive lichen species undergo structural changes in response to air pollution, including reduced photosynthetic activity and discoloration. Lichens require pure, unpolluted air to survive. Lichens may be completely destroyed by pollution, and the lichen fungus may also change color and develop more slowly as a result. It's possible that species that can tolerate pollution may gradually replace sensitive ones. As a result, a lot may be learned about the quality of the air from the sorts of lichens that grow there and the quantity of pollutants they contain. Nitrogen and sulfur dioxide are the two principal air contaminants that have an impact on lichen formation. Even in the harshest and most arid parts of the globe, lichens can live. They can tolerate extremely cold and dry circumstances because of dormancy and the ability to recover fast when the conditions are right. Although they only require a tiny quantity of moisture to exist, lichens can flourish in locations with significant rainfall. This could be significant when global temperature extremes increase due to climate change. Lichens can simply go dormant when the environment is too severe, and the fungi's threads will shield the organism until the environment is more hospitable. Lichens are frequently referred to as "extremists" or "pioneer organisms," which are a monument to their capacity for adaptation and survival in many of the challenging conditions on Earth. Due to this lenience, it has even been hypothesized that lichens may have formerly existed on Mars and that they may still do so in some extremely dry deserts and arid valleys in Antarctica. Lichens can photosynthesize in frigid climes even when they are frozen, collecting water from snow and the moist air. They perform entirely distinct ecological functions and have quite different needs than microorganisms. Additionally, microorganisms frequently face a threat to their existence without lichens. Lichens are crucial for the sequestration of carbon. They possess metabolic processes that are capable of trapping extra Carbon in secondary molecules. The Pedogenesis process, where topsoil formation and remediation take place, involves cycling of minerals in lithophytic lichen thalli. The defense mechanisms of lichens' photosynthetic systems in wet environments are less well understood. The preservation of lichen variety is not a priority. Conservers are of the opinion that conserving macro flora and fauna also implies conservation of lichen diversity. Determining specific dangers to various lichen communities, such as soil lichens, and putting effective countermeasures in place. Raise public awareness of the value of conserving lichens at the proper levels. The preservation of habitat quality, connectivity, and size should be the core goal of lichen conservation. The single most significant factor contributing to the population reduction of lichens in the future will likely be land-use change, which includes habitat loss and degradation. In contrast to the conservation of dynamic ecosystems like forests, grasslands, and gravel fields, the conservation of permanent habitats, such as mountain ridges, may best be followed by a static protection and is very simple to accomplish. Lichen conservation biology will only be effective if the species is viewed as a component of a community that is the result of intricate relationships between climatic conditions, quality of the environment, and anthropogenic management, similar to other species that are connected in an ecological network.

References

1 Negi, H.R. (2003). Lichens: a valuable bioresource for environmental monitoring and sustainable development. *Resonance* 8: 51–58.

2 Mishra, R., Chand, P.K., and Satapathy, K.B. (2020). Some new additions to the lichen flora of Odisha. *Plant Archives* 20 (1): 1979–1986.

3 Muci, L., Valacho, M., Dimopoulos, P. et al. (2000). Epiphytic lichen and moss vegetation along an altitude gradient on Mount Aenos (Kefallinia, Greece). *Biologia* 55 (1): 43–48.

4 Pinokiyo, A., Singh, K.P., and Singh, J.S. (2008). Diversity and distribution of lichens in relation to altitude within a protected biodiversity hot spot, north-east India. *The Lichenologist* 40 (1): 47–62. https://doi.org/10.1017/S0024282908007214.

5 Shukla, V., Upreti, D., and Bajpai, R. (2013). Lichen diversity in different lichenogeographical regions of India. In: *Lichens to Biomonitor the Environment* (ed. V. Shukla, D.K. Upreti, and R. Bajpai), 61–94. Springer. https://doi.org/10.1007/978-81-322-1503-5.

6 Sethy, P. (2012). Lichens on mangrove plants in Andaman Islands, India. *Mycosphere* 3 (4): 476–484. https://doi.org/10.5943/mycosphere/3/4/11.

7 Lisci, M., Monte, M., and Pacini, E. (2003). Lichens and higher plants on stone: a review. *International Biodeterioration and Biodegradation* 51 (1): 1–17. https://doi.org/10.1016/S0964-8305(02)00071-9.

8 Spribille, T., Thor, G., Bunnell, F.L. et al. (2008). Lichens on dead wood: species-substrate relationships in the epiphytic lichen floras of the Pacific Northwest and Fennoscandia. *Ecography* 31 (6): 741–750. https://doi.org/10.1111/j.1600-0587.2008.05503.x.

9 Knudsen, K. and Lendemer, J.C. (2006). Changes and additions to the North American lichen mycota - V. *Mycotaxon* 95 (November): 309–313. https://doi.org/10.1635/0097-3157(2007)156[55:caattn]2.0.co;2.

10 Aptroot, A. and Da Silva Cáceres, M.E. (2018). New lichen species from Chapada Diamantina, Bahia, Brazil. *Bryologist* 121 (1): 67–79. https://doi.org/10.1639/0007-2745-121.1.067.

11 Asplund, J. and Wardle, D.A. (2017). How lichens impact on terrestrial community and ecosystem properties. *Biological Reviews* 92 (3): 1720–1738. https://doi.org/10.1111/brv.12305.

12 Jackson, T.A. (2015). Weathering, secondary mineral genesis, and soil formation caused by lichens and mosses growing on granitic gneiss in a boreal forest environment. *Geoderma* 251–252: 78–91. https://doi.org/10.1016/j.geoderma.2015.03.012.

13 De La Coba, F., Aguilera, J., Figueroa, F.L. et al. (2009). Antioxidant activity of mycosporine-like amino acids isolated from three red macroalgae and one marine lichen. *Journal of Applied Phycology* 21 (2): 161–169. https://doi.org/10.1007/s10811-008-9345-1.

14 Randall, R.E. (2004). Management of coastal vegetated shingle in the United Kingdom. *Journal of Coastal Conservation* 10 (1–2): 159–168. https://doi.org/10.1652/1400-0350(2004)010[0159:MOCVSI]2.0.CO;2.

15 Büdel, B. (2003). Biological soil crusts in European temperate. In: *Biological Soil Crusts: Structure, Function, and Management* (ed. J. Belnap and O.L. Lange), 75–86. Berlin: Springer.

16 Hauck, M. (2011). Site factors controlling epiphytic lichen abundance in northern coniferous forests. *Flora: Morphology, Distribution, Functional Ecology of Plants* 206 (2): 81–90. https://doi.org/10.1016/j.flora.2010.02.001.

17 Chen, J., Blume, H.-P., and Beyer, L. (2000). Weathering of rocks induced by lichen colonization—a review. *Catena* 39: 121–146.

18 Armstrong, R. and Bradwell, T. (2010). Growth of crustose lichens: a review. *Geografiska Annaler, Series A: Physical Geography* 92 (1): 3–17. https://doi.org/10.1111/j.1468-0459.2010.00374.x.

19 Bradwell, T. (2010). Studies on the growth of *Rhizocarpon geographicum* in NW Scotland, and some implications for lichenometry. *Geografiska Annaler, Series A: Physical Geography* 92 (1): 41–52. https://doi.org/10.1111/j.1468-0459.2010.00376.x.

20 Wangmo, S. and Dorji, P. (2017). Effect of feeding *Dolichousnea longissima* (Ach.) Articus as supplement during winter on milk yield and its composition in Yaks. *Bhutan Journal of Natural Resources and Development* 4 (2): 39–45. https://doi.org/10.17102/cnr.2017.08.

21 Armstrong, R.A. (2014). Within-site variation in lichen growth rates and its implications for direct lichenometry. *Geografiska Annaler, Series A: Physical Geography* 96 (2): 217–226. https://doi.org/10.1111/geoa.12043.

22 Bidussi, M., Goward, T., and Gauslaa, Y. (2013). Growth and secondary compound investments in the epiphytic lichens *Lobaria pulmonaria* and *Hypogymnia* occidentalis transplanted along an altitudinal gradient in British Columbia. *Botany* 91 (9): 621–630. https://doi.org/10.1139/cjb-2013-0088.

23 Kidron, G.J., Temina, M., and Starinsky, A. (2011). An investigation of the role of water (rain and dew) in controlling the growth form of lichens on cobbles in the negev desert. *Geomicrobiology Journal* 28 (4): 335–346. https://doi.org/10.1080/01490451.2010.501707.

24 Garty, J. (2010). Critical reviews in plant sciences biomonitoring atmospheric heavy metals with lichens: theory and application biomonitoring atmospheric heavy metals with lichens: theory and application. *Critical Reviews in Plant Sciences* 2011: 37–41.

25 Upreti, D.K., Divakar, P.K., Shukla, V., and Bajpai, R. (2015). *Recent Advances in Lichenology: Modern Methods and Approaches in Biomonitoring and Bioprospection*, vol. 1. https://doi.org/10.1007/978-81-322-2181-4.

26 Sancho, L.G., Palacios, D., De Marcos, J., and Valladares, F. (2001). Geomorphological significance of lichen colonization in a present snow hollow: Hoya del cuchillar de las navajas, sierra de gredos (Spain). *Catena* 43 (4): 323–340. https://doi.org/10.1016/S0341-8162(00)00131-4.

27 Després, V.R., Alex Huffman, J., Burrows, S.M. et al. (2012). Primary biological aerosol particles in the atmosphere: a review. *Tellus, Series B: Chemical and Physical Meteorology* 64 (1). https://doi.org/10.3402/tellusb.v64i0.15598.

28 Coxson, D.S. and Stevenson, S.K. (2007). Growth rate responses of *Lobaria pulmonaria* to canopy structure in even-aged and old-growth cedar-hemlock forests of central-interior British Columbia, Canada. *Forest Ecology and Management* 242 (1): 5–16. https://doi.org/10.1016/j.foreco.2007.01.031.

29 Stevenson, S.K. and Coxson, D.S. (2008). Growth responses of *Lobaria retigera* to forest edge and canopy structure in the inland temperate rainforest, British Columbia. *Forest Ecology and Management* 256 (4): 618–623. https://doi.org/10.1016/j.foreco.2008.05.025.

30 Armstrong, R. and Bradwell, T. (2011). Growth of foliose lichens: a review. *Symbiosis* 53: 1–16.

31 Gauslaa, Y. and McEvoy, M. (2005). Seasonal changes in solar radiation drive acclimation of the sun-screening compound parietin in the lichen *Xanthoria parietina*. *Basic and Applied Ecology* 6: 75–82. https://doi.org/10.1016/j.baae.2004.10.003.

32 Allen, A. (2008). Lichen specialities of lundy: an overview. *Journal of the Lundy Field Society* 1 (4): 33–40.

33 Gisela, G.O., Dahlman, L., and Palmqvist, K. (2005). Responses of the lichen *Xanthoria parietina* (L.) Th. Fr. to varying thallus nitrogen concentrations. *The Lichenologist* 37 (2): 171–179. https://doi.org/10.1017/S0024282904014598.

34 Dahlman, L., Näsholm, T., and Palmqvist, K. (2002). Growth, nitrogen uptake, and resource allocation in the two tripartite lichens *Nephroma arcticum* and *Peltigera aphthosa* during nitrogen stress. *New Phytologist* 153 (2): 307–315.

35 Gunde-Cimerman, N., Oren, A., and Plemenitas, A. (2005). *Adaptations to Life at High Salt Concentrations in Archaea, Bacteria, and Eukarya*. Dordrecht: Springer.

36 Godinho, R.M., Verburg, T.G., Freitas, M.C., and Wolterbeek, H.T. (2009). Accumulation of trace elements in the peripheral and central parts of two species of epiphytic lichens transplanted to a polluted site in Portugal. *Environmental Pollution* 157 (1): 102–109. https://doi.org/10.1016/j.envpol.2008.07.021.

37 Van Dobben, H.F., Wolterbeek, H.T., Wamelink, G.W.W., and Ter Braak, C.J.F. (2001). Relationship between epiphytic lichens, trace elements and gaseous atmospheric pollutants. *Environmental Pollution* 112 (2): 163–169.

38 Kumar, J., Dhar, P., Tayade, A.B. et al. (2014). Antioxidant capacities, phenolic profile and cytotoxic effects of saxicolous lichens from trans-Himalayan cold desert of Ladakh. *PLoS One* 9 (6): e98696. https://doi.org/10.1371/journal.pone.0098696.

39 Guvenc, S. and Kula, H. (2021). The effects of picnic areas on epiphytic lichen diversity at Abies forests in Uludag National Park, Turkey. *Fresenius Environmental Bulletin* 30 (9): 10568–10575.

40 Sonina, A. and Androsova, V. (2021). Coastal lichens: distribution patterns and adaptive strategies. In: *Handbook of Halophytes* (ed. M. Grigore), 1–22. Springer Cham. https://doi.org/10.1007/978-3-030-17854-3

41 Krzewicka, B., Matura, N., Adamska, E., and Osyczka, P. (2020). Species composition of freshwater lichens in temperate mountain streams: the effect of site, habitat and local spatial isolation. *Preslia* 92 (3): 235–254. https://doi.org/10.23855/preslia.2020.235.

42 Matura, N. and Krzewicka, B. (2015). Verrucaria species and other rare amphibious lichens in the Beskid Sądecki Mts. *Acta Mycologica* 50 (1): 1057. https://doi.org/10.5586/am.1057.

43 Forbes, B.C. and Kumpula, T. (2009). The ecological role and geography of reindeer (*Rangifer tarandus*) in Northern Eurasia. *Geography Compass* 3 (4): 1356–1380.

44 Jung, P., Emrich, D., Briegel, L. et al. (2019). Ecophysiology and phylogeny of new terricolous and epiphytic chlorolichens in a fog oasis of the Atacama Desert. *MicrobiologyOpen* 8: e894. https://doi.org/10.1002/mbo3.894.

45 Lalegerie, F., Gager, L., and Stiger-Pouvreau, V. (2020). The stressful life of red and brown seaweeds on the temperate intertidal zone: effect of abiotic and biotic parameters on the physiology of macroalgae and content variability of particular metabolites. *Advances in Botanical Research* 95: 247–287. https://doi.org/10.1016/bs.abr.2019.11.007.

46 Sonina, A.V. (2012). Epilithic lichens and their morphological adaptations to the conditions of the White and Barents Seas coast (Russian Arctic). *Czech Polar Reports* 2 (2): 109–116.

47 Taylor, T.N., Klavins, S.D., Krings, M. et al. (2004). Fungi from the Rhynie chert: a view from the dark side. *Transactions of the Royal Society of Edinburgh: Earth Sciences* 94 (1): 457–473.

48 Delmail, D., Grube, M., Parrot, D. et al. (2013). Halotolerance in lichens: symbiotic coalition against salt stress. In: *Ecophysiology and Responses of Plants Under Salt Stress* (ed. P. Ahmad, M. Azooz, and M. Prasad), 115–148. New York: Springer. https://doi.org/10.1007/978-1-4614-4747-4.

49 Lakatos, M., Rascher, U., and Budel, B. (2006). Functional characteristics of corticolous lichens in the understory of a tropical lowland rain forest. *New Phytologist* 172 (4): 679–695.

50 Mcgenity, T.J., Gessesse, A., Hallsworth, J.E. et al. (2020). Editorial visualizing the invisible: class excursions to ignite children's enthusiasm for microbes. *Microbial Biotechnology* 13 (4): 844–887. https://doi.org/10.1111/1751-7915.13576.

51 Bhattarai, H., Paudel, B., Hong, S. et al. (2008). Thin layer chromatography analysis of antioxidant constituents of lichens from Antarctica. *Journal of Natural Medicines* 62: 481–484. https://doi.org/10.1007/s11418-008-0257-9.

52 Armstrong, R.A. (2019). The lichen symbiosis: lichen "Extremophiles" and survival on Mars. *Journal of Astrobiology and Space Science Reviews* 1: 378–399.

53 Roach, P., Shirtcliffe, N.J., Newton, M.I. et al. (2008). Progress in superhydrophobic surface development. *Soft Matter* 4 (2): 224–240. https://doi.org/10.1039/b712575p.

54 Hauck, M. (2010). Ammonium and nitrate tolerance in lichens. *Environmental Pollution* 158 (5): 1127–1133. https://doi.org/10.1016/j.envpol.2009.12.036.

55 Edwards, H.G.M., Moody, C.D., Jorge, S.E., and Wynn-williams, D.D. (2005). Raman spectroscopic detection of key biomarkers of cyanobacteria and lichen symbiosis in extreme Antarctic habitats: evaluation for Mars Lander missions. *Icarus* 174: 560–571. https://doi.org/10.1016/j.icarus.2004.07.029.

56 Brandt, A., De Vera, J., Onofri, S., and Ott, S. (2015). Viability of the lichen *Xanthoria elegans* and its symbionts after 18 months of space exposure and simulated Mars conditions on the ISS. *International Journal of Astrobiology* 14 (3): 411–425. https://doi.org/10.1017/S1473550414000214.

57 Raggio, J., Pintado, A., Ascaso, C. et al. (2011). Whole lichen thalli survive exposure to space conditions: results of lithopanspermia experiment with *Aspicilia fruticulosa*. *Astrobiology* 11 (4): 281–192. https://doi.org/10.1089/ast.2010.0588.

58 Onofri, S., De Vera, J.P., Zucconi, L., and Selbmann, L. (2015). Survival of Antarctic cryptoendolithic fungi in simulated martian conditions on board the international space station. *Astrobiology 15* (12): 1052–1059. https://doi.org/10.1089/ast.2015.1324.

59 Pyakurel, D., Smith-Hall, C., Bhattarai-Sharma, I., and Ghimire, S.K. (2019). Trade and conservation of Nepalese medicinal plants, fungi, and lichen. *Economic Botany* 73: 505–521.

60 Fremstad, E.L.I., Paal, J., and Möls, T. (2005). Impacts of increased nitrogen supply on Norwegian lichen-rich alpine communities: a 10-year experiment. *Journal of Ecology* 93: 471–481. https://doi.org/10.1111/j.1365-2745.2005.00995.x.

61 Joly, K., Jandt, R.R., and Klein, D.R. (2009). Decrease of lichens in Arctic ecosystems: the role of wildfire, caribou, reindeer, competition and climate in north-western Alaska. *Polar Research* 28 (3): 433–442. https://doi.org/10.1111/j.1751-8369.2009.00113.x.

62 Armstrong, R.A. (2011). The biology of the crustose lichen *Rhizocarpon geographicum*. *Symbiosis* 55 (2): 53–67. https://doi.org/10.1007/s13199-011-0147-x.

63 Bungartz, F., Garvie, L.A.J., and Nash, T.H. (2004). Anatomy of the endolithic Sonoran Desert lichen *Verrucaria rubrocincta* Breuss: implications for biodeterioration and biomineralization. *The Lichenologist* 36 (1): 55–73. https://doi.org/10.1017/S0024282904013854.

64 Crews, T.E., Kurina, L.M., and Vitousek, P.M. (2001). Organic matter and nitrogen accumulation and nitrogen fixation during early ecosystem development in Hawaii. *Biogeochemistry* 52: 259–279.

65 Stretch, R.C. and Viles, H.A. (2002). The nature and rate of weathering by lichens on lava flows on Lanzarote. *Geomorphology* 47 (1): 87–94.

6

Physiology of Lichen

Farishta Yasmin and Rosni Jabin

Department of Botany, Nowgong College (Autonomous), Nagaon, Assam

6.1 Introduction

Lichen is a composite organism with mutualistic relationships and with highly adaptive capabilities and is well recognized as a pioneering entity on Earth. The body of lichen is a thallus (pl. thalii) composed of mainly photobiont and mycobiont as inhabitant and exhabitant, respectively, and is attached to the substratum by rhizines.

Anatomically, lichen thallus usually consists of the upper cortex, photosynthetic layer, medulla, and lower cortex. The thickness of the layers can vary in different species, which is a response to different environmental conditions [1]. Based on the ratio of photobiont cells to mycobiont cells, the thallus of lichen can be classified as homoiomerous type, where the number of algal cells is more than fungal cells and heteromerous type and where fungal cells are dominant over algal cells. Lichen symbiosis occurs due to dynamic evolution that is reflected in the various growth forms of the thalli [2].

Morphologically, the thallus structures of lichen are categorized as crustose, squamulose, foliose, fruticose, leprose, filamentous, and gelatinous. Crustose lichens are tightly attached to the substrate by whole thalli and are difficult to remove without any physical damage. The physiological studies of these lichens remain inadequate due to the complicated extraction process from the substrate and low biomass production, making routine analysis difficult. Foliose lichens are leaf-like structures that are partially attached to the substrate and spread horizontally over the surface. The thallus is usually divided into lobes and is easy to collect compared with crustose forms. Foliose lichens are mainly used for biochemical and ecophysiological studies because of their biomass and easy collection. Fruticose lichens are shrubby and a preferential choice of lichenologists for ease of removal from the substratum. Squamulose lichens are an intermediate form between crustose and foliose lichens. Leprose lichens are lack of fruiting bodies. The structure of filamentous lichens comprises fungal hyphae enveloping chains of algal cells. In gelatinous lichen, cyanobacteria produce polysaccharides that absorb and retain water giving them a gel-like appearance.

Chemistry, Biology and Pharmacology of Lichen, First Edition. Edited by Ashoke Kumar Das, Ajay Sharma, Deepika Kathuria, Mohammad Javed Ansari, and Garima Bhardwaj.

6.1.1 Mycobiont

The lichen-forming fungi or lichenized fungi are taxonomically a distinct group capable of acquiring carbon from its photoautotrophic partner. The fungus controls the exposure of the algal cells to sunlight to its physiological working stages, as vigorous light intensity can damage the algal partner [3]. Fungal cells around the algal partner are highly hydrophobic in nature; they seal the apoplastic cells and allow metabolic flux among symbiotic partners. The hydrophobic coat is mainly composed of proteins, lipids, and phenolic secondary compounds of fungal origin [4–6]. The fungal partner of mycobiont has the ability to shift the location of algal cells for sufficient illumination and optimum photosynthesis [7, 8]. Lichenized fungi are ecologically obligate but physiologically facultative. Lichenized fungi differ from non-lichenized fungi by nutritional strategies. They are adapted to live with symbiotic photobiont partners. The central portion of the mycobiont without a symbiotic partner is typically made up of clumped masses of cartilaginous cells that thrive in low-oxygen conditions, while aerial hyphae with filamentous growth occur at the outer region. The type of photobionts determines the transfer of metabolites from photobionts to mycobionts. In lichen, carboxylation occurs within the photobiont, and carbon is then transformed into the mycobiont as glucose, ribitol, sorbitol, or erythritol [9].

6.1.2 Photobiont: Phycobiont and Cyanobiont (Cyanolichen)

Photobionts, the photosynthetic partner of lichen, provide nutrition to mycobionts by reducing CO_2 into organic carbon sugars (carbohydrates). Photobionts produce glucose, ribitol, sorbitol, or erythritol. Most studies have shown that mycobionts are not restrictive in their choice of photobionts and that one type of photobiont can occur widely in different lichens. Lichen thalli can also be classified by the ratio of phycobiont cells to mycobiont cells. The homoiomerous type of thallus consists of numerous algal cells distributed among a lesser number of fungal cells, while the heteromerous thallus has a predominance of fungal cells. The primary function of the photobiont is to provide the fungus with photo-assimilated carbon [8]. Therefore, from the fungal perspective, it then appears to be particularly advantageous to choose a cyanobacterial photobiont, which is capable of both CO_2 and N_2 fixation and helps them in growth and mineral cycling [10]. Nearly 100 species from 40 genera of algae and cyanobacteria have been reported as photobionts in lichens [11, 12]. Three genera *Trebouxia*, *Trentepohlia*, and *Nostoc*, are the most frequent photobionts. *Trebouxia* and *Trentepohlia* are eukaryotic and belong to green algae; the genus *Nostoc* belongs to the oxygenic photosynthetic bacteria (cyanobacteria). The transfer of metabolites between the autotrophic photobiont and the heterotrophic mycobiont is influenced by the type of photobiont involved. As mentioned earlier, phycobionts produce sugar alcohols (carbohydrates) and cyanobionts produce glucose. The mode of activation of CO_2 uptake varies depending on cell types of photobionts, i.e. prokaryotic or eukaryotic. Positive net photosynthesis can occur in phycobiont lichens through water uptake alone, whereas in cyanobacterial lichens, there is no detectable gas exchange because photosynthetic processes necessitate higher water content levels, which can only be achieved with liquid water. Photosynthesis in lichens with cyanobacterial photobionts always requires the addition of liquid water to recover from desiccation [13].

6.2 Physiological Interaction

Lichens are widely distributed on Earth from sea level to a very high altitude. They can grow on almost any surface in many environmental conditions, including some of the most extreme environments, such as rocky coasts, hot deserts, Arctic Tundra, Antarctica, objects of metals and glasses, toxic slag heaps, etc. Lichen is combination of alga or cyanobacterium and fungus. But lichen has a different physiology and biochemistry other than the individual identity of its partners. Different types of associations of partners of lichen symbionts might enable the lichen to exist in a wider range of environmental conditions and habitats. The mycobiont part of lichen reproduces through the release of different outgrowth and spores produced in asexual and sexual structures like soredia, isidia, pycnidia, apothecia, and perithecia. These spores require a finding of compatible photobiont for survival as a symbiont [14]. The lectins produced by mycobionts play a significant role in attracting and binding photobionts [15, 16]. Lectins are proteins with sugar-binding sites that mediate interactions between cells and the outside environment [16]. It has been observed that lectin isolated from various mycobionts binds both cyanobionts and phycobionts [16–19]. It has been established that lectin released by *Peltigera canina* acts as a chemoattractant to *Nostoc* cells [20, 21]. Lectin genes can also be increased by the presence of the photobiont, proving that the mycobiont is also able to recognize the signals that the photobiont emits [22, 23]. Lectin might be crucial in determining photobiont selectivity. Different lichen metabolites generated from mycobionts, like usnic acid and vulpinic acid show an inhibitory effect on the proliferation of photobionts which indicates that secondary metabolites regulate the cell division of algal partners. Secondary metabolites may be functioning in the lichen thallus as allelochemicals [24].

6.3 Metabolism

Lichen produces different types of metabolites categorically and can be divided into primary metabolites and secondary metabolites, which are intracellular and extracellular, respectively. Huneck and Yoshimura [25] discuss the isolation, identification, and structural analysis of lichen substances. Amino acids, amines, peptides, proteins, polyols, saccharides (mono-, oligo-, poly-), carotenoids, and vitamins bound in cell walls and protoplasts could be considered as products of primary metabolism. Many of these primary metabolites are produced by higher plants, free-living fungi, and algae. Lichen thallus contains 1.5–24 mg/g of dry weight of carotenoids, 3.5% dry weight of polysaccharides, and 1.6–11.4% dry weight of nitrogen compounds [26, 27].

A considerable chemical diversity of lichen secondary metabolites, also known as lichen acids, is recorded. They are gaining importance as a natural source for preclinical studies and pharmaceutical use. All of the secondary metabolites seen in lichens are derived from mycobiont [28]. However, it has also been demonstrated that the production of secondary metabolites varies throughout seasons by location and by environmental variables including UV radiation [29–32]. This might justify the enhancement of the capacity of lichens to adjust to the environment.

The secondary metabolites of lichen symbiont are derived from three biochemical pathways: the acetate–polymalonate pathway, the shikimic acid pathway, and the mevalonic

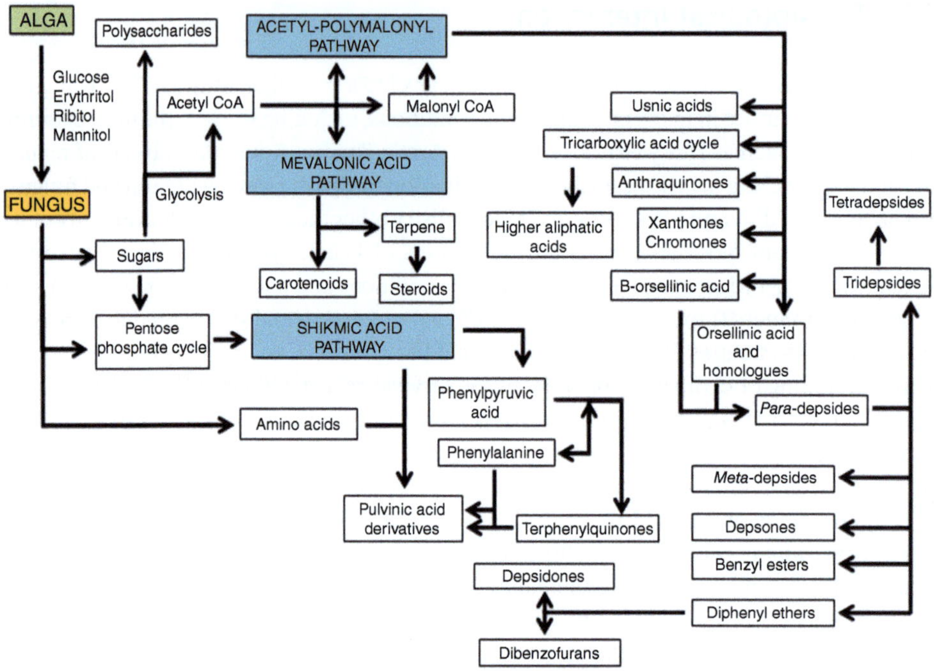

Figure 6.1 Biosynthetic pathways of lichen secondary metabolites.

acid pathway. A representative diagram adopted from Ranković, B. [28] of these pathways is given in Figure 6.1.

It has been seen that most of the secondary metabolites are produced through the Acetate-polymalonate pathway rather than the Mevalonic acid and Shikimic acid pathways. A summary of secondary metabolites produced by lichens through different pathways (Table 6.1) and the chemical structures of some of them (Figure 6.2) are given below.

6.4 Physiological Peculiarities

Lichens are poikilohydric organisms and can withstand dehydrated conditions for a long time. In a lichen thallus, photobiont is properly hydrated through photosynthesis. Lichens undergo anhydrobiosis when they totally lose their water and remain there until they are rehydrated [33]. The absence of moisture for a long period of time has led to desiccation for the majority of vascular plants and organisms. Lichens are desiccation tolerant and can remain in a suspended state until water becomes available again, after which they revive and resume normal metabolism [34]. Usually, lichen thallus absorbs moisture contents of dew drops and fog, which triggers dark respiration and the continuing exchange of CO_2 below the zero line throughout the night. The exchange of CO_2 above the zero line is started with sunlight in the morning, and the photosynthesis process is activated for the day. The carbohydrate produced through photosynthesis is used for the growth and development of photobionts as well as mycobionts. The net photosynthetic rate of lichen is also significantly influenced by the water content of their thalli.

Table 6.1 Metabolic pathways and their related compounds.

Metabolic pathways	Secondary metabolites
Acetate–polymalonate pathway	Secondary aliphatic acids, esters and related derivates
	Mononuclear phenolic compounds
	Depsides, tridepsides and benzyl esters
	Depsidones and diphenyl esters
	Depsones
	Dibenzofurans, usnic acids and derivates
	Anthraquinones and biogenetically related xanthones
	Chromones
	Naphthoquinones
	Xanthones
Mevalonic acid pathway	Di-, sester- and triterpenes
	Steroids
Shikimic acid pathway	Terphenylquinones
	Pulvinic acid derivates

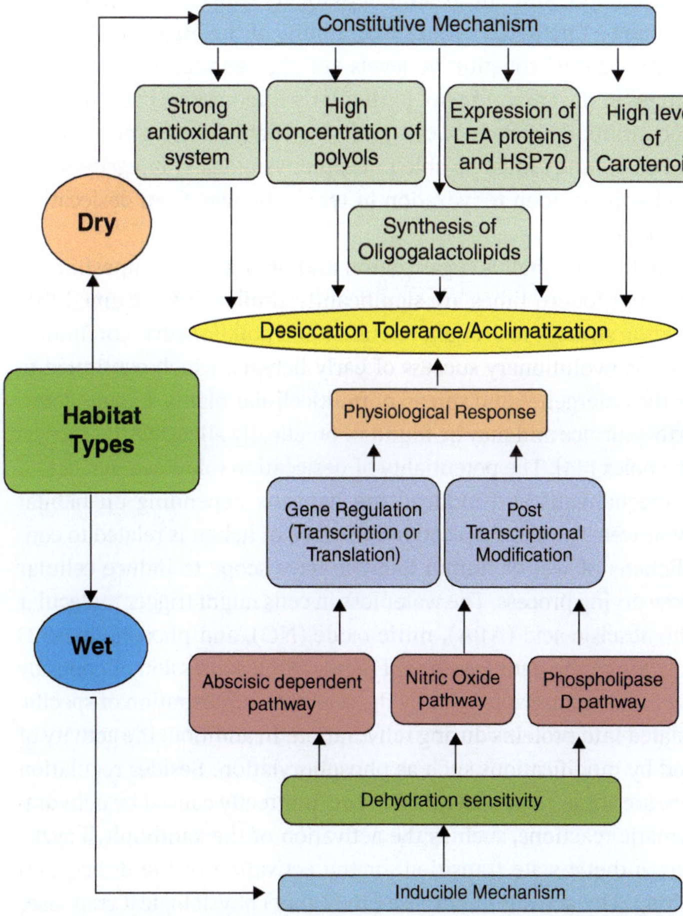

Figure 6.2 Diagrammatic representation of adaptive mechanisms of lichens toward habitat types (*Source:* Adopted from [33]).

Diffusion of CO_2 to the phycobiont is hampered in thallus with saturated water, which results in the decline of the CO_2 assimilation rate. Moreover, the photobiont may suddenly die at maximal water saturation in continuous light since all of its products are transferred to the mycobiont. Again, peak photosynthesis does not happen until the thallus dries to between 65% and 90% of its maximal water content. After that, when temperatures and light levels rise, net photosynthesis and water content both decrease [35–37]. Desiccation occurs reasonably slowly, over hours rather than minutes [38]. Lichen photobionts are able to maintain maximum photosynthetic activity until the water content reaches 20% [39]; thus, during this period lichens photosynthesize at rates that are sufficient to allow a net positive carbon gain over the year. Also, thalline growth rates depend on the frequency and length of this period per day and year [40].

Lichens have evolved unique mechanisms to prevent the formation of oxidative damage during the desiccation and rehydration processes at the photobiont cellular level. Lichens fulfill their carbon assimilation requirements by the procedures used to distribute additional light energy as heat and through antioxidant systems considered protective tolerance mechanisms against oxidative stress. Glutathione appears to be the primary antioxidant found in lichens. Antioxidant levels and desiccation tolerance, however, do not necessarily correlate. Glutathione, ascorbic acid, and antioxidant activity, like superoxide dismutase (SOD) or peroxidase (POX), can be fluctuated based susceptibility of the organism to desiccation. Most likely, these antioxidants' constitutive levels are high enough to shield cells from the effect of rapid changes in the environment, particularly fluctuations in light intensity and humidity. Lichens continually cycle between periods of dehydration and rehydration throughout their lifespan. So, they must be able to keep physiological integrity while dried out and mobilize mechanisms upon rehydration to repair damage from desiccation and subsequent rehydration [41–43].

It is very interesting to note that the thallus organization and photobiont composition of lichen thallus of Devonian and Modern times are significantly similar. It is assumed that early lichens set their adaptation strategy to drought tolerance and poikilohydric conditions. It is enormously helpful for the evolutionary success of early lichens, which continued to diversify on land even after the emergence and spread of multicellular plants. Lichen dominates roughly 8% of the earth's surface and may be found in practically all terrestrial ecosystems, from the tropics to the poles [44]. The potentiality of desiccation tolerance in a lichen is governed by constitute mechanisms and induced mechanisms depending on habitat types, whether they are dry or wet. Though desiccation tolerance of lichen is related to constitutive mechanism, the lichens of wet or humid habitats have scope to induce cellular responses because of the slow drying process. The water loss in cells might trigger molecular signal cascades, such as the abscisic acid (ABA), nitric oxide (NO), and phospholipase D (PLD) pathways (Figure 6.2). Signaling pathways might regulate the expression of genes by the induction or inhibition of their transcription, or by the selective sequestration of specific mRNA that would be translated into proteins during rehydration. In addition, the activity of enzymes might be regulated by modifications such as phosphorylation. Besides regulation by signaling pathways, there are some physiochemical events indirectly caused by dehydration that can activate enzymatic reactions, such as the activation of the xanthophyll cycle, the regulation of the photosynthetic state transition, or the activation of the desiccation non-photochemical quenching. The activation of these (and other) physiological responses would allow the cells to face a desiccation event in a better position and to recover quickly upon rehydration. In this sense, priming of physiological responses could help lichens to acclimate to environmental conditions [33]. The growth and distribution of lichen thallus are also governed by the ecophysiology of microhabitats. Microhabitat affects the physiological state of individual thallus that can be location-specific. The growth-rate responses depend on species-specific abilities to assimilate nutrients (e.g. increasing photobiont abundance and tolerating higher light levels for their functioning) and to tolerate increased parasitism and plant competition [45, 46]. The impact of micro-environmental fluctuations (microclimatic extremes) on lichen habitat use varies along with species-specific protection mechanisms and speed in recovering vital functions, such as cell turgor [47], photosystem II activity [48, 49] and (in cyanolichens) nitrogen fixation [10]. In such a way, through the physiological process, lichens develop the ability to colonize in extremely dry environmental conditions and become ubiquitous.

Lichens are highly successful self-sustaining ecosystems involving lichen-forming fungi, phototrophic algae, and/or cyanobacteria [50].

6.5 Conclusion

The capacity to thrive in extreme environments makes lichen physiology interesting for researchers. Though many researchers have used their best efforts to understand the physiology of lichens and their metabolites, further investigations are needed to understand the exact mechanisms for stress conditions that occur due to water relations and pollution. The impact of emerging pollutants from a wide variety of sources like organic sources, artificial light, and newly recognized air pollutants is reflected in ultrastructural changes and physiological processes. It widens the scope of research on lichen physiology. Moreover, in-depth studies are needed to understand ecophysiology in respond to microclimate as well as to discover lichen-derived bioactive secondary metabolites to develop new sustainable products.

References

1 Goga, M., Elečko, J., Marcinčinová, M. et al. (2020). Lichen metabolites: an overview of some secondary metabolites and their biological potential. In: *Co-Evolution of Secondary Metabolites* (ed. J.M. Mérillon and K. Ramawat), 175–209. Cham: Springer.

2 Margulis, L. and Fester, R. (ed.) (1991). *Symbiosis as a Source of Evolutionary Innovation: Speciation and Morphogenesis*. MIT press.

3 Nash, T.H. III, Kappen, L., Lösch, R. et al. (1987). Cold resistance of lichens with *Trentepohlia*-or *Trebouxia*-photobionts from the North American west coast. *Flora* 179 (4): 241–251.

4 Honegger, R. (1984). Cytological aspects of the mycobiont–phycobiont relationship in lichens: haustorial types, phycobiont cell wall types, and the ultrastructure of the cell surface layers in some cultured and symbiotic myco-and phycobionts. *The Lichenologist* 16 (2): 111–127.

5 Honegger, R. (1991). 22 fungal evolution: symbiosis and morphogenesis. In: *Symbiosis as a Source of Evolutionary Innovation: Speciation and Morphogenesis* (ed. L. Margulis and R. Fester), 319–340. Cambridge: MIT Press, Massachusetts.

6 Scherrer, S., De Vries, O.M., Dudler, R. et al. (2000). Interfacial self-assembly of fungal hydrophobins of the lichen-forming ascomycetes *Xanthoria parietina* and *X. ectaneoides*. *Fungal Genetics and Biology* 30 (1): 81–93.

7 Honegger, R. (1990). Surface interactions in lichens. In: *Cell Walls and Surfaces, Reproduction, Photosynthesis* (ed. W. Wiessner, D.G. Robinson, and R.C. Starr), 40–54. Berlin, Heidelberg: Springer.

8 Honegger, R. (1996). Experimental studies of growth and regenerative capacity in the foliose lichen *Xanthoria parietina*. *New Phytologist* 133 (4): 573–581.

9 Nash, T.H. (1996). Photosynthesis, respiration, productivity and growth. In: *Lichen Biology* (ed. T.H. Nash), 88–120. Cambridge: Cambridge University Press.

10 Nash, T.H. III (2008). Nitrogen, its metabolism and potential contribution to ecosystems. In: *Lichen Biology* (ed. T.H. Nash III), 216–233. Cambridge, UK: Cambridge University Press.

11 Friedl, T. and Büdel, B. (2008). Photobionts. In: *Lichen Biology* (ed. T.H. Nash III), 9–26. Cambridge, UK: Cambridge University Press.

12 Büdel, B. (1992). Taxonomy of lichenized procaryotic blue-green algae. In: *Algae and Symbioses* (ed. W. Reisser). Bristol: Biopress Limited.

13 Lange, O.L., Kilian, E., and Ziegler, H. (1986). Water vapor uptake and photosynthesis of lichens: performance differences in species with green and blue-green algae as phycobionts. *Oecologia* 71: 104–110.

14 Meeßen, J. and Ott, S. (2013). Recognition mechanisms during the pre-contact state of lichens: I. Mycobiont-photobiont interactions of the mycobiont of *Fulgensia bracteata*. *Symbiosis* 59: 121–130.

15 Insarova, I.D. and Blagoveshchenskaya, E.Y. (2016). Lichen symbiosis: search and recognition of partners. *Biology Bulletin* 43: 408–418.

16 Singh, R.S. and Walia, A.K. (2014). Characteristics of lichen lectins and their role in symbiosis. *Symbiosis* 62: 123–134.

17 Legaz, M.E., Fontaniella, B., Millanes, A.M., and Vicente, C. (2004). Secreted arginases from phylogenetically far-related lichen species act as cross-recognition factors for two different algal cells. *European Journal of Cell Biology* 83 (8): 435–446.

18 Sacristán, M., Millanes, A.M., Legaz, M.E., and Vicente, C. (2006). A lichen lectin specifically binds to the α-1, 4-polygalactoside moiety of urease located in the cell wall of homologous algae. *Plant Signaling & Behavior* 1 (1): 23–27.

19 Lockhart, C.M., Rowell, P., and Stewart, W.D.P. (1978). Phytohaemagglutinins from the nitrogen-fixing lichens *Peltigera canina* and *P. polydactyla*. *FEMS Microbiology Letters* 3 (3): 127–130.

20 Díaz, E.M., Sacristán, M., Legaz, M.E., and Vicente, C. (2009). Isolation and characterization of a cyanobacterium-binding protein and its cell wall receptor in the lichen *Peltigera canina*. *Plant Signaling & Behavior* 4 (7): 598–603.

21 Díaz, E.M., Vicente-Manzanares, M., Sacristan, M. et al. (2011). Fungal lectin of *Peltigera canina* induces chemotropism of compatible Nostoc cells by constriction-relaxation pulses of cyanobiont cytoskeleton. *Plant Signaling & Behavior* 6 (10): 1525–1536.

22 Wang, Y.Y., Liu, B., Zhang, X.Y. et al. (2014). Genome characteristics reveal the impact of lichenization on lichen-forming fungus *Endocarpon pusillum* Hedwig (Verrucariales, Ascomycota). *BMC Genomics* 15: 1–18.

23 Miao, V.P., Manoharan, S.S., Snæbjarnarson, V., and Andrésson, Ó.S. (2012). Expression of lec-1, a mycobiont gene encoding a galectin-like protein in the lichen *Peltigera membranacea*. *Symbiosis* 57: 23–31.

24 Bačkor, M., Ivanova, V., Laatsch, H. et al. (2013). Allelopathic effects of lichen secondary metabolites on lichen photobiont Trebouxia erici. *Allelopathy Journal* 31 (1): 189–198.

25 Huneck, S. and Yoshimura, I. (1996). *Identification of Lichen Substances*, 11–123. Berlin Heidelberg: Springer.

26 Hale, M.E. (1983). *The Biology of Lichens*, 3e. London: Edward Arnold.

27 Culberson, W.L. (1970). Chemosystematics and ecology of lichen-forming fungi. *Annual Review of Ecology and Systematics* 1: 153–170.

28 Kosanić, M. and Ranković, B. (2015). Lichen secondary metabolites as potential antibiotic agents. In: *Lichen Secondary Metabolites Bioactive Properties and Pharmaceutical Potential* (ed. B. Ranković). Springer Cham Heidelberg New York Dordrecht London: Springer International Publishing.

29 BeGora, M.D. and Fahselt, D. (2001). Usnic acid and atranorin concentrations in lichens in relation to bands of UV irradiance. *The Bryologist* 104 (1): 134–140.

30 McEvoy, M., Nybakken, L., Solhaug, K.A., and Gauslaa, Y. (2006). UV triggers the synthesis of the widely distributed secondary lichen compound usnic acid. *Mycological Progress* 5: 221–229.

31 McEvoy, M., Gauslaa, Y., and Solhaug, K.A. (2007). Changes in pools of depsidones and melanins, and their function, during growth and acclimation under contrasting natural light in the lichen Lobaria pulmonaria. *New Phytologist* 175 (2): 271–282.

32 Solhaug, K.A., Lind, M., Nybakken, L., and Gauslaa, Y. (2009). Possible functional roles of cortical depsides and medullary depsidones in the foliose lichen Hypogymnia physodes. *Flora-Morphology, Distribution, Functional Ecology of Plants* 204 (1): 40–48.

33 Gasulla, F., Del Campo, E.M., Casano, L.M., and Guéra, A. (2021). Advances in understanding of desiccation tolerance of lichens and lichen-forming algae. *Plants* 10 (4): 807.

34 Kappen, L. and Valladares, F. (2007). Opportunistic growth and desiccation tolerance: the ecological success of poikilohydrous autotrophs. In: *Functional Plant Ecology* (ed. F. Pugnaire and F. Valladares), 7–66. CRC Press.

35 Green, T.G.A., Nash, T.H., and Lange, O.L. (2008). Physiological ecology of carbon dioxide exchange. In: *Lichen Biology* (ed. T.H. Nash), 152–181. Cambridge: Cambridge University Press.

36 Harris, G.P. and Kershaw, K.A. (1971). Thallus growth and the distribution of stored metabolites in the phycobionts of the lichens *Parmelia sulcata* and *P. physodes*. *Canadian Journal of Botany* 49 (8): 1367–1372.

37 Lange, O.L. and Tenhunen, J.D. (1981). Moisture content and CO 2 exchange of lichens. II. Depression of net photosynthesis in Ramalina maciformis at high water content is caused by increased thallus carbon dioxide diffusion resistance. *Oecologia* 51: 426–429.

38 Kappen, L. (1973). Response to extreme environments. In: *The Lichens* (ed. V. Ahmadjian and M.E. Hale), 311–380. Academic Press.

39 Gasulla, F., de Nova, P.G., Esteban-Carrasco, A. et al. (2009). Dehydration rate and time of desiccation affect recovery of the lichenic algae Trebouxia erici: alternative and classical protective mechanisms. *Planta* 231: 195–208.
40 Lange, O.L. and Matthes, U. (1981). Moisture-dependent CO_2 exchange of lichens. *Photosynthetica* 15 (4): 555–574.
41 Bewley, D.J. (1979). Physiological aspects of desiccation tolerance. *Annual Reviews of Plant Physiology* 30 (1): 195–238.
42 Oliver, M.J. and Derek Bewley, J. (1996). Desiccation-tolerance of plant tissues: a mechanistic overview. *Horticultural Reviews* 18: 171–213.
43 Oliver, M.J., Velten, J., and Wood, A.J. (2000). Bryophytes as experimental models for the study of environmental stress tolerance: *Tortula ruralis* and desiccation-tolerance in mosses. *Plant Ecology* 151: 73–84.
44 Larson, D.W. (1987). The absorption and release of water by lichens. *Bibliotheca Lichenologica* 25 (35): 351–360.
45 Hauck, M. and Wirth, V. (2010). Preference of lichens for shady habitats is correlated with intolerance to high nitrogen levels. *Lichenologist* 42: 475–484.
46 Carter, T.S., Clark, C.M., Fenn, M.E. et al. (2017). Mechanisms of nitrogen deposition effects on temperate forest lichens and trees. *Ecosphere* 8: e01717.
47 Nardini, A., Marchetto, A., and Tretiach, M. (2013). Water relation parameters of six Peltigera species correlate with their habitat preferences. *Fungal Ecology* 6: 397–407.
48 Jonsson Cabrajić, A.V., Lidén, M., Lundmark, T. et al. (2010). Modelling hydration and photosystem II activation in relation to in situ rain and humidity patterns: a tool to compare performance of rare and generalist epiphytic lichens. *Plant Cell Environment* 33: 840–850.
49 Lidén, M., Jonsson Cabrajić, A.V., Ottosson-Löfvenius, M. et al. (2010). Species-specific activation time-lags can ˇ explain habitat restrictions in hydrophilic lichens. *Plant Cell Environment* 33: 851–862.
50 Hawksworth, D.L. and Grube, M. (2020). Lichens redefined as complex ecosystems. *The New Phytologist* 227 (5): 1281.

7

Lichen as Pioneer of Natural Ecosystem

Bhaskor Kolita[1] *and Ridip Kumar Gogoi*[2]

[1] *Department of Botany, Jorhat Kendriya Mahavidyalaya, Jorhat, Assam, India*
[2] *Bioinformatics Infrastructure Facility Centre, Rajiv Gandhi University, Doimukh, Arunachal Pradesh, India*

7.1 Introduction

Lichens are symbiotic organisms made up of filaments of various fungal species known as mycobiont and algae or cyanobacteria as the photobiont. Algal or cyanobacterial cells that are capable of photosynthesis transform ambient CO_2 into organic sugars of carbon that support both the symbionts. The fungus filaments then defend the photobiont and deliver nutrients and moisture from the surrounding environment [1]. As pioneers, lichens colonize wood from ancient trees, terricolous lichens soil, rocks from saxicolous lichens, old wood mosses, and evergreen leaves. Even some lichen species will grow on other lichens like parasites. Lichens show their presence on a number of substrate surfaces, including old walls, metal or wooden fences, posts, railway rails, worn-out leather, glass, plastic, and so forth.

Apothecia, the fruiting bodies developed by the fungal partner, which have to germinate and find an algal partner previous to forming a new lichen thallus, or may develop into minute fragments of finger-like outgrowths called as isidia, or sugar-like granules known as soredia, which contains the both partners, have the capacity to disperse quickly and colonize the available habitats. This way, they are responsible for the development and establishment of lichen on a substrate. Lichens frequently establish themselves on the tree trunks, twigs, and branches because the bark offers a stable environment where they may gather the necessary sunlight, rainfall, and materials from the environment. Lichens are an important part of the ecosystem as they can establish life on barren, rock, and disturbed areas because they are pioneers on rock surfaces. As lichens grow on rocks, they trap silt, dust, and water, causing the rock surface to wither biogeophysically and biogeochemically, resulting in the development of soil. Another crucial substrate for lichens is soil. It offers shelter as well as moisture, nutrients, and space for growth, depending on the environment. Lichens can grow on any substrate and in any climate, although the diversity of lichens found in a particular location or substratum is greatly influenced by the local microclimate [2]. Lichens provide a way for algae to convert atmospheric carbon dioxide into oxygen, which is necessary for all life, and

Chemistry, Biology and Pharmacology of Lichen, First Edition. Edited by Ashoke Kumar Das, Ajay Sharma, Deepika Kathuria, Mohammad Javed Ansari, and Garima Bhardwaj.

to allow them to thrive in a variety of climates around the planet. By modifying soil qualities, including porosity, pH regulation, water holding capacity, mineralization, and converting atmospheric carbon dioxide into oxygen through photosynthesis, lichens have a profound impact on ecosystems [3]. The impact of lichens on natural ecosystems and their distinctive traits as pioneer species in ecological succession are covered in this chapter.

7.2 Lichens as Pioneer of Ecological Succession

Because they were the first to settle on bare rocks, mountains, and cliffs, lichens are considered pioneer creatures. They gather minerals and organic substances while corroding rocks. Later in the succession, plants like mosses and grasses emerge, taking advantage of the initial soil created by lichens. Lichens can thereby turn a desolate region into one that can sustain vegetation. Tree trunks can also be overgrown with lichens.

7.2.1 Lichens as a Pioneer on Rock Surfaces

Lichens are multicellular complex organisms with symbiotic relationships that interact with exhabitant fungi (sometimes more than one) and an arrangement of extracellular photosynthetic partners with unspecified number of additional microscopic organisms to form a self-sustaining ecosystem. Since lichens offer a number of benefits over other species and are among the first microbes to colonize surfaces like rocks, they are regarded as pioneers in this field. Saxicolous lichens, which make up the bulk of lichens, like to colonize the surfaces of rocks, but endolithic lichens, which interact with the lithomatrix, prefer to do so [4].

According to research on how lichens affect historical sites, their mycelial filaments can pierce granitic substrates up to 2 mm deep [5]. The silica can then be effectively "digested" and incorporated by the lichen thalli. Lava released during eruptions is swiftly colonized by lichens. Let's look at two examples: *Dictyonema pavonia*, which grows on the side of the Guadeloupe Soufriere volcano, and *Stereocaulon vesuvianum*, a type of cyanobacteria and green algae that was noticed on the deposits of volcanic eruptions of Vesuvius. Researchers have been able to comprehend *Rhizocarpon geographicum*'s invasion strategy better thanks to study on quartzite colonization in alpine regions [6]. The compartments that make up the crustose thallus of this species include both fungus and algae (or areolae). The earliest areolae are produced when the germinated spores of fungus forms a black mycelium (hypothallus), which expands radially. The hypothallus and confluent both have additional areolae. The initial thalli can then confluence and grow into complete thalli with a maximum diameter of 15 cm [5].

7.2.2 Lichens as Pioneers on Soil Surfaces

True soil pioneers include lichens like *Cladonia* and *Baeomyces*. Two species, a lichen named *Baeomyces rufus* and a moss named *Pogonatum urnigerum*, were seen colonizing slopes that had been left barren for the construction of a ski slope, according to observations made in the Alps. In order to help stabilize the surface soil, *Baeomyces* starts to grow on the bare substrate's surface and produces a significant mycelial felting [7]. In contrast to the lichen, the *Pogonatum* moss adopts a diverse strategy and develops deeper. The development of an actual lichen-moss complex encourages the development of a more complex soil [7].

In *Cladonia*, the base of the podetions lyses, releasing mycelial filaments that progressively bind to the earthy particles. As in the preceding instance, the development of a true lichen-soil complex, frequently rich in bacterial colonies and promotes plant colonization [8]. In various places of the world, lichens are an essential component of ecosystems. In the Arctic, the ratio of lichenic to phanerogamic flora can approach 100, compared to the global norm of 0.1–1. In the tundra regions or northern temperate woodlands, various species of lichens, including reindeer lichen, *Cladonia rangiferina*, are favored foods for many animals, including Reindeer, Moose, Mouflon, and infrequently certain domestic animals.

In dry and semi-arid regions, soil lichens help in the formation of "biological crusts" that colonize desert soils. Lichens, cyanobacteria, fungus, bryophytes, and algae, which make up these organisms, cohabit in the upper centimeters of the soil surfaces. These crusts span an area of five times the Europe's surface area of about 50 million km^2. They are crucial in the defense against erosion and fertilization of the soil, as well as the fight against desertification. They alone are thought to be responsible for fixing 7% of the carbon dioxide that humans emit on a yearly basis when they burn fossil fuels [9].

7.2.3 Lichens as Pioneers for Colonizers on Tree

Because the bark of trees provides a stable environment for gathering necessary sunshine, moisture, and materials from the air, lichens typically colonize tree trunks, branches, and twigs. Both healthy and ill trees—those under stress or in other ways—have them grow. Homeowners may be alarmed by the appearance of colored creatures mounting on the bark of nearby shrubs or trees. Homeowners may misidentify lichens as a sort of moss or believe that they are the source of plant problems due to their mystery.

A "cause-and-effect" relationship between disease and deterioration is suggested by the fact that certain lichens are extra noticeable on stressed or aged tree branches and trunks. The higher amount of sunlight that is available on branches and trees with less or partial foliation are the main cause of their enhanced possibility. A healthy tree's bark keeps growing and shedding as it becomes bigger. An aged or stressed tree's bark may turn into more fragile, with additional fissures and irregular surfaces, which makes it simpler for lichens to cling to. The kind of lichen that can thrive there changes as bark matures due to changes in its capacity, chemistry, and texture to hold water.

While certain fungi outside of lichens can and do damage by penetration and start the disintegration of the tree, but the lichen-dwelling fungus do not harm trees. Fungi play a significant part in nature by removing fallen tree trunks and dumping the residual material into the forest floor. They are among one of the limited living organisms that can degrade all of the compounds in wood tissue. On the same dead branch or area of a diseased or stressed tree, lichen and other fungi can coexist. If so, the homeowner or a tree specialist may need to remove the dead tissue in order to preserve the health of the surviving plant. When making these kinds of choices, homeowners need to be aware of the special characteristics of lichen as well as the distinctions between fungi that are symbiotic with lichen and those that are acting independently [10, 11].

7.3 Impact of Lichen on Natural Ecosystem

Although lichens only make up a small portion of the biomass in most terrestrial habitats, they do so in some, including drylands, forests, and tundras. In environments like dry, cold, poor nutrient, or a continuous or full plant cover, lichens usually predominate. On average, lichens cover 8% of the terrestrial area on Earth. With more than 18,000 species globally, lichens outnumber vascular plant species at higher latitudes [12].

7.3.1 Impacts of Lichens on Biogeochemical Cycles of Ecosystem

Despite their potential influence on biogeochemistry, ecosystem health, and social processes, these topics have received relatively less attention from the scientific community. Lichen species work with cyanobacteria to fix atmospheric N_2 and ingest nutrients from the air, which drives ecosystem biogeochemistry. As a result, through animal ingestion, decomposition, and leaching, the nutrients retained by lichens reach other ecological components. In addition, lichens provide habitats for a variety of invertebrates, which may or may not use them as food. Rarely the significance of lichen functional features for driving ecological processes in further ecologically significant autotrophs recognized [13–15]. Regardless of this, they have a unique set of functional qualities that are comparable to the kinds of functional traits commonly researched in vascular plants, offering a potential mechanistic framework to comprehend their role in community and ecosystem processes [16]. The effect of lichens on communities and ecological features depends on species-specific changes in lichen growth form, functional traits, kind of photobiont, capacity of water-holding, color, and secondary chemistry. Along various lichen growth forms, lichenized fungus also differs in their relationships with their photobionts, which have significant levels of ecosystem effects. For instance, cephalolichens, cyanolichens, and chlorolichens all contain green algae as their sole photobiont. Cyanolichens also have cyanobacteria as their sole photobiont. The main apparent dissimilarity among these groups is that those lichens which include cyanobacterial symbionts can normally fix N_2 and have elevated concentrations of N. Because of their ability to accumulate and take up nutrients, lichens in a number of ecosystems store a considerable proportion of the whole nutrients present in the ecosystem. Roughly 10% of every lichen species enclose N_2-fixing cyanobacteria, which can deeply enhance the inputs of N to the nutrient-poor ecosystems. There is not much difference in the release of N during decomposition amid N_2-fixing and non-N_2-fixing lichens. Thereby, lichens growth form also plays a significant job in the release of N to the atmosphere [14].

7.3.2 Impact of Lichens on Maintaining Humidity and as a Source of Food and Shelter in the Ecosystem

Contrarily, cyanolichens require liquid water to activate photosynthesis [17], which explains why they are most plentiful in open and rainforest areas with recurrent intense dewfall [18]. Cephalolichens and Chlorolichens readily turn on their photosynthesis in elevated ambient air humidity and favor habitats that are scarce in liquid water [17, 19]. Lichens have a wide range of functional characteristics related to consuming and retaining resources [14, 16], which may be crucial in defining how they affect ecological processes [13] and coupled invertebrate communities [20]. Some of the properties, which are similar to the functional characteristics of vascular plant leaves and are widely acknowledged as significant ecological drivers, are defensive chemicals, thallus nutrition content, water-holding capacity, and specific thallus mass (STM). Due to their considerable intraspecific variability, lichens can soak up and build up nutrients from atmospheric sources differently depending on the species [21]. The fact that lichens diverge significantly in their ability to retain moisture is one of the significant ecological consequences of lichens. Lichens come in a variety of shapes and have varying effects on the ecosystem. For instance, thin, pendulous lichens have a limited capacity to store water [22], despite the fact that they may swiftly absorb it from humid air. Long-lasting water retention is possible in foliose cyanobacterial lichens that are thick or gel-like [23–26]. Lichens' ability to retain water is positively correlated with both their inter- and intra-species STM [22, 27, 28]. Thin chlorolichens use damp air every night, which is a trade-off between the technique and the water-storage conservative method used by cyanolichens, which limits their photosynthesis during less frequent wet periods [29]. Epiphytic lichens are important in the distribution of moisture from precipitation, which helps to keep the interior forest's humidity levels high [30]. Fruticose epiphytic lichens in low-rainfall areas take up moisture from the fog and release it into the underlying soils, which improve soil moisture for plant growth [31, 32]. Lichens come in a wide range of colors, from white to black. Wide variations in thallus surface temperatures are caused by this variation in spectral distinctiveness [33, 34]. Variations in lichen color may also have an impact on the soil's surface microclimate [35]. For instance, the albedo of the ground surface can be increased by around 1% by mat-forming light-colored lichens [36].

Fog and dew, which frequently contain more nutrients than rain, are primarily captured by the growth form of hair-like fruticose lichens [12]. They are also better at capturing elements than foliose lichens due to their relatively large surface area [37], while foliose lichens are often rich in N, Ca, and P than fruticose lichens [38]. Various microorganisms can also find microhabitats in lichens, depending on the lichen species [39–41]. The classic notion of lichens as being just a symbiosis among fungus or photobionts is challenged by the idea that lichen is an integral constituent of the lichen meta-organism. Lichens play a significant function in lichen-derived food webs as food for protozoa and nematodes, with bacterial cell densities exceeding those on vascular plant leaves [40, 42–44]. The diversity and abundance of bacteria in different lichen species vary greatly, and this variation is mostly caused by variations in lichen photobiont type and growth form [45]. Numerous aquatic microfauna, including nematodes, rotifers, protozoa, and tardigrades, are supported by lichen thallus, which also serves as the base of a complex food web. Lichens are a supply of food and shelter for a number of other terrestrial invertebrates, including moth larvae, mites, mite larvae, springtail larvae, beetles, and woodlice [46]. A small number of snail species are solely dependent on lichens for their whole diet. Lichens provide gastropods with a food source as well as protection from predators and desiccation [47]. Lichens are used by many different types of vertebrate animals, including birds, as adornment or nesting material [48]. Additionally, flying squirrels build their nests out of lichens, primarily the fruit-eating *Bryoria* genus. Mammals like deer (*Cervidae*), ibex (*Capra* spp.), elk (*Cervus elaphus*), gazelle (*Gazella* spp.), mountain goat (*Oreamnos americanus*), musk ox (*Ovibos moschatus*), polar bear, vole (*Myodes* spp.), *Lemmus* spp., marmot (*Marmota* spp.), squirrel (*Sciuridae*) and monkeys also feed on lichens to varying extents in various regions of the world [49].

Non-N-fixing lichens also increase the availability of soil nutrients. For instance, in dry environments, biological soil crust lichens can change the chemistry of the soil by storing nutrients in their thalli [50]. Early lichen colonization also accelerates rock weathering by releasing minerals into the soil that plants can use [51–53]. Additionally, depending on the neighboring environment and the kinds of plants and lichen species present, they have been shown to either decrease or increase the establishment of vascular plant seedlings [54–57]. By changing water availability, nutrient fluxes, and erosion processes in dry land habitats, biological soil crust lichens have an impact on plant communities [58]. Thus, the functional characteristics of lichens are highly variable, and this variation affects how lichens interact with other primary producers and higher trophic level producers, with implications for community and ecosystem properties. As a result, lichens have an impact on ecosystem processes that primarily involve carbon fluxes, nitrogen fixation, and nutrients.

7.4 Conclusion

Lichens have a significant impact on ecosystem processes, particularly those related to carbon fluxes, nitrogen fixation, and nutrient fluxes. Lichen interactions with other primary producers and higher tropic levels have a diverse impact on the ecosystem, resulting in proper ecosystem balance. More than 18,000 different types of lichens add to the planet's biological variety. The availability of suitable substrates in climate-friendly locales, continuity of their habitats, and high air quality all contribute to their variety. They frequently live in communities with different species. The distribution of microbes in soil, leaf, and aquatic habitats and invertebrates can be shaped according to lichen-dominated areas with their unique chemical compounds. Cyanobacterial lichens can convert atmospheric nitrogen into forms that lichens and other plants and animals can use. They capture and release nutrients from fine particles in the air, rain, fog, dew, and gases into the natural ecosystem, enriching it with natural properties. They are also effective ecological indicators of natural ecosystems. Lichens aid in forest continuity by detecting the distribution of specialized microhabitats and microclimates, as well as biological diversity. They also aid in the assessment of air and water quality. Sensitive lichens are initially impacted by air pollutants that comprise nitrogen and sulfur due to their increased ecological effects. Because of how deeply entwined they are with the ecosystem, damage to them also has a negative impact on more tolerant or commercially useful species. They provide food, cover, and nesting material for a wide range of birds, insects, and mammals, as well as contributing to the water and mineral cycles of forests and rangelands. Lichens thus play a significant part in the planet's biological diversity.

References

1 Adenubi, O.T., Famuyide, I.M., McGaw, L.J., and Eloff, J.N. (2022). Lichens: an update on their ethnopharmacological uses and potential as sources of drug leads. *Journal of Ethnopharmacology* 298: 115657.

2 Shukla, V., Upreti, D.K., and Bajpai, R. (2014). Introduction. In: *Lichens to Biomonitor the Environment*. New Delhi: Springer. https://doi.org/10.1007/978-81-322-1503-5_1.

3 Antony, A.C., George, S., Chandran, S., and Sebastian, A. (2020). Chapter 10 – Lichens as a source and indicator of agrochemicals. In: *Agrochemicals Detection, Treatment and Remediation* (ed. M.N.V. Prasad), 241–261. Cambridge, MA: Butterworth-Heinemann.

4 Jung, P. and Büdel, B. (2021). 2.3 Lichens as pioneers on rock surfaces. In: *Life at Rock Surfaces: Challenged by Extreme Light, Temperature and Hydration Fluctuations* (ed. B. Büdel and T. Friedl), 141–160. Berlin: De Gruyter.

5 Galsomiès, L., Robett, M., and Orial, G. (1999). Interaction lichens-roche sur monument historique en granite. *Bulletin d'Informations de l'Association Française de Lichénologie, Mémoires* 3: 35–42.

6 Asta, J. and Letrouit, M.A. (1995). Observations on the early growth of *Rhizocarpon geographicum* thalli. *Herzogia* 11: 229–252.

7 Asta, A. and Souchier, B. (1999). Lichens et pédogenèse: dynamique de la végétation et étudesmicromorphologiques de l'interface-lichen-sol. *Bulletin d'Informations de l'Association Française de Lichénologie, Mémoires* 3: 29–34.

8 Asta, J., Orry, F., Toutain, F. et al. (2001). Micromorphological and ultastructural investigations of the lichen-soil interface. *Soil Biology and Biochemistry* 33: 323–337.

9 ASTA Juliette (2019). Lichens, surprising pioneering organisms, *Encyclopedia of the Environment* (21 August), [online ISSN 2555-0950]. https://www.encyclopedie-environnement.org/en/life/lichens-pioneering-organisms/ (accessed 1 February 2024).

10 Sandoval, S. (2008). *What About the Lichen on My Tree?* Las Cruces, NM: New Mexico State University, Cooperative Extension Service, College of Agriculture and Home Economics Guide H-167: 9 pars, June 13; aces. nmsu.edu/pubs/_h/h-167.pdf (PDF).

11 Stephenson, S.L. (2010). *The Kingdom Fungi: The Biology of Mushrooms, Molds, and Lichens*. Portland: Timber Press.

12 Nash, T.H. (2008). *Lichen Biology*. Cambridge: Cambridge University Press.

13 Lang, S.I., Cornelissen, J.H.C., Klahn, T. et al. (2009). An experimental comparison of chemical traits and litter decomposition rates in a diverse range of subarctic bryophyte, lichen and vascular plant species. *Journal of Ecology* 97: 886–900.

14 Asplund, J. and Wardle, D.A. (2013). The impact of secondary compounds and functional characteristics on lichen palatability and decomposition. *Journal of Ecology* 101: 689–700.

15 Zedda, L. and Rambold, G. (2015). The diversity of lichenised fungi: ecosystem functions and ecosystem services. In: *Recent Advances in Lichenology* (ed. D.K. Upreti, P.K. Divakar, V. Shukla, and R. Bajpai), 121–145. New Delhi: Springer.

16 Cornelissen, J.H.C., Lang, S.I., Soudzilovskaia, N.A., and During, H.J. (2007). Comparative cryptogam ecology: a review of bryophyte and lichen traits that drive biogeochemistry. *Annals of Botany* 99: 987–1001.

17 Lange, O.L., Kilian, E., and Ziegler, H. (1986). Water vapor uptake and photosynthesis of lichens: performance differences in species with green and blue-green algae as phycobionts. *Oecologia* 71: 104–110.

18 Gauslaa, Y. (2014). Rain, dew, and humid air as drivers of morphology, function and spatial distribution in epiphytic lichens. *The Lichenologist* 46: 1–16.

19 Smith, C.W., Aptroot, A., Coppins, B.J. et al. (2009). *The Lichens of Great Britain and Ireland*, 2e. London: British Lichen Society.

20 Bokhorst, S., Asplund, J., Kardol, P., and Wardle, D.A. (2015). Lichen physiological traits and growth forms affect communities of associated invertebrates. *Ecology* 96: 2394–2407.

21 Asplund, J., Sandling, A., and Wardle, D.A. (2012). Lichen specific thallus mass and secondary compounds change across a retrogressive fire-driven chronosequence. *PLoS ONE* 7: e49081.

22 Esseen, P.-A., Olsson, T., Coxson, D., and Gauslaa, Y. (2015). Morphology influences water storage in hair lichens from boreal forest canopies. *Fungal Ecology* 18: 26–35.

23 Lange, O.L., Büdel, B., Heber, U. et al. (1993). Temperate rainforest lichens in New Zealand: high thallus water content can severely limit photosynthetic CO_2 exchange. *Oecologia* 95: 303–313.

24 Gauslaa, Y. and Solhaug, K.A. (1998). The significance of thallus size for the water economy of the cyanobacterial old forest lichen *Degelia plumbea*. *Oecologia* 116: 76–84.

25 Lange, O.L., Belnap, J., and Reichenberger, H. (1998). Photosynthesis of the cyanobacterial soil-crust lichen *Collema tenax* from arid lands in southern Utah, USA: role of water content on light and temperature responses of CO_2 exchange. *Functional Ecology* 12: 195–202.

26 Lange, O.L. (2000). Photosynthetic performance of a gelatinous lichen under temperate habitat conditions: long-term monitoring of CO_2 exchange of *Collema cristatum*. *Bibliotheca Lichenologica* 75: 307–332.

27 Gauslaa, Y. and Coxson, D. (2011). Interspecific and intraspecific variations in water storage in epiphytic old forest foliose lichens. *Botany* 89: 787–798.

28 Merinero, S., Hilmo, O., and Gauslaa, Y. (2014). Size is a main driver for hydration traits in cyano- and cephalolichens of boreal rainforest canopies. *Fungal Ecology* 7: 59–66.

29 Gauslaa, Y., Coxson, D.S., and Solhaug, K.A. (2012). The paradox of higher light tolerance during desiccation in rare old forest cyanolichens than in more widespread co-occurring chloro- and cephalolichens. *New Phytologist* 195: 812–822.

30 Van Stan, J.T. II and Pypker, T.G. (2015). A review and evaluation of forest canopy epiphyte roles in the partitioning and chemical alteration of precipitation. *Science of the Total Environment* 536: 813–824.

31 Stanton, D.E. and Horn, H.S. (2013). Epiphytes as 'filter-drinkers': life-form changes across a fog gradient. *The Bryologist* 116: 34–42.

32 Stanton, D.E., Armesto, J.J., and Hedin, L.O. (2014). Ecosystem properties self-organize in response to a directional fog–vegetation interaction. *Ecology* 95: 1203–1212.

33 Kershaw, K.A. (1975). Studies on lichen-dominated systems. XII. The ecological significance of thallus color. *Canadian Journal of Botany* 53: 660–667.

34 Gauslaa, Y. (1984). Heat resistance and energy budget in different Scandinavian plants. *Holarctic Ecology* 7: 1–78.

35 Kershaw, K.A. (1978). The role of lichens in boreal tundra transition areas. *The Bryologist* 81: 294–306.

36 Stoy, P.C., Street, L.E., Johnson, A.V. et al. (2012). Temperature, heat flux, and reflectance of common subarctic mosses and lichens under field conditions: might changes to community composition impact climate-relevant surface fluxes? *Arctic, Antarctic, and Alpine Research* 44: 500–508.

37 Yemets, O.A., Solhaug, K.A., and Gauslaa, Y. (2014). Spatial dispersal of airborne pollutants and their effects on growth and viability of lichen transplants along a rural highway in Norway. *The Lichenologist* 46: 809–823.

38 Mangelson, N.F., Argyle, D.C., Kelly, R. et al. (2002). Elemental analysis of lichens from the western United States: distribution of phosphorus and calcium from a large data set. *International Journal of PIXE* 12: 167–173.

39 Lawrey, J.D. and Diederich, P. (2003). Lichenicolous fungi: interactions, evolution, and biodiversity. *Bryologist* 106: 80–120.

40 Grube, M. and Berg, G. (2009). Microbial consortia of bacteria and fungi with focus on the lichen symbiosis. *Fungal Biology Reviews* 23: 72–85.

41 Bates, S.T., Cropsey, G.W.G., Caporaso, J.G. et al. (2011). Bacterial communities associated with the lichen symbiosis. *Applied and Environmental Microbiology* 77: 1309–1314.

42 Aschenbrenner, I.A., Cernava, T., Berg, G., and Grube, M. (2016). Understanding microbial multi-species symbioses. *Microbial Symbioses* 7: 1–9.

43 Cardinale, M. Jr., Vieira de Castro, J., Müller, H. et al. (2008). In situ analysis of the bacterial community associated with the reindeer lichen *Cladonia arbuscula* reveals predominance of *Alphaproteobacteria*. *FEMS Microbiology Ecology* 66: 63–71.

44 Saleem, M. (2015). Microbiome ecosystem ecology: unseen majority in an anthropogenic ecosystem. In: *Microbiome Community Ecology*, 1–11. Cham: Springer International Publishing. https://doi.org/10.1007/978-3-319-11665-5_1.

45 Hodkinson, B.P., Gottel, N.R., Schadt, C.W., and Lutzoni, F. (2012). Photoautotrophic symbiont and geography are major factors affecting highly structured and diverse bacterial communities in the lichen microbiome. *Environmental Microbiology* 14: 147–161.

46 Gerson, U. and Seaward, M.R.D. (1977). Lichen-invertebrate associations. In: *Lichen Ecology* (ed. M.R.D. Seaward), 69–119. London: Academic Press.

47 Kerney, M.P. (1999). *Atlas of the Land and Freshwater Molluscs of Britain and Ireland*. Colchester: Harley.

48 Richardson, D.H.S. and Young, C.M. (1977). Lichens and vertebrates. In: *Lichen Ecology* (ed. M.R.D. Seaward), 121–144. London: Academic Press.

49 Seaward, M.R.D. (2008). Environmental role of lichens. In: *Lichen Biology*, 2e (ed. T.H. Nash), 274–298. Cambridge: Cambridge University Press.

50 Delgado-Baquerizo, M., Gallardo, A., Covelo, F. et al. (2015). Differences in thallus chemistry are related to species-specific effects of biocrust-forming lichens on soil nutrients and microbial communities. *Functional Ecology* 29: 1087–1098.

51 Viles, H. (1995). Ecological perspectives on rock surface weathering: towards a conceptual model. *Geomorphology* 13: 21–35.

52 Adamo, P. and Violante, P. (2000). Weathering of rocks and neogenesis of minerals associated with lichen activity. *Applied Clay Science* 16: 229–256.

53 Chen, J., Blume, H.-P., and Beyer, L. (2000). Weathering of rocks induced by lichen colonization – a review. *CATENA* 39: 121–146.

54 Deines, L., Rosentreter, R., Eldridge, D.J., and Serpe, M.D. (2007). Germination and seedling establishment of two annual grasses on lichen-dominated biological soil crusts. *Plant and Soil* 295: 23–35.

55 Zamfir, M. (2000). Effects of bryophytes and lichens on seedling emergence of alvar plants: evidence from greenhouse experiments. *Oikos* 88: 603–611.

56 Houle, G. and Filion, L. (2003). The effects of lichens on white spruce seedling establishment and juvenile growth in a spruce-lichen woodland of subarctic Quebec. *Ecoscience* 10: 80–84.

57 Escudero, A., Martínez, I., de la Cruz, A. et al. (2007). Soil lichens have species-specific effects on the seedling emergence of three gypsophile plant species. *Journal of Arid Environments* 70: 18–28.

58 Eldridge, D.J., Zaady, E., and Shachak, M. (2002). Microphytic crusts, shrub patches and water harvesting in the Negev Desert: the *Shikim* system. *Landscape Ecology* 17: 587–597.

8

Conservation of Lichens

Beena Kumari, Girish Kumar Sharma, Amit Vaish, Prashant Kumar, and Mohammad Javed Ansari

Department of Botany, Hindu College Moradabad (Affiliated to MJP Rohilkhand University), Bareilly, India

8.1 Introduction

8.1.1 History

Theophrastus, who is regarded as the inventor of botany (370–285 BC), used the name "lichen," a word with Greek roots, to describe the superficial growth on olive tree bark. This is when the field of lichenology first began. All lichen species are included in the genus Lichen, which Linnaeus designated for future research and assigned to one of his final students, Erik Acharis. Erik Acharius, a Swedish botanist, was born in the Swedish town of Gävle in 1757 (Acharius 1803, 1810, and 1814). He moved to Uppsala at the age of 17 to study botany as one of Linnaeus' final pupils. In his three monumental books Methodica Lichenum, Lichenographia Universalis, and Synopsis Methodica Lichenum, Acharis—known as the father of lichenology—described numerous new genera and numerous new species on the basis of external morphology and coined several terms for the structures unique to lichens.

8.1.2 What are Lichens?

A fungus and an alga grow together in intimate symbiotic cooperation to form lichens, a particular type of plant. These two separate species have no known biological relationship. Even though lichen and mosses combined make up the dominating organism in an ecosystem that covers over 10% of all terrestrial environments on Earth, lichen research is still largely underfunded globally [1, 2]. Lichens with cyanobacterial blue-green symbionts are important for the fixation of nitrogen in forests [3]. Lichens are employed for a variety of purposes, including monitoring pollution [4, 5]. These are the plants that grow in the most unfavorable climatic and soil circumstances. As a result, this group has a very high level of significance within an ecosystem [6]. According to Fleischner [7], lichens act like

Chemistry, Biology and Pharmacology of Lichen, First Edition. Edited by Ashoke Kumar Das, Ajay Sharma, Deepika Kathuria, Mohammad Javed Ansari, and Garima Bhardwaj.

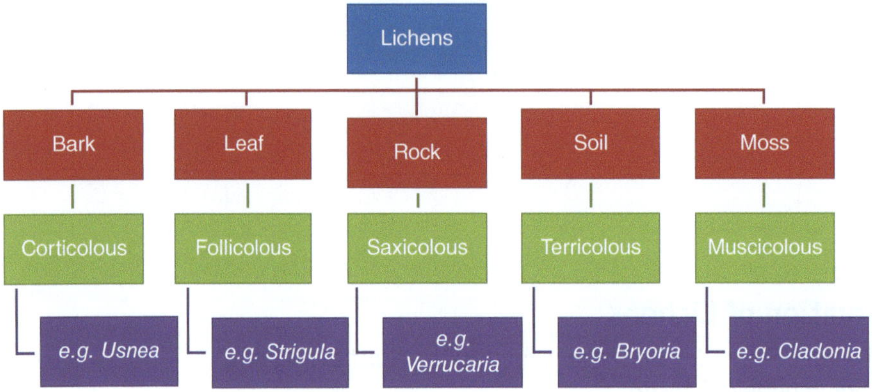

Figure 8.1 Different habitats of lichens.

tiny sponges that absorb everything that comes their way, including air pollution. Lichens can grow in one of three ways: The lichens that lack a distinct upper and lower surface are the most three-dimensional. Fruticose lichens can either be small and shrubby or long and stringy. Foliose lichens have rather flat lobes that can be divided into an upper and lower surface. A foliose lichen can have broad, leaf-like lobes or long, narrow ones. Some of these lobes stick closely to their substrate, making it challenging to tell them apart from the following category. Crustose lichens lack a lower surface that can be distinguished from the substrate and appear to the human eye to be practically two-dimensional, growing on the surface of their substrate almost as if they were painted on (Figure 8.1).

Other advanced forms like "Leprose, Placodioid, Squamulose and Dimorphic," are also recognized in addition to these three primary varieties [8, 9]. A recent study from the University of Alberta revealed that the rainforests of Alaska are a hotspot for lichen worldwide. In Glacier Bay National Park in southern Alaska, biologists Toby Spribille and Willis [10] discovered more than 1300 types of lichen, including 27 species that had never been recorded before.

In only 0.18% of the world's land area, New Zealand, according to Hutchison and Marley [11], is a global hotspot for lichen biodiversity. It is home to about 10% of all currently known lichen species. More than 2030 different species of lichen are currently recognized in New Zealand.

8.1.3 Substrate and Habitat

Lichens, which cover up to 8% of the earth's surface [12, 13], are able to withstand extreme climatic conditions. Additionally, they rank among the most significant markers of both ecosystem health and air pollution [14, 15], in addition to having many practical medical uses. They react strongly to changes in the microclimate. Therefore, lichen populations will be impacted by any natural or man-made disruptions. Lichens often benefit from the ideal conditions of moisture, light, clean air, and an undisturbed substrate for optimum growth and abundance. Lichens are a varied group of symbiotic organisms that play a crucial role in ecosystems all over the world. There are 13,000 known species of lichens worldwide.

According to Sipman and Aptroot [2], there are 13,000 known species of lichens worldwide, and this number grows to 20,000 if the orphaned species are taken into account.

Since Linnaeus indicated that Lichen fuciformis (L.) DC. (*Roccella montagnei* Bél.) occurs in India, lichen research scenarios in India are being researched. Eric Acharius, the founder of lichenology, wrote about lichens in his classic works "Lichenographia Universalis" and "Synopsis Methodica Lichenum" (1810, 1814), which are included in his masterpiece "Species Plantarum". Following that, several lichenologists described the Indian lichens. Bélanger [16], who described 40 taxa from Pondicherry and the Coromandel Coast, six of which were brand-new to science, was the first person to study the lichens of the Western Ghats. Important European lichenologists Jatta [17], Smith [18], Choisy [19], Santesson [20], Moreau and Mme [21] all studied the lichens of the Western Ghats. The characteristics of lichens and how they deteriorate on stone have been studied by a large number of authors from throughout the world [11, 22–27], found a variety of lichen colonies on British Isles hard limestone. Allen et al. [28] evaluated the diversity of lichen species in a region with around 6000 species in Canada and the United States.

According to environmental factors and the kind of ecosystem, lichen diversity and species composition vary [29]. Lichens can be found thriving everywhere, from low tide to mountain tops, and from desert and cold climates to tropical ones [14, 30]. There are three typical distribution patterns for the many lichens that are restricted to the tropics [31]. All of the tropics contain the pantropical taxa. The Old-World tropics (from Africa east to Australasia and the Pacific) are home to the paleotropical taxa, but the Americas are not one of them. Substrates include things like buildings, gravestones, cars, old farm machinery, trees, rocks, and soil.

Trees, rocks, and soil are the three most prevalent natural substrates. Lichens, moss, and ferns all naturally grow on rocks [32]. Numerous studies [33–36] have listed the distribution of lichens growing on historical sites in various parts of India. The variety of lichens found on the monuments in central India has previously been the subject of some research [37–40], but more work is still needed. The main areas of variety and occurrence for lichens in India include the Eastern Himalayas, North-Eastern India, Western Ghats, Western Himalayas, and the Andaman and Nicobar Islands [41].

In the country, they are represented by 2532 species in 324 genera and 78 families, 541 of which are endemic [42]. These can reach heights of up to 5000 m above sea level and thrive in tropical to temperate environments. Rout et al. [43] described two growth patterns, crustose and foliose, under 55 species under 26 genera and 15 families of lichen variation in a reserve forest in southern Assam, northeast India, which is situated in the Indo-Burma biodiversity Hotspot. According to Satya et al. [44], *Shorea robusta* trees are an ideal phorophyte for the luxuriant growth and development of crustose lichens. The category includes 42 of the 64 lichen species discovered on *Shorea robusta* trees in various Indian forest settings.

In the Kumaun Himalayan districts of Pithoragarh and Almora, Upreti and Chatterjee [45] investigated the distribution of epiphytic lichens on Quercus and Pinus trees. The Uttarakhand regions of Chopta-Tunganath, Saryu River Valley in Pindari Glacier (Bageshwar district), Nain Singh top en route to Milam Glacier, Chaubatia to Ranikhet (Almora district), and Mussoorie Hills (Dehra Dun district) have been proposed as the "Lichen Sanctuary" because of their rich, distinctive, and some endemic lichen taxa.

Acroscyphus sphaerophoroides Lév., a rare, non-endemic species, was only found in the bare alpine rocks of West Sikkim in India, according to Sinha and Ram [46]. In the Uttar Pradesh Katarniaghat Wildlife Sanctuary, lichen diversity and distribution patterns were examined by Nayaka et al. in [47] 2011. Uppadhyay et al. [48] observed a relationship between rock porosity and lichen growth from the monuments of the Gwalior division, Madhya Pradesh.

China boasts the highest diversity of lichens, with over 2400 taxa, second only to India. It's noteworthy to notice that 518 taxa from the Himalayas share traits with China, or 30% of their total taxonomy. About 650 lichens are found in Thailand, one of the tropical nations that have been well-researched for lichens. Of these, 190 (30%) share characteristics with lichens from South India. There are roughly 1100 lichen taxa in Sri Lanka, of which 210 (20%) are shared with the Indian portion of the Western Ghats. It is evident that the Western Ghats, which include Sri Lanka, have the highest diversity of lichens, with about 2100 different species [49]. Lichens from numerous nations are listed on the Red List. The "Red List" of endangered species, however, does not include lichens in India. *Campylothelium superbum* is probably the oldest species that has not been collected since 1825. Such lichens are either extinct or critically endangered.

The goal of conservation biology is to keep uncommon and endangered species' populations self-sustaining. This goal can be accomplished by a combination of techniques aimed at specific species and habitat conservation initiatives. By comprehending and documenting the needs of rare or endangered species in dynamic landscapes, identifying critical stages in the target species' life cycle [50–54], and other factors, population biology can significantly contribute to the development of appropriate conservation strategies for lichens.

To increase total and threatened lichen species richness, Boch et al. [55] suggested that

1) Unmanaged woods should be preserved;
2) silvicultural practices should be encouraged;
3) ancient trees in managed forests should be safeguarded and preserved;
4) Promoting natural deciduous tree species rather than coniferous plantings is necessary, and
5) There is more deadwood now.

"Habitat destruction throughout the world is causing widespread extinction of lichen species, due in large part to loss of trees, especially in the tropics, Urban and industrial expansion, including mining, resulting in major changes to the land surface, and pollution of air, soil, and water," claims Seaward [27]. The destruction of the usual habitat is a method that is accelerating globally and is one of the main dangers to biotic multiplicity, according to Giovanna et al. [56]. Landscapes are significantly impacted by habitat loss and fragmentation, which may also have an impact on ecosystems, populations, and species.

Comparing lichens to other types of species (such as plants and vertebrates), they are typically underrepresented in conservation evaluations and implementation. However, improvements in lichen conservation have been made in recent years. Lichens favorably develop on mangrove trees in inland streams as opposed to those growing near river mouths and open water, according to Savillo [57]. Lichen was observed growing on trees in bays or on islands that are partially surrounded by islets. Mangroves in India offer rich and distinctive lichen flora, yet they are among the least investigated areas, according to research done over the years [58].

Joshi et al. [59] examined the distribution of lichens in the "Pindari Glacier valley's various habitats" and discovered an intriguing growth pattern along the riverbank and in regions damaged by forest fires and human intervention. Growth in environmental or anthropogenic challenges led to several adaptive and self-preservation methods in lichens. Lichens are much more significant than just the biomass they produce in the valley, which is constantly in danger. To adopt some conservation techniques against rising unregulated harvesting that is putting the lichens on the edge of extinction, documentation of the lichen flora in the "*Pindari Glacier valley*" is required.

In the southern Appalachian Mountains of North Carolina, the United States, Allen [60] employed transplant techniques for the conservation of "*Graphis sterlingiana*, *Hypotrachyna virginica*, and *Lepraria lanata*." Smaller gaps and closer proximity to edges encourage the recolonization of epiphytic lichens in Boreal and temperate rainforests, according to Paula et al. [61] recommendation for Silvicultural techniques for lichen conservation. Lichens are vital to the ecology because they permit algae to flourish in various temperatures around the world. They also offer a means to convert atmospheric CO_2 into O_2, which is necessary for human existence.

The Uttarakhand Forest Department (UFD) in 2019 created "the country's first lichen park" near Munsiyari (Pithoragarh district) with the intention of preserving, protecting, and cultivating lichen. "*Parmotrema perlatum*, *Usnea longissima*, *Lecanora subfuseescens*, *Sarcogyne privigna*, *Arthonia impolitella*, *Acarospora fusca*, *Acarospora oxytona*, *Polysporina dubia*," and many other lichen species that can be found in the state. Around 150 of the nearly 600 lichen species reported in the state of Uttarakhand are found in Munsiyari [49].

8.1.4 Licheno-geographic Regions of India

Based on the 10 prominent families and genera in each region, Singh and Sinha [62] categorized India into eight lichen-geographical regions. With the exception of the Lakshadweep Islands, these regions covered the majority of the territory within India's political borders: (i) the Western Himalayas; (ii) the Eastern Himalayas and North-east India; (iii) the Western Dry Region; (iv) the Gangetic Plain; (v) central India; (vi) the Western Ghats; (vii) the Eastern Ghats and Deccan Plateau; and (viii) the Andaman and Nicobar Islands.

8.1.5 Lichens in Indian Languages

Lichens are referred to in Sanskrit as "Shailaya" and "Shila Pushp" in Indian contexts, respectively, where Shila is a rock and Pushp is a flower. The English name "Stone Flower" for lichen species is also known in Hindi as "Pathar Ka Phul," Marathi as "Dagad Phul," Kannada as "Kalahu," Tamil as "Kalpasi," and Urdu as "Rihamkarmani." Locally, it is referred to as "Mukku," "Shewal," and "Jhoola" in Garhwal; "Chaarila" in Kumaun; and "Chaai" and "Mendi" in Himachal. The lichens are sometimes referred to as "Shaik" in Hindi (Shai = Shewal and K = Kawak).

8.1.6 Threats and Conservation of Lichens

Fluctuations in ecological circumstances, woodland cover, habitat forfeiture, and an upsurge in urban and industrial areas are the causes of the loss of lichen variety in the studied area. The anthropogenic activities in hilly areas including tourism, mineral mining, road building projects, hydroelectric projects as agriculture, "Jhoom" cultivation, etc., are hastening the decline of lichen-growing environments, overuse, and choosy removal by locals of commercially significant lichens. The quantity of raw materials needed for lichens' varied applications is substantial. Since dried lichens are so light, a significant amount of these plants is collected for usage.

Evidence from around the world indicates that many lichen-rich places are under threat from human activities such as air pollution, habitat degradation, haphazard logging, forest fires, tourism, overexploitation, and illegal collections. According to Kathiresan [63], the three main issues that affect the majority of India's mangrove ecosystem are (i) excessive fishing activity, which hinders the regeneration of mangrove seedlings and causes unnecessary tree cutting, (ii) damage to trees caused by firewood and cattle feed, as well as rehabilitation, reclamation, and conversion activities, and (iii) illegal logging. (iii) Lack of public awareness of and involvement in conservation efforts.

In the Himalayas, lichens are painstakingly gathered and are frequently utilized as ingredients in spices, Ayurvedic medicines, and other products. In India's Central Himalayan region, lichen is collected on an annual basis in amounts of roughly 10,000 tons. However, the exploitation of lichen is completely prohibited in protected regions like KWLS. The lichens are marketed under the names "Charrila" and "Jhoola" [64].

8.2 Important Roles of Lichens

8.2.1 Role of Lichen as a Food Source

Lichens provide a food source for a wide range of species, and in areas with little plant life, they may be the only available source of nutrition. For example, the name "reindeer lichen" comes from the fact that reindeer rely heavily on it for nourishment throughout the winter. Other creatures that eat lichens include pronghorn, deer, bighorn sheep, elk, rats, moose, mountain goats, and insects.

Even organisms that do not consume lichens directly gain from their availability as a food source. Lichens, for instance, provide indirect benefits to species like stoats, owls, and hawks as well as creatures that eat various insects and mammals that consume lichen. Lichens are used as nesting material and shelter by a variety of animals, demonstrating the broad distribution of these unusual symbiotic organisms.

8.2.2 Role of Lichen in Nitrogen Fixation

Lichens take in nutrients and minerals from the air around them. By doing this, these unusual organisms enable other living things to access these nutrients. Cyanobacteria-containing lichens can "fix" ordinarily inaccessible atmospheric nitrogen so that it is useful to the lichen and consequently to other organisms that consume the lichen. *Leptogium* sp. works with cyanobacteria (Nostoc sp.), which are capable of fixing atmospheric N_2.

8.2.3 Lichen's Role as an Environmental Indicator

Researchers use these unusual organisms as monitors for different contaminants and heavy metals because they ingest nutrients from the air around them. Scientists can use samples of the lichen to test the area for pollution because, if pollutants are present in the air nearby the lichen, they absorb those contaminants.

Researchers can also use lichen growth and death to assess the overall health of an ecosystem. Many lichens cannot endure high amounts of air pollution, despite the fact that they can survive in areas that are otherwise unsuitable. Researchers can presume that pollution levels are rising in a certain place when lichens start to die there. They can then look into this more. On the other hand, lichen-rich places suggest robust and healthy ecosystems. Healthy lichen populations in habitats and ecosystems can support a wide variety of animals. Lichens offer nutrients to pests and herbivores, fix nitrogen to replenish it in a form that may be used by other animals in the ecosystem, and indirectly sustain predators that also prey on those critters.

8.3 Biogeography (Geographic Distribution of Lichen)

From the Arctic to the Antarctic, deserts to the tropics, and coastal zones to mountain peaks, lichens can be found. While some are ubiquitous and grow on various trees, soils, and rocks, others favor a specific habitat type, such as limestone or sand dunes. Nearly every terrestrial ecosystem is inhabited by lichens, and they also have a presence in aquatic habitats. Some tumbleweed, or vagrant, lichens grow free and drift over soil in arid environments. They can grow on practically any firm surface. Epiphytic plants are those that grow on wood or bark, including organic fence materials. Saxicolous lichens are those that grow on rocks, and they can be further divided into those that occupy acidic rock (siliceous) or those that grow on basic rocks that are calcium-rich (calcareous), including limestone, cement, and even pavements! Terricolous lichens also develop directly on the ground. Terricolous lichens, along with moss and free-living cyanobacteria, create biological soil crusts, also known as biocrusts, which are crucial soil improvers and stabilizers in desert ecosystems. Lichens can be found on rocks as far south as Antarctica, in tropical rainforests, and in deserts with no predictable annual rainfall. Additionally, freshwater lakes, mountain streams, and marine tidal zones all have semi-aquatic lichen habitats. Even when completely immersed in spring-fed mountain streams, *Peltigera gowardii*, previously known as the western populations of *Peltigera hydrothyria* (or *Hydrothyria venosa*), thrives [65].

Within many terrestrial settings, lichens and mosses coexist in a gradient, with lichens dominating the driest sections and mosses the wettest (although certain mosses do coexist with lichens). Gradients can often be seen within the lichens themselves. In the Pacific Northwest and California's north-coastal mountains, conifer forests are initially inhabited by common chlorolichen species with a somewhat mossy zone close to the ground; as they mature, long fruticose lichens known as alectorioid lichens (primarily Alectoria and Bryoria) colonize the mid-canopy; then as old growth conditions develop, cyanolichens colonize a zone in the lower canopy, just above a mossy understory [66]. Even more compact microhabitats are specialized by some lichens. Gradients can be seen going around a tree trunk. Mosses, followed by bigger fruticose and foliose lichens, may predominate in regions where canopy drip produces the most moisture. These lichens then transition to protected areas that receive no direct liquid water, where a variety of powdery crustose leprarioid lichens and tiny pin-lichens with tiny stalked fruiting bodies may be present. Old-growth forests are the finest places to develop these protected microhabitats. Gradients in conifer forests may be among the best characterized in the literature, but similar gradients are frequently seen elsewhere, even running the length of the face of a single big rock [67].

Dirinaria applanata, *Lecanora* sp., *Parmotrema grayanum*, and *Leptogium* sp. all represented saxicolous lichens. Some macro-lichen species displayed host tree specialization, which might be related to different ecological circumstances. *Mangifera indica* (Mango tree) serves as the exclusive host for *R. montagnei*. The only corticolous species discovered in deciduous woodlands is *Usnea stigmatoides*. *Tectona grandis* and *Syzygium* tree species supported the branches of *Coccocarpia erythroxyli* and *Heterodermia speciosa*, which are hosted by *Litsea floribunda* and several *Usnea* species. Lichens such as *Pyxine coccifera* and *D. applanata* are both tolerant of pollution and can be found growing in the research area's almost open spaces, scrubby forests, and arid regions. In deciduous and semi-evergreen forests, it is common to find *Parmotrema tinctorum*, *Parmotremacristiferum*, *Parmelinella wallichiana*, *Haterodermia diademata*, *Hymenothrix dissecta*, and *Leptogium burnetiae*.

In both deciduous and semi-evergreen woods, the most prevalent genera are *P. tinctorum* and *L. burnetiae*. Deciduous woods are plenty in *Parmotrema reticulatum*, *Ramalina pacifica*, *Ramalina conduplicans*, and *Usnea galbinifera*, whereas Thelotremataceae and Graphidaceaeous species dominated semi-evergreen forests. The corticolous taxa in Southeast Asia's seasonal tropics can be utilized to identify regions with long ecological continuity, areas with high biodiversity, and regions where forest degradation is taking place [68]. Most foliose species, including *Parmotrema*, *Heterodermia*, *Leptogium*, and *Pyxine* species, are found in lowland environments. With sufficient canopy openings, Parmeliaceae and Physciaceae are either attached to the higher canopy branches or are growing on rocks, where they would receive more light and wind. Negi [69] discovered that in two Indian settings in the "Chopta-Tungnath" and "Nanda Devi Biosphere Reserve," over 64% of the lichen species were located on woody components. Five hundred and sixty-five lichen species were recorded from the Himalayas of Uttarakhand by Upreti et al. [70]. The largest diversity is found in evergreen forests, where foliose species are on the decline, and crustose species are on the rise [71]. The diversity of crustose species in dry deciduous forests varies with fire and forest history, while the variety of foliose species is lowest in regularly burned plots that vary with tree type [68]. Even when the same host trees are present, epiphytic lichens switch hosts in various climate regimes [43].

Similar host specificity patterns of lichens were also documented by Hale [72] in the highland woods of Wisconsin. Depending on the types of forests and altitude, there are differences in macro and micro climate conditions as well as tree bark features. Even though the availability of light in evergreen forests is lower than it is in deciduous forests, it nevertheless plays a significant role in lichen dispersal. According to Bruteig [73], the frequency, length, and kind of precipitation all play a role in how lichens are distributed. Even where precipitation is light, mist and fog can nevertheless result in humid conditions.

8.3.1 Prominent Areas of Research in Conservation Biogeography

The biogeography of degradation (including impacts from urbanization, homogeneity, and habitat fragmentation)

The following processes are involved:

a) Colonization, climate as a chief determinant of distribution, dispersal, disturbance, extinction, persistence, range extension, resilience, and speciation.
b) Issues with data, inventory, and mapping (atlas data, breeding bird surveys, citizen science, detectability/discovery probability, herbaria, and other collections, sampling intensity, and biases).
c) Bioclimatic modeling, habitat suitability analysis, model performance, niche-based models, presence-only vs. presence-absence data, and dispersion kernel analysis are a few of the techniques used to simulate species distribution.
d) Identification of alternate baselines from long-term ecological data, characterizing biotas (conservation status, diversity indices and patterns, ecoregions, endemism, rarity, range size, species-area connections, threatened species, and so on).
e) Methods (molecular methods, palaeoecology, remote sensing, scenario development).
f) Reintroduction ecology, invasion ecology, bioinformatics, molecular phylogenetics, network analysis, risk analysis, behavioral ecology, and population viability analysis are related fields.
g) Themes that run throughout the paper include the following: imperiled species, the discovery of alternative baselines from long-term ecological data, niche (basic vs. realized), unique climates and ecosystems, scale difficulties, uncertainty, Linnean shortfall, and Wallacean shortfall.

8.3.2 Why Lichens Require Special Conservation Measures

- The majority of lichens are very small (Micro-lichens), making it difficult to detect the extinction of a species or changes in the pattern of their distribution. Lichens are Eukaryotic microorganisms that have stable genetic makeup and exhibit microbe-like characteristics.
- They react fast to changes in their habitat and air pollution.
- Lichens depend heavily on habitats that macro-organisms do not often colonize (rocky terrains, harsh conditions, man-made buildings).
- Many of their succession patterns and community dynamics have not been thoroughly examined.
- They perform entirely distinct ecological functions and have quite different needs than macro-organisms. Additionally, macro-organisms frequently face danger of extinction in the absence of lichens.
- Lichens are very well known for their ability to adapt to harsh conditions and to colonize freshly exposed substrata like those created by lava flow and glacial retreat, but they are not strong enough to survive risks caused by the environment or by humans.
- Lichens play an important part in the ecosystem's ability to regulate the water cycle and nitrogen cycling.

8.4 Conservation of Lichen Diversity

8.4.1 International Initiatives on the Conservation of Lichens

- The Field Museum in Chicago, United States, established and maintains MYCONET, a database on the classification of fungi. Forest Health Monitoring Program—Lichens as Indicators of Air Pollution/United States Department of Agriculture/USA http://www.fieldmuseum.org/myconet/ www.fs.fed.us/r6/aq/lichen
- European mapping recommendations for lichen diversity as a measure of environmental stress. http://www.thebls.org.uk/content/documents/eumap.pdf
- IUCN Red Database/Lichens: Books on Endangered Species http://www.iucnredlist.org
- A global database of information about the diversity of lichens and lichen-forming fungus.
- American Lichen Society International Association of Lichenologists, http://www.thebls.org.uk http://lichenology.org
- (http://www.sbf.c.se/slf/slf.htm) Swedish Lichen Society
- Middle European Bryological and Lichenological Society, http://www.blam.privat.t-online.de/Czech Botanical Society's Bryological and Lichenological Section, http://botanika.bf.jcu.cz/BLS/english/online_en.html
- The Netherlands' lichenology is described at http://www.biodiv.nl/lichens/index.htm. Swiss Association of Bryology and Lichenology (http://www.bryolich.ch/english/index_engl.html)
- Italian Society for Lichenology (http://dbiodbs.univ.trieste.it/sli/home.html#sli)
- Hattori Botanical Laboratory of Japan (http://www7.ocn.ne.jp/%7Ehattorib)
- Association Française de Lichénologie (http://www.lichenologie.org/)

8.4.2 Lichen Conservation Initiatives in India

- There is no particular effort to preserve the diversity of lichens.
- Counting and recording the lichen-rich settings in the Western and Eastern Ghats, mangrove swamps, and other forested places.
- About 5% of the entire Indian land area has been studied for lichens using a qualitative methodology, according to conservationists, who believe that conserving macroflora and fauna also involves conserving lichen variety.

8.5 Conservation Challenges of Lichens

Due to their extreme sensitivity, lichens may alter their variety, abundance, morphology, physiology, accumulation of contaminants, and other characteristics in reaction to environmental changes. For example, habitat loss and degradation [74], habitat fragmentation [56, 75], overexploitation [76], species invasions [77], air pollution, and climate change [78] are all threats to biodiversity that also apply to lichens. There has been a noticeable loss in the wide expanses of forested areas in various parts of the world as a result of the rapid pace of urbanization and industrialization in recent years. The many endeavors, including agricultural, mineral mining, tourism, hydropower projects, and road construction, contribute to the quick loss of lichens.

The main dangers to the lichen flora have been identified which can be environmental pollution, urbanization, forest fires, and deforestation [79]. Lichens are becoming extinct due to reindeer and other animals overgrazing. The unobstructed reaping of lichens has become a severe menace to biodiversity in the Himalayas and Western Ghats because of the sharp rise in demand for herbal items and non-timber forest products from the forests. Lichens are slow-growing creatures, so if they are ever removed from their natural environment, it will take them a while to re-establish. Due to their sensitivity to various levels of pollution, air pollution represents yet another significant hazard to lichens. Bangalore [80] and Kolkata cities clearly showed the loss of lichen diversity and changes to their community due to urbanization and changing climate.

Luca Paoli et al. [81] investigated the effects of forest management on threatened epiphytic macro lichens in a Mediterranean mixed oak forest in Italy. They found that this management has a significant impact on biodiversity, the structure and dynamics of the forest, and environmental factors that are important for the dispersal, establishment, and maintenance of epiphytic lichens. As a result of climate change and forestry practices, *Lobaria pulmonaria* Hoffm. has generally declined throughout Europe [82]. *L. pulmonaria* is currently an endangered species in a number of European nations. In the Paneveggio Pale di San Martino Natural Park, Italy, one of the best-known lichenological regions in the world, species like *Nephroma laevigatum*, *Sticta fuliginosa,* and *Usnea longissima* actually vanished as a result of global changes (i.e. climate, land-use, and forest management) [13].

According to Esseen et al. [83], industrial forestry, along with nitrogen, is the primary cause of lichen decreases in Sweden. Logging in densely forested areas with long stretches of continuous tree cover, short rotation cycles, substrate limitations, and low light levels is bad for lichens. At four locations in northeast Scotland, Mark and Christopher [84] utilized predictive niche modeling to look at how *Vulpicida pinastri* responded to various current climate variables. At small scales, *V. pinastri* appeared to be dispersion-restricted, and it was substantially more prevalent on aged and deteriorated juniper plants. According to a study, churchyards are crucial for the preservation of lichen. Churchyards all over Great Britain and Ireland are a fantastic source of lichens. They take over the church, memorials, boundary walls, and trees in the churchyard.

Juriado and Liira [85] recommended that *L. pulmonaria*-containing overgrown woody meadows in Estonia be restored carefully because sudden changes in the habitat's light and moisture management are likely to have an impact on the species' numbers. Even the existing habitat situations should be better, which calls for moderate habitat management, such as tree removal to minimize canopy closure. Conifers should be removed first during thinning, whereas *L. pulmonaria* host trees should be left intact. To aid the lichen's ability to colonize the new habitat, potential host trees for *L. pulmonaria* should be retained next to old trees.

According to James and Allen [23], the Atlantic Coastal Plain is home to a hitherto underappreciated lichen biodiversity hotspot that is immediately threatened by sea-level rise. According to conservative sea-level rise forecasts, all of the region's highest-diversity locations will be submerged within the next century. Important ramifications for conservation strategy result from the removal of less well-known but varied, biologically significant groups from global biodiversity surveys.

The Pindari Glacier Valley's various climatic conditions provide habitats for several lichen species. On the river bank and in areas where there had been a forest fire or other human intervention, the valley's lichens displayed an intriguing growth pattern. Lichens developed some adaptation and self-preservation methods resulting in an increase in natural or artificial adversities [59]. *Pseudocyphellaria aurata* (Ach.) Vain was discovered by Gintaras [86] and is currently endangered locally and scarce. The other lichen, such as *Cladia ferdinandii* (Müll.Arg.) Filson, in South Australia, appears to be extinct.

8.6 Recommendation for Conservation of Lichens

Lichens are a varied group of symbiotic organisms that play a crucial role in ecosystems all over the world. Comparing lichens to other types of species (such as plants and vertebrates), they are typically underrepresented in conservation evaluations and implementation. In the Garhwal Himalaya's wet temperate zones, where oak and other trees sustain luxuriant lichen development, lichen exploitation is a frequent activity among neighbors and rivals. According to Kumar and Upreti [87], two fruticose genera, *Ramalina* and *Usne*, as well as the parmelioid lichens *Everniastrum, Parmotrema, Cetrariopsis, Bulbothrix, Hypotrachyna,* and *Rimelia* are economically traded lichens from the Garhwal Himalaya.

8.6.1 Species-based Conservation

Maintaining genetic variety is really what conservation of biodiversity is all about. The genetic variety as categorized by evolutionary history is conveniently represented by species. The goal of conservation of these units is to ensure that these evolutionary groups can endure over time. We have known for a long time that species are vulnerable when their genetic diversity is low. Thus, conservation efforts should support a wide range of genetic variation across the evolutionary groups that we commonly refer to as species. Despite the fact that genetic research is currently making it more difficult to define lichen species, species-based conservation is still the most effective strategy to preserve genetic variation.

8.6.2 Habitat-based Conservation

Given the difficulties in currently classifying taxa and the incomplete knowledge of lichen distributions worldwide, "*habitat-based conservation*" might provide an effective method for preserving lichen variety lacking necessitating a high level of taxonomic knowledge for surveys. Lichen "*habitat-based conservation*" in the western hemisphere is still not well understood. But for those who routinely study lichens, many of the areas that contribute the most to lichen biodiversity are obvious. These habitats should be preserved wherever possible because they contribute significantly to the landscape ecology. The ecosystems of interest are likely to include old-growth conifer forests, unspoiled oak savannas, rock outcrops with groundwater seeps, and other habitats that are frequently of conservation value for a variety of animals. A method to more precisely pinpoint biodiversity hotspots over the landscape and at sizes adequate for designating zones of preservation may involve intersecting these habitats with more general biogeographic patterns in lichens.

8.6.3 Issue-based Conservation

8.6.3.1 Urban/Sub-urban Development

It would be ideal to preserve what we can while humans encroach on the natural world, and perhaps even to urge nature to encroach on our urbanized regions. A great place to start would be by simply maintaining trees and rock outcrops while planning for open space. Xeriscaping is a different concept that can enhance current development. Eliminating landscape irrigation can promote the colonization and survival of local lichens by preventing the disruption of lichens' natural moisture regimes caused by landscape watering. Increased usage of natural materials is a third suggestion. Consider using unprocessed redwood or cedar for fencing instead of plastic, metal, or treated wood.

8.6.3.2 Air Pollution

Most lichens are susceptible to air pollution, which has been shown to degrade lichen communities at local sizes in European Countries This has already been mentioned. The Central Valley's agricultural activities are probably to blame for similar symptoms of air pollution problems that are being observed in the Sierra Nevada Mountains [88]. Therefore, Upreti et al. [89] advocated that we must resolutely keep moving forward in the direction of cleaner air.

8.6.3.3 Invasive Exotic Plants

Plant incursions have unknowable effects on lichen populations by changing the habitats' ecology or even whole biomes. Invading annual grasses have nearly completely destroyed the last open regions in the Central Valley. Similar changes are occurring in the desert biomes, particularly in the Great Basin [90]. According to evidence, annual grasses and soil biocrusts cannot coexist [91–93]. As a result, recovery after loss will probably be challenging, but annual grasses might be kept in control where unbroken crusts still exist. This implies that a two-pronged strategy may be possible: first, retaining what is left to give biocrusts a chance to colonize; second, finding a technique to lessen annual grasses to allow biocrust recolonization to flourish.

8.6.3.4 Forestry Practices

Over the past century, forestry has advanced significantly, but there is still more work to be done in this area. More study is needed on the biodiversity that old-growth coniferous and hardwood stands support [94]. Thinning forests rather than clear-cutting should be encouraged in newly managed stands. As homogenization of forests occurs as a result of uniform thinning, communities of lichens are pushed toward the most prevalent species [95]. Miniature hotspots of biodiversity can be created through variable thinning, which leaves pockets of older trees among young stands [96]. Partial-cut Harvesting techniques to Conserve Terrestrial Lichens in Managed Landscapes should be encouraged, as suggested by Darwyn [97].

8.6.3.5 Wildfire

There is little research on how wildfire affects lichens. The majority of research focuses on ground-dwelling lichens in other regions' woods or biocrust groups in cool deserts [98, 99]. The few studies that do exist either come from different systems [100] or only tangentially deal with wildfire [88] in terms of effects on epiphytic lichens. The main issue that negatively impacts habitat management and lichen conservation is wildfire. Natural regeneration is either lost or delayed as a result of forest fires, some of which are intentionally started, and the forest floor is lacking of rich humus, which is crucial for the formation of lichen. In addition to impacting many species' regular existence, growth, propagation, and mobility, fires harm the microfauna irreparably.

Many lichenologists hold the belief that a brief exposure to smoke while lichens are dormant will only have very modest impacts, although further research is undoubtedly needed on this issue [101]. Accordingly, especially where directly burned, naturally occurring low-intensity wildfires may cause slight effects on lichen. However, because lichens may be physiologically active and hence more sensitive, controlled fires carried out while moisture levels are high may result in higher harm to lichens. Therefore, when developing the ecosystem and conserving the lichen in the sanctuary, fire control methods must be given top attention. During high-fire risk days, the Fire Protection Force and fire-extinguishing trucks could be deployed in advance at the world's most populous fire-prone regions.

8.6.3.6 Smuggling and Poaching or Hunting

Smuggling is a constant threat to the natural habitation of nearly most valuable timber tree species, including "*T. grandis*, *Dalbergia latifolia*, *Lagerstroemia microcarpa*, *Santalum album*," and others. The variety of epiphytic lichens is being lost as a result of these tree smuggling activities. Illegal lichen harvesting and commercial export to local markets should be outlawed in some areas.

8.6.3.7 Illegal Grazing

The villagers' livestock is present in and around the sanctuary in great numbers. The cattle access the sanctuary's core zones in addition to the settlements for grazing. The sanctuary's terricolous lichen diversity will be impacted by the trampling effects of their foot in several spots.

8.6.3.8 Tourism

- In 25% of lichen locations, tourism has been classified as a danger.
- Coastal locations that have abundant lichen communities because of a combination of the tidal regime and the rock type are also in danger because of the growth of tourism.
- Lichen communities are frequently abundant in exposed locations above the tree line on mountains, and these habitats may be under more pressure as a result of the growth of ski resorts and ski areas.

8.6.4 *In Vitro* Techniques for Lichen Conservation

- The *in vitro* culture of Lichen symbionts or re-synthesis of whole thallus seems to be an alternative, especially for lichens under threat in natural conditions.
- In India, some highly efficient methods have been developed for the cultivation of isolated symbionts as well as the whole lichen tissue [102].

- In India, many different lichen species have been cultured *in vitro* and standardized for their physiological conditions [103].
- However, the tissue culture technique of lichens is un-economical, time-consuming and has also proved to be unsuccessful in certain cases [104].

8.6.5 Awareness Creation Activities

- The human activities involving the maximum utility in agriculture, tourism, transport, etc., have led to severe destruction and depletion of natural resources.
- Hence, for developing biodiversity conservation measures, it is imperative to be aware of different sections of societies through communication, education, and public awareness activities.
- It involves generating interactions, enthusiasm, and ensure participation of people rather than just making scientific information available to the public.

8.7 Conclusion

The proposals for a strategy to effectively conserve the lichens' biodiversity is summed up here. Forest fires ought to be avoided at all costs. Effective restoration of old-growth forest structure through the management of associated species invasion. To stop the gathering of firewood, substitute fuel sources should be made accessible to the inhabitants. Outlaw the illicit gathering of lichens for profit. Prior to implementation, the viability of development projects that could upset the equilibrium of a protected area should be examined. Work on lower plants like lichens should be promoted and supported by inviting the researchers. Create a long-term plot for lichen biomonitoring research. It is proposed that meteorological stations be created in order to have regular weather data because meteorological data is frequently lacking in the districts' areas. Set up some lichen gardens in the regions with a lot of biodiversity. Specialist macro lichen taxa found on rocks, soil, or wood should be conserved.

References

1 Nash, T.H. and Egan, R.S. (1988). The Biodiversity of Lichens and Bryophytes. In: *Lichens, Bryophytes and Air Quality*, vol. 30 (ed. T. Nash III and V. Wirth), 11–22. Borntraeger.

2 Sipman, H.J.M. and Aptroot, A. (2001). Where are the missing lichens? *Mycology Research* 105: 1433–1439.

3 Slack, N.G. (1988). The ecological importance of lichens & bryophytes. *Bibliotheca Lichenologica* 30: 23–53.

4 Giordani, P. (2019). Lichen diversity and biomonitoring: a special issue. *Diversity* 11: 171–181.

5 Moxham, T.H. (1986). The commercial exploitation of lichens for perfume industry. In: *Progress in Essential Oil Research* (ed. E.J. Brunke). Berlin: Walter de gruyter.

6 Richardson, D.H.S. (1992). *Pollution Monitoring with Lichens*. England: Richmond Publishing Co. Ltd. 1992.

7 Fleischner, T.L. (1994). Ecological costs of livestock grazing in western North America. *Conservation Biology* 8 (3): 629–644.

8 Feuerer, T. and Hawksworth, D.L. (2007). Biodiversity of lichens, including a world-wide analysis of checklist data based on Takhtajan's floristic regions. *Biodiversity and Conservation* 16 (1): 85–98. https://doi.org/10.1007/s10531-006-9142-6.

9 Nayaka, S. (2014). Methods, techniques in collection, preservation and identification of lichens. In: *Plant Taxonomy and Biosystematics: Classical and Modern Methods* (ed. T.S. Rana, K.N. Nair, and D.K. Upreti), 101–105. New India Publishing Agency.

10 Toby, S. and Willis, K. (2020). Alaskan rainforests are a global lichen hotspot. www.ualberta.ca/science/news/2020/may/alaska-rainforest-lichen.html (accessed 12 May 2020).

11 Hutchison, M.A.S. and Marley, F. (2021). New Zealand: a global hotspot for lichen diversity. International Association for Lichenology 9th Symposium (1–6 August 2021) - https://doity.com.br/ial9/blog/artigos.

12 Ahmadjian, V. (1995). Lichens are more important than you think. *Bio Science* 45 (3): 124.

13 Nascimbene, G.G., Peter, O.B., Luana, F. et al. (2022). A hotspot of lichen diversity and lichenological research in the Alps: the Paneveggio-Pale di San Martino Natural Park (Italy). *Myco Keys* 94: 37–50. https://doi.org/10.3897/mycokeys.94.95858.

14 Nash, T.H. and Thomas, H. (2008). Lichen sensitivity to air pollution. In: *Lichen Biology*, 2e (ed. T.H. Nash III), 299–314. Cambridge: Cambridge University Press. https://doi.org/10.1017/CBO9780511790478.016.

15 Rose, C.I. and Hawksworth, D.L. (1981). Lichen recolonization in London's cleaner air. *Nature* 289 (5795): 289–292. https://doi.org/10.1038/289289a0.

16 Bêlanger, M.C. (1834). Voyage aux bides Orientales Annees 1825–1829. *Botanique II. Partie. Cryptogamie par ch. Belanger et Bory de St. Vince* 1838: 113–114.

17 Jatta, A. (1903). *Licheni esotici dell'erbario Levier raccolti nell'Asia meridionale, nell'Oceania, nel Brasile e nel Madagascar, serie 1: determinati*. Ciminago.

18 Smith, A.L. (1926). Cryptotheciaceae: a family of primitive lichens. *Transactions of the British Mycological Society* 11 (3-4): 189-IN1.

19 Choisy, M. (1931). Lichens nouveaux. *Bulletin de la Société Botanique de France* 78 (4): 453–460.

20 Santesson, R. (1952). *Foliicolous Lichens I. A Revision of the Taxonomy of the Obligately Foliicolous, Lichenized Fungi*. Lundequistska bokhandeln.

21 Moreau, M. and Fernaud, M. (1952). Lichens del Inde. Rev. Bryolog. *Lichenlogist, NS* 21: 135–143.

22 Giacobini, C., Nugari, M.P., Micheli, M.P. et al. (1985). Lichenology and the conservation of ancient monuments: an interdisciplinary study. In: *Biodeterioration*, 6e (ed. S. Barry and D.R. Houghton), 386–392. Slough: International Mycological Institute and The Biodeterioration Society.

23 James, C.L. and Allen, J.L. (2014). Lichen biodiversity under threat from sea-level rise in the Atlantic Coastal Plain. *Bio Science* 64 (10): 923–931. https://doi.org/10.1093/biosci/biu136.

24 Jones, D. and Wilson, M.J. (1985). Chemical activity of lichens on mineral surfaces-a review. *International Biodeterioration* 21: 99–104.

25 Paine, S.G., Linggood, F.V., Shimmer, F., and Throupp, T.C. (1933). The relationship of microorganism to the decay of stone. *Philosophical Transactions of the Royal Society* 8: 97–127.

26 Sabbioni, C. and Monte del, M. (1987). A study of patina called “Scialbatura” on imperial Roman marbles. *Studies in Conservation* 32: 114–121.

27 Seaward, M.R.D. (1982). Principles and priorities of lichen conservation. *The Journal of the Hattori Botanical Laboratory* 52: 401–406.

28 Allen, J.L., McMullin, R.T., Tripp, E.A., and Lendemer, J.C. (2019). Lichen conservation in North America: a review of current practices and research in Canada and the US. *Biodiversity and Conservation* 28 (12): 3103–3138.

29 Nimis, P.L., Scheidegger, C., and Wolseley, P.A. (2002). *Monitoring with Lichens*. Netherland: Springer.

30 Abas, A., Awang, A., and Aiyub, K. (2018). Lichen as bio-indicator for air pollution in Klang Selangor. *Pollution Research* 37 (4): 35–39.

31 Rosentreter, R. and DeBolt, A. (2021). Comparing local, state, and global lichen distribution patterns at Jupiter Inlet Lighthouse Outstanding Natural Area Florida. *Evansia* 37 (4): 125–138. https://doi.org/10.1639/0747-9859-37.4.125.

32 Lalremruata, P.C., Laha, R.C., and Lalhriatpuia, R.V. (2018). Diversity of lichens; its threat need conservation in Mizoram. *International Journal of Life Sciences Research* 6 (2): 69–75.

33 Chatterjee, S., Acharya, A., Jha, A.B. et al. (1995). *Setting Standards for Sustainable Harvest of Wild Medicinal Plants in Uttarakhand: A Case Study of Lichens*. TERI Publication.

34 Saxena, S., Upreti, D.K., Singh, A., and Singh, K.P. (2004). Observation on lichens growing on artifacts in the Indian subcontinents. In: *Biodeterioration of Stone Surface* (ed. L. Clair St and M. Seaward), 181–193. Netherland: Kluwer Academic Publishers.

35 Singh, A. and Upreti, D.K. (1991). Lichen flora of Lucknow with special reference to its historical monuments. In: *Biodeterioration of Cultural Property* (ed. O.P. Agarwal and S. Dhawan), 219–231. New Delhi: Mac Millan.

36 Singh, A., Chatterjee, S., and Sinha, G.P. (1999). Lichens of Indian monuments. In: *Biology of Lichens* (ed. K.G. Mukerjee, B.P. Chamola, D.K. Upreti, and R.K. Upadhyaya), 115–151. New Delhi: Aravali Book International.

37 Bajpai, R., Upreti, D.K., and Dwivedi, S.K. (2008). Diversity and distribution of lichens on some major monuments of Madhya Pradesh, India. *Geophytology* 37: 23–29.

38 Jain, A.K. (2001). Biodeterioration of rock cut images at Gwalior fort. In: *Studies in Biodeterioration of Material* (ed. O.P. Agarwal, S. Dhawan, and N. Pathak), 59–68. Lucknow: ICCI & ICBCP.

39 Upreti, D.K., Nayaka, S. and Joshim Y. (2004). Lichen activity over rock shelter of Bhimbetka world Heritage zone, Madhya Pradesh. Rock Art research: changing paradigms. *The tenth Congress of the International Federation of Rock Art Organization (IFRAO) & Federation of Rock Art*, pp. 30–31, (28 November–2 December 2004). Agra, India: Rock Art Society of India.

40 Upreti, D.K., Divakar, P.K., and Nayaka, S. (2002). Notes on species of the lichen genus Phyllopsora in India. *Bibliotheca Lichenologica* 86: 185–191.

41 Awasthi, D.D. (2000). *Lichenology in Indian Subcontinent – A Supplement to “A Handbook of Lichens”*. Dehra Dun: Bishen Singh Mahendra Pal Singh.

42 Sinha, G.P. (2020). *Lichens of India*. Botanical Survey of India.

43 Rout, J., Das, P., and Upreti, D.K. (2010). Epiphytic lichen diversity in a Reserve Forest in southern Assam India. *Tropical Ecology* 51 (2): 281–288.

44 Upreti, D.K. and Nayaka, S. (2005). *Shorea robusta* -an excellent host tree for lichen growth in India. *Current Science* 89 (4): 594–595.

45 Upreti, D.K. and Chatterjee, S. (1999). Distribution of lichens on Quercus and Pinus trees in Almora District Kumaon Himalayas, India. *Geophytology* 28: 41–49.

46 Sinha, G.P. and Ram, J.T.A. (2018). *Lichen Diversity in Sikkim*. Sikkim Himalayan Regional Centre, Gangtok: Botanical Survey of India.

47 Nayaka, S., Upreti, D.K., and Khare, R. (2011). Diversity and distribution of lichens in Katarniaghat Wildlife Sanctuary, Uttar Pradesh. *Journal of Indian Botanical Society* 90 (3&4): 360–366.

48 Uppadhyay, V., Ingle, K.K., Trivedi, S., and Upreti, D.K. (2016). Diversity and distribution of lichens from the monuments of Gwalior division, Madhya Pradesh with special reference to rock porosity and lichen growth. *Tropical Plant Research* 3 (2): 384–389.

49 Srivastava, K. and Sirohi, S. (2021). Role of lichen diversity in India. *IJESC* 11 (7): 28467–28471.

50 Hilmo, O. and Såstad, S. (2001). Colonization of old-forest lichens in a young and an old boreal Picea abies forest: an experimental approach. *Biological Conservation* 102: 251–259.

51 Scheidegger, C., Schroeter, B., and Fre, Y.B. (1995). Structural and functional processes during water-vapor uptake and desiccation in selected lichens with green algal photobionts. *Planta* 197: 399–409.

52 Werth, S., Wagner, H.H., Gugerli, F. et al. (2006). Quantifying dispersal and establishment limitation in a population of an epiphytic lichen. *Ecology* 87: 2037–2046.

53 Sillett, S.C., McCune, B., Peck, J.E. et al. (2000). Dispersal limitations of epiphytic lichens result in species dependent on old-growth forests. *Ecological Applications* 10 (3): 789–799.

54 Zoller, S., Frey, B., and Scheidegger, A.C. (2000). Juvenile development and diaspore-survival in the threatened epiphytic lichen species *Sticta fuliginosa, Leptogium saturninum* and *Menegazzia terebrata*: conclusions for in situ conservation. *Plant Biology* 2 (4): 496–504.

55 Boch, S., Prati, D., Hessenmöller, D. et al. (2013). Richness of lichen species, especially of threatened ones, is promoted by management methods furthering stand continuity. *PLoS One* 8 (1): 55461. https://doi.org/10.1371/journal.pone.0055461.

56 Giovanna, P., Gianluca, G., Simonetta, F., and Leonardo, R. (2022). Habitat fragmentation and lichen diversity in peri-urban woodlands: a case study in the municipality of potenza (Southern Italy). *Plants* 11: 6–11. https://doi.org/10.3390/plants11141858.

57 Savillo, I. (2009). *Lichens in Mangroves*, 189. Madison, WI: In SWS Annual Meeting Abstracts.

58 Nayaka, S., Upreti, D. K., and Ingle, K. K. (2012). Mangroves of India: Potential Habitats for Unique Lichen Flora. International Day for Biological Diversity, *National Conference on Marine Biodiversity*, Lucknow, India (22 May 2012), pp. 68–74, UP State Biodiversity Board, Lucknow.

59 Joshi, S., Yadav, N.K., Kaur, A., and Khan, S. (2022). Lichen adaptation and self-conservation strategies in Pindari Glacier Valley. *Invertis Journal of Renewable Energy* 12 (1): 14–23. https://doi.org/10.5958/2454-7611.2022.00002.

60 Allen, J.L. (2017). Testing lichen transplant methods for conservation applications in the southern Appalachian Mountains, North Carolina, USA. *The Bryologist* 120 (3): 311–319. https://doi.org/10.1639/0007-2745-120.3.311.

61 Paula, B., Lilles, E., and Yngvar, G. (2022). Silvicultural strategies for lichen conservation: smaller gaps and shorter distances to edges promote recolonization. *Ecosphere* 13 (1): 1–34.

62 Singh, K.P. and Sinha, G.P. (1997). Lichens. In: *In Floristic Diversity and Conservation Strategies In India*, vol. 1 (ed. V. Mudgal and P.K. Hajra), 195–234. Calcutta: Botanical survey of India.

63 Kathiresan, K. (2003). Conservation strategies for mangroves in India. *The Botanica* 53: 61–75.

64 Kumar, B. (2009). Lichen resource use pattern and its socioeconomic status in temperate region of Garhwal Himalaya India. *Nature and Science* 7 (2): 101–106.

65 Peterson, E. B. (2010). Conservation assessment with management guidelines for *Peltigera hydrothyria* Miadlikowska & Lutzoni (a.k.a. *Hydrothyria venosa* J. L. Russell). Report from the California Native Plant Society to the U. S. Forest Service. http://californialichens.org/wp-content/uploads/2012/04/Peterson2010-PEHY-Assessment-in-CA.pdf (accessed 23 May 2010).

66 Neitlich, P.N. and McCune, B. (1997). Hotspots of Epiphytic Lichen Diversity in Two Young Managed Forests: Sitios Críticos de Diversidad de Líquenes Epífitos en Dos Bosques Jóvenes Bajo Manejo. *Conservation Biology* 11 (1): 172–182.

67 Young, A. and Clifton, K. (2016). Tardigrades inhabit lichen and moss in Smith Rock State Park, Oregon. *Bulletin of the California Lichen Society* 22 (2): 48–53.

68 Wolseley, P. A. & Aguirre-Hudson, B. (2003). The ecology and distribution of lichens in tropical deciduous and evergreen forests of northern Thailand. *Journal of Biogeography* 24 (3): 327–343. https://doi.org/10.1046/j.1365-2699.1997.00124.x

69 Negi, H.R. (2000). On the patterns of abundance and diversity of macrolichens of Chopta-Tunganath in the Garhwal Himalaya. *Journal of Biosciences* 25 (4): 367–378.

70 Upreti, D.K., Nayaka, S., and Chatterjee, S. (2010). Lichen diversity of Uttarakhand Himalayas. In: *The Plant Wealth of Uttara Khand* (ed. P.L. Uniyal, B.P. Chamola, and D.P. Semwal), 79–193. New Delhi: Jagdamba Publishing Co.

71 Negi, H.R. and Upreti, D.K. (2000). Species diversity and relative abundance of lichens in Rumbak catchment of Hemis National Park in Ladakh. *Current Science* 78 (9): 1105–1112.

72 Hale, M.E. (1955). Phytosociology of corticolous crytogams in the upland forests of southern Wisconsin. *Ecology* 36 (1): 45–63.

73 Bruteig, I.E. (1993). Large-scale survey of the distribution and ecology of common epiphytic lichens on Pinus sylvestris in Norway. In: *Annales Botanici Fennici*, 161–179. The Finnish Botanical Publishing Board.

74 Groom, M.J., Meffe, G.K., and Carroll, C.R. (2006). *Principles of Conservation Biology*. Sunderland: Sinauer Associates.

75 Bergamini, A., Scheidegger, C., Silvia, S., and Palmira, C. (2005). Performance of macrolichens and lichen genera as indicators of lichen species richness and composition. *Conservation Biology* 19 (4): 1051–1062. https://doi.org/10.1111/j.1523-1739.2005.00086.x.

76 Upreti, D.K. and Nayaka, S. (2008). Need for creation of lichen gardens and sanctuaries in India. *Current Science* 94 (8): 976–978.

77 La Greca, S. and Stutzman, B.W. (2006). Distribution and ecology of Lecanora conizaeoides (Lecanoraceae) in eastern Massachusetts. *The Bryologist* 109 (3): 335–347.

78 Giordani, P., Benesperi, R., and Nascimbene, J. (2022). Lung lichens face challenges from climate change. *Frontiers for Young Minds* 10: 659341. https://doi.org/10.3389/frym.2022.659341.

79 Negi, H.R. (2003). Lichens: A valuable bioresource for environmental monitoring and sustainable development. *Resonance* 8 (1): 51–58.

80 Nayaka, S., Upreti, D.K., Gadgil, M., and Pandey, V. (2003). Distribution pattern and heavy metal accumulation in lichens of Bangalore city with special reference to Lalbagh garden. *Current Science* 674–680.

81 Paoli, L., Renato, B., Zuzana Fačkovcová, J.N. et al. (2019). Impact of forest management on threatened epiphytic macrolichens: evidence from a Mediterranean mixed oak forest (Italy). *iForest-Biogeosciences and Forestry* 12 (4): 383–388. https://doi.org/10.3832/ifor2951-012.

82 Johnson, C.J., Ehlers, L.P.W., and Seip, D.R. (2015). Witnessing extinction – cumulative impacts across landscapes and the future loss of an evolutionarily significant unit of woodland caribou in Canada. *Biological Conservation* 186: 176–186.

83 Esseen, P.A., Ekström, M., Grafström, A. et al. (2022). Multiple drivers of large-scale lichen decline in boreal forest canopies. *Global Change Biology* 28: 3293–3309. https://doi.org/10.1111/gcb.16128.

84 Mark, D.B. and Christopher, J.E. (2008). Conservation of the rare British lichen *Vulpicida pinastri*: changing climate, habitat loss and strategies for mitigation. *The Lichenologist* 40 (1): 63–79. https://doi.org/10.1017/S0024282908007275.

85 Juriado, I. and Liira, J. (2011). Threatened forest lichen *Lobaria pulmonaria* - its past, present and future in Estonia. *Forestry Studies* 53: 15–24. https://doi.org/10.2478/v10132-011-0086-6.

86 Gintaras, K. (2018). South Australian lichens — a Kangaroo Island case study. *Swainsona* 30: 17–24.

87 Kumar, B. and Upreti, D.K. (2008). An account of lichens on fallen twigs of three Quercus species in Chopta forest of Garhwal Himalayas, India. *Annals of Forestry* 15 (1): 92–98.

88 McCune, B., Grenon, J., Mutch, L.S., and Martin, E.P. (2007). Lichens in relation to management issues in the Sierra Nevada national parks. *Pacific Northwest Fungi* 2 (3): 1–139.

89 Upreti, D.K., Nayaka, S., and Dubey, U. (2009). Air pollution monitoring studies with lichens in India. *Applied Botany Abstracts* 29 (2): 174–186.

90 Pellant, M. (1996). *Cheatgrass: The Invader That Won the West*. Idaho State Office, Boise: Interior Columbia Basin Ecosystem Management Project, US Bureau of Land Management http://www.icbemp.gov/science/pellant.pdf.

91 Belnap, J., Prasse, R., and Harper, K.T. (2001b). Influence of biological soil crusts on soil environments and vascular plants. In: *Biological Soil Crusts: Structure, Function, and Management, Ecological Studies*, vol. 150 (ed. J. Belnap and O.L. Lange), 281–300. New York: Springer.

92 Deines, L., Rosentreter, R., Eldridge, D.J., and Serpe, M.D. (2007). Germination and seedling establishment of two annual grasses on lichen-dominated biological soil crusts. *Plant and Soil* 295: 23–35.

93 Peterson, E.B. (2013). Regional-scale relationship among biological soil crusts, invasive annual grasses, and disturbance. *Ecological Processes* 2: 2. https://doi.org/10.1186/2192-1709-2-2.

94 Williams, C.B. and Sillett, S.C. (2007). Epiphyte communities on redwood (Sequoia sempervirens) in north-western California. *Bryologist* 110 (3): 420–452.

95 Peterson, E.B. and McCune, B. (2001). Diversity and succession of epiphytic macrolichen communities in low-elevation managed conifer forests in Western Oregon. *Journal of Vegetation Science* 12 (4): 511–524.

96 Peterson, E.B. and McCune, B. (2003). The importance of hotspots for lichen diversity in forests of western Oregon. *Bryologist* 106: 246–256.

97 Darwyn, S.C. (2015). Using partial-cut harvesting to conserve terrestrial lichens in managed landscapes. *Canadian Wildlife Biology & Management* 4 (2): 151–162.

98 Debolt, A.M., Rosentreter, R., and Martin, E.P. (2007). Macrolichen diversity in subtropical forests of North-Central Florida. *Bryologist* 110: 254–265.

99 Hilty, J.H., Eldridge, D.J., Rosentreter, R. et al. (2004). Recovery of biological soil crusts following wildfire in Idaho. *Journal of Range Management* 57: 89–96.

100 Mistry, J. and Berardi, A. (2005). Effects of phorophyte determinants on lichen abundance in the Cerrado of central Brazil. *Plant Ecology* 178: 61–76. https://doi.org/10.1007/s11258-004-2493-8.

101 Collins, W.B., Dale, B.W., Adams, L.G. et al. (2011). Fire, grazing history, lichen abundance, and winter distribution of Caribou in Alaska's taiga. *Journal of Wildlife Management* 75: 369–377.

102 Yamamoto, Y., Kinoshita, Y., and Yoshimura, I. (2001). Culture of thallus fragments and redifferentiation of lichens. In: *Methods in Lichenology* (ed. I. Kranner, R. Becknett, and A. Varma), 34–46. Heidelberg: Springer-Verlag.

103 Behera, B.C., Verma, N., Sonone, A., and Makhija, U. (2005). Evaluation of antioxidant potential of the cultured mycobiont of a lichen *Usnea ghattensis*. *Phytotherapy Research* 19: 58–64. https://doi.org/10.1002/ptr.1607.

104 Ellis, C.J., Coppins, B.J., Dawson, T.P., and Seaward, M.R.D. (2007). Response of British lichens to climate change scenarios: trends and uncertainties in the projected impact for contrasting biogeographic groups. *Biological Conservation* 140: 217–235.

9

Lichen at the Age of Climate Change

Barsha Devi

Department of Botany, Pandit Deendayal Upadhyaya Adarsha Mahavidyalaya, Tulungia, Assam, India

9.1 Introduction

Climate change caused by global warming and anthropogenic activities is one of the greatest environmental threats around the World. From 1990 to 2100, global mean surface air temperature is projected to increase by 1–3.5 °C. It is anticipated that the sea level will rise 15–95 cm by 2100 from its present level [1]. As global temperatures rise, heat waves and extremely hot days are expected to become more frequent and intense. Conversely, colder regions may experience fewer extremely cold days as the average temperature rises. The changing climate can cause variations in precipitation patterns, leading to both an increase in drought severity and an increase in the intensity of rainfall that can trigger flooding [2]. Climate change impacts are increasingly affecting flora and ecosystems in terms of breeding, flowering, nesting, abundance, and diversity of biota [3, 4].

All regions of the earth, from the poles to the tropics are colonized by lichen. However, they become a prominent part of the vegetation in harsh climates such as polar regions and arid and semi-arid deserts [5]). Lichen diversity is sensitive to climate change along with several associated factors such as the type of substratum, nutritional status, etc. Lichens are thought to have a stress tolerance capacity and hence are kept under the category of "stress-tolerant" organisms. Such organisms are attributed to low nutritional requirements, a long life span, sluggish growth, and adaptation to a specific environment [6]. Being diverse and functionally significant in forest ecosystems and well-known pollution indicators, lichens are a crucial category for observing environmental change [7]. Lichens can act as early indicators of environmental degradation and offer important information about how climate change affects ecosystems. In the face of mounting environmental challenges such as pollution and climate change, the study of indicator species has become increasingly important in determining the health of ecosystems and their response to these global issues.

Chemistry, Biology and Pharmacology of Lichen, First Edition. Edited by Ashoke Kumar Das, Ajay Sharma, Deepika Kathuria, Mohammad Javed Ansari, and Garima Bhardwaj.

9.2 Adaptation of Lichen to the Harsh Environment

Lichens can withstand a wide range of environmental conditions like varying temperature regimes, and Ultraviolet radiation (Table 9.1). During their life span, they spend most of their time in a metabolically inactive and parched state. Lichens likely possess a variety of mechanisms capable of coping with stress, such as an antioxidant system [20]. For example, Antarctic lichens have high levels of carotenes and xanthophyll, which help them resist oxidative stress [21]. A widespread biochemical adaptation involves the accumulation of compatible solutes, such as sugars and polyols, under stressed conditions [22]. The ability of lichens to maintain pools of polyols indicates their stress tolerance. Different parts of the thallus have different concentrations of these polyols, which give lichens a clue about their stress response [23]. A defensive approach of a lichen consists of chemical substances. The substances, known as allelochemicals, are thought to be an adaptive strategy in lichens [24, 25]. Due to their poikilohydrous nature, lichens can resist low temperatures, deficits in water, and light irradiance. Due to their simple water uptake system, lichens have resilience to dehydration stress and other adverse environmental conditions [26]. The exceptional symbiotic relationship of the lichen allows them to thrive in extreme climate conditions. For instance, the lichen species *Usnea* of Antarctica and *Umbilicaria* of Antarctic Maritime have limited growth of up to 20 cm in height [27]. Lichen species from Arctic habitats such as *Xanthoria elegans* and *Cetraria islandica* maintain a high protective mechanism against UV-B irradiance Nybakken et al. [9]. Usnic acid, a characteristic yellow pigment found in *Usnea* species, protects the lichen against UV radiation [28]. Similarly, *Caloplaca coralligera*, a lichen species from the maritime Antarctic, which is thought to be endemic to the place, exhibits unique morphological and anatomical adaptations. A study conducted by Henri [29] suggested that lichen populations from different continents exhibit intra-species genetic variation. Research has shown that despite freezing temperatures lichens can perform positive net photosynthesis [30]. The Antarctic continent, which has an extreme environment and supports a sparse vegetation cover is a home for several lichen species [31, 32]. As the driest continent in the world, Antarctica presents special challenges to lichens, which include special adaptations to survive severe dehydration, subzero temperatures, and strong radiations throughout the year. According to an old report, two lichen species namely, *Psorotrichianum idella* and *Collema tenax* of Negev desert of southern Israel shows xerophytic adaptation just like succulent higher plants [11].

9.3 Impact of Climate Change on Lichen Flora

Lichens are increasingly under threat from climate change, which may have an adverse effect on lichen habitats. It is challenging to pinpoint common responses because many of the seen reactions seem to be habitat- or species-specific. Around the world, research is being done on lichen communities to determine how climate change and other anthropogenic disturbances have an impact. Several studies have shown that climate change will have significant impacts on lichens and their communities [33, 34]. It has been demonstrated that many lichen species are very dependent on the climate and are impacted by even the smallest changes in climatic variables. Changes in climate have had a significant

Table 9.1 Adaptive strategies of lichen to the harsh environment.

Study area/ location	Lichen species	Adaptive features	References
White and Barents Seas coast (Russian Arctic)	*Lecanora polytropa, Lecanora intricata*	Thallus is adapted to dehydration, the formation of small-sized ascospores as a reproductive approach, presence of a high amount of usnic acid that enables them to thrive in direct sunlight.	Sonina [8]
Svalbard, Finse, S Norway, Col du Lautaret (Europe)	*Xanthoria elegans, Cetraria islandica*	Maintain a high level of screening pigments like Parietin and melanin as a UV-B radiation.	Nybakken et al. [9]
Maritime Antarctic	*Caloplaca coralligera*	Presence of air spaces in the thallus formed by a well-developed hyphal network; the frutescent growth form of the thallus increases surface area as well as the photosynthetic active area.	Ott and Sancho [10]
Negev desert (Israel)	*Psorotrichia Numidella, Collema tenax*	Presence of xerophytic adaptations like huge water storage capacity.	Kappen [11]
Kaski (Nepal)	*Parmelia flexilis*	A high level of usnic acid in the lichen thallus collected from lower level of altitude while a low level of usnic acid was found in lichens receiving highest precipitation.	Neupane et al. [12]
Maritime Antarctica	*Circinaria gyrosa, Rhizocarpon geographicum, X. elegans, Buellia frigida, Pleopsidium chlorophanum*	Presence of diverse morphological and anatomical traits such as mucilaginous thalli, pigmented cortices.	Meeßen et al. [13]
South Norway	*Xenylla maritima and Anurophorus laricis*	A supercooling ability is achieved during autumn because of gut evacuation and buildup of cryoprotective substances.	Leinaas and Sømme [14]
Europe and North America	*Flavoparmelia caperata*	Presence of enzymatic and nonenzymatic defense system against ozone induced oxidative stress	Pellegrini et al. [15]
Turkey	*Pseudevernia furfuracea*	Accumulation of glutamate decarboxylase (GAD) and glutamate dehydrogenase (GDH) as acclimation to high temperature	Sağlam et al. [16]
South Australia	*Chondropsis semiviridis (F. Muell. Ex. Nyl.) Nyl.*	The algal partner has specific adaptations to increased temperature, as evidenced by the optimum temperature for net photosynthesis of approximately 30 °C.	Rogers [17]

(Continued)

Table 9.1 (Continued)

Study area/ location	Lichen species	Adaptive features	References
North Wales, UK	*R. geographicum (L.)*	Increased trehalose accumulation in the hypothallus as an adaptive strategy to withstand desiccation.	Armstrong and Smith [18]
Northern Norway	*Flavocetraria nivalis, Ophioparma ventosa*	Accumulation of usnic acid, a UV absorbing phenolic compound at higher altitudes.	Bjerke et al. [19]

impact on lichen distribution [35]. Long-term lichen monitoring has revealed the effects of climate change. A study by Crabtree et al. found a link between projected temperature changes and species in the higher mountains of Scotland. The study found that distant arctic and alpine species are more at risk from climate change [33]. The recent emergence of several sub-tropical species in a temperate zone demonstrates that many lichen species can alter their environments [34]. Lichens are one of the dominant vegetation types of the Antarctic Peninsula and adjacent islands. It is estimated that the Antarctic peninsula is experiencing some of the fastest warming on Earth. There are varying responses among species to increased temperature in cold environments, such as Antarctica, with some responding negatively and others positively [36, 37]. Growth rates may also be positively influenced by temperature in such environments, however. Experimental manipulations using open-top chambers (OTCs) have demonstrated that global warming negatively affects *Usnea antarctica* [36]. To track climate change, Insarov et al. [38] developed a trend detection index (TDI) using epilithic lichens over flat calcareous rock transects. According to the report of Escolar and colleagues, biological soil crust communities will be affected greatly by climate change, and their recovery from disturbances will also be impacted [39]. Rai et al. [40] used a multivariate strategy to validate lichen as a bioindicator of shifting planets. These techniques are increasingly frequently employed to analyze the effects of anthropogenic pressures on habitats, such as global warming, pollution, altered land use patterns, and others. Exposure to temperatures higher than 40 °C adversely affected the photosynthetic pigments of the lichen *Evernia prunastri* [41]. Moreover, an increase in global temperature led to a strong loss of photobiont activity in *Evernia mesomorpha* [42]. Lichens with *Trentepohlia* as their photobiont has been undergoing rapid growth [43]. According to them, this expansion of the photobiont might be attributed to global warming, and lichens with this specific alga may benefit from it. However, it has been predicted that climate change will be less severe for saxicolous or epiphyte species, lichens having green algae as a photobiont, and smaller species [44]. Because lichen species spread quickly and disperse well, global warming may shift lichen species from tropical to higher temperate areas [45].

9.4 Sensitivity of Lichen to Climate Change

Lichens have been found to be sensitive indicators of environmental changes, including global warming [3]. Monitoring programs have confirmed the sensitivity of lichens to global warming over a long period of time [19, 34]. A change in the photosynthetic process is indicative of lichen stress response [46]. Lichens are well-known for their capacity to withstand harsh environmental circumstances, such as high-temperature stress, particularly while they are in their dry state [11]. However, even in their dry state, lichens are susceptible to the negative effects of high temperatures as elevated temperatures can lead to the degradation of photosynthetic pigments, such as chlorophyll [41]. Long-term exposure to temperatures exceeding 35 °C can have an adverse impact on lichen in its dry state leading to a decline in net photosynthesis [47]. Studies confirmed that Lichen diversity has declined in Alaska as a result of 16 years of experimental warming and in Sweden, which has experienced nine years of experimental warming [48]. Alatalo et al. [49] studied the responses of lichen communities to 18 years of experimental warming in sub-arctic Sweden. According to the study, climate change may adversely affect arctic and high-alpine lichen communities. A study reported that endemic lichens found in high elevations face significant threats from climate change [50]. In the sub-arctic and Arctic, lichen communities have declined dramatically in response to elevated temperatures [48]. In the future climate change scenario, reduced precipitation will affect endemic lichens' carbon balance by reducing photosynthetic activity time and increasing desiccation damage, increasing their vulnerability [51]. The recent changes in the lichen flora of the Netherlands may also be attributed to global warming, alongside decreasing SO_2 and increasing NH_3 levels [34].

The growth of many lichen species is largely determined by climate conditions, which are often influenced by minor changes. This group of organisms is highly sensitive to the effects of climate change. For the first time in 1979, a long-term monitoring program was conducted to monitor the effect of climate warming on terrestrial ecosystems in the Netherlands. The distribution of lichen epiphytes changed significantly during the study period, which the researchers relate to climate change. A significant increase has been seen in species characteristic of warm temperate environments, while a decrease in species of cold climate has been observed [34].

9.5 Lichen as an Indicator of Climate Change

A climate change indicator species is particularly sensitive to particular climatic variables (temperature, rainfall) and has therefore been identified as a target for monitoring its impacts [52]. Because of their sensitivity to environmental conditions and unique physiology, lichens have been considered important climate change indicators [53]. Surveys of lichen communities are mostly conducted to monitor air quality and forest health. When climate changes occur, lichens adapt quickly, enabling early detection of shifts in vegetation before they are apparent to other plants [53]. The growth of lichen thallus is a reliable parameter for studying the response of lichens to various environmental factors, including both microclimatic and macroclimatic variables [46]. Observing lichen thalli growth allows researchers to determine how lichens respond to microclimate and macroclimate changes. In a study, the growth rates of six different lichens from different regions, namely, maritime Antarctica, Livingston Island, and South Shetland Islands were monitored and compared between 1991 and 2002. The study indicated that the lichen species exhibited significantly higher growth rates than those previously reported for crustose lichens in the region [54]. This suggests that an increased growth rate was observed in these lichens during the study period. A study conducted in Central Europe reports that only one-third of lichen species saw an increase in abundance in recent years from the study period, while the other two-thirds declined in numbers compared to the historical records [55]. Changes in lichen species composition can provide valuable information about changes in climate and air quality [35]. In the US and Europe, lichens have been employed for climate research [34, 56]. According to the findings of the study by Giordani et al., numerous epiphytic lichen species in the study region may be effective bioclimatic indicators that may be used to track climatic changes over the long term [57].

9.6 Transplant Experiment on Lichen

Worldwide, anthropogenic activities and climate change are increasingly threatening species with extinction. As a means of ensuring the conservation status of threatened and endangered species, translocations are increasingly considered and accepted as a means of establishing, re-establishing, or bolstering populations [58]. Lichens are special because of the symbiotic relationship between a mycobiont and a photobiont, which is also one of the factors that makes them challenging to conserve. The methods used to conserve flora and fauna sometimes conflict with those used to conserve threatened lichens. Therefore, lichens need special conservation strategies that take into consideration their complexity. Conservation translocations are a technique for lichen conservation that is gaining popularity [59]. "Transplant" and "translocation" are terms that are used synonymously in the literature [60], which means the intentional moving of lichens from one place to another. Many lichen translocation programs have been conducted, however, most of them concentrate on lichens as bioindicators rather than using them in conservation translocations [59]. It has been found that lichen translocations can be successful for a wide variety of species of lichen. Lichen transplantation experiments have been carried out to address a variety of questions pertaining to lichen morphology, physiology, development, seasonality, survival, and growth rates [61–63]. Research involving lichen translocations for nonconservation goals however may aid conservational benefits because it investigates the growth and development of lichens, ecophysiology, ecology, and the influence of the environment on lichens. Experiments with transplanting lichen have also been done explicitly for conservation. [60, 64–66]. Epiphytic lichen species were transplanted and induced climate change during a two-year period in a highland moist evergreen broad-leaf forest in southwest China. The research has shown that even seemingly insignificant climatic changes have a significant detrimental effect on the well-being and rate of growth of these transplanted epiphytes [67].

As many species are in danger due to global environmental change, there is a definite need for efficient transplantation techniques. Utilizing transplantation technology, degraded habitats can be restored, aiding in global conservation [66]. The threat of global climate change is putting many species at risk, and effective translocation methods are needed to save lichen diversity around the world. In areas that have been significantly depleted of lichens or are threatened with extinction, lichen transplant experiments are indeed an effective conservation strategy. It is possible to protect and conserve endangered lichen species through transplant experiments. It is probable to ensure the survival and recovery of threatened lichen populations by either translocating individuals to suitable

Table 9.2 Lichen transplantation experiments for conservation.

Lichen species	Experiment and findings	References
Lobaria amplissima	A total of 14 transplants were monitored for 10 years from which 70% survived. Failures were observed only during one year of transplantation.	Gilbert [60]
Parmelia saxatilis	New colonies were formed successfully. Although a small fraction of the lichen fragments was trapped in cobwebs or rugose surfaces and some were anchored by rhizinal growth after just four weeks, the bulk of the lichen fragments were washed into dark crevices or blown away, where they disintegrated.	Gilbert [64]
Lobaria pulmonaria (L.) Hoffm	vegetative diaspores and adult thalli were transplanted as a method of re-establishment and population reinforcement	Scheidegger et al. [65]
Lepraria lanata, Graphis sterlingiana, Hypotrachyna virginica	Discs of surgical gauze were used as an artificial substrate which did not show positive response and gauze fell from the trees within six months. In the next trial burlap and plastic air filter were used as artificial substrates which showed 80% and 60% success rate, respectively.	Allen [68]
Usnea angulata	By suspending fragments on monofilament, transplants of *U. angulata* were established to determine their suitability for transplantation. The experiment showed the growth rate of the species higher than the previously reported growth rates for the same species.	Allen [68]

habitats or creating special conservation sites. By adopting this strategy, lichens that are rare or endemic can be protected, and their ecological functions can be maintained. Table 9.2 summarizes some of the lichen transplantation experiments conducted for conservation purposes.

9.7 Carbon Sequestration by Lichen

Carbon sequestration is the process of capturing and storing carbon dioxide (CO_2) from the atmosphere. Photosynthesis is nature's method for capturing carbon. Carbon dioxide and nitrogen from the atmosphere are taken up by cryptogams and fixed at the surface of the earth [69]. Since autotrophs, lichens play a crucial role in carbon sequestration [70]. Approximately 0.34–3.3 Gt of carbon are absorbed each year by lichens from the atmosphere due to their carbon fixation capacity [71]. Both carbon and nitrogen can be fixed by cyanolichens [72]. They possess metabolic processes that are capable of trapping extra Carbon in secondary molecules. Carbon can be retained up to 30% of its weight through the production and storage of secondary compounds in lichens [73]. In habitats where other vegetation struggles to survive, lichens can be crucial participants even if they may not be able to store as much carbon as a giant tree.

9.8 Conclusion

Extreme weather events like drought, heat waves, wind disturbances, and changes in the precipitation pattern will become more common and severe as a result of climate change. Numerous studies on the potential effects of climate change predict that most plants will likely become more stressed and less prolific in the future. Climate and weather changes could drive certain species to relocate. As bioindicators, lichens are among the first species to adapt to environmental changes. Because of their sensitivity, they are crucial to scientists in measuring climatic changes. However, there are limited studies and historical data that confirm the usefulness of different lichen species as climate change indicators. The findings of this type of study may lead to better conservation strategies to prevent species from extinction due to climate change. Lichen populations should be regularly monitored because this can provide insights into local climate changes. According to scientists, certain lichens will be able to survive the effects of climate change by shifting to more favorable habitats. However, certain species are more likely to perish from warmer temperatures, particularly cold-adapted species that are already close to their climate limit. Translocation attempts may be taken into consideration when lichens are threatened with extinction as a result of climate change effects. A key conservation objective should be the protection and restoration of habitats that support diverse lichen communities. Scientists can gain a better understanding of how climate change is affecting various locations by monitoring changes in the distribution and abundance of these species. Insights into the ecological implications of climate change can be gained from tracking the distribution and health of lichen species, which can also help with the creation of effective conservation tactics.

References

1 Allen, J.L., McMullin, R.T., Tripp, E.A., and Lendemer, J.C. (2019). Lichen conservation in North America: a review of current practices and research in Canada and the United States. *Biodiversity and Conservation* 28: 3103–3138.

2 Lee, D., Min, S.K., Fischer, E. et al. (2018). Impacts of half a degree additional warming on the Asian summer monsoon rainfall characteristics. *Environmental Research Letters* 13: 044033.

3 Parmesan, C. and Yohe, G. (2003). A globally coherent fingerprint of climate change impacts across natural systems. *Nature* 421: 37–42.

4 Root, T.L., Price, J.T., Hall, K.R. et al. (2003). Fingerprints of global warming on wild animals and plants. *Nature* 421 (6918): 57–60.

5 Armstrong, R. A. (2017). A study of fragmentation rates in lichen populations on rock surfaces using the Kaplan-Meir estimator and Cox regression. In *Annales Botanici Fennici* (Vol. 54, No. 1–3, pp. 169-178). Finnish Zoological and Botanical Publishing Board.

6 Matthews, J.A. and Trenbirth, H.E. (2011). Growth rate of a very large crustose lichen (*Rhizocarpon* subgenus) and its implications for lichenometry. *Geografiska Annaler: Series A, Physical Geography* 93: 27–39.

7 Ellis, C.J., Eaton, S., Theodoropoulos, M. et al. (2014). Response of epiphytic lichens to 21st Century climate change and tree disease scenarios. *Biological Conservation* 180: 153–164.

8 Sonina, A.V. (2012). Epilithic lichens and their morphological adaptations to the conditions of the White and Barents Seas coast (Russian Arctic). *Czech Polar Reports* 2: 109–116.

9 Nybakken, L., Solhaug, K.A., Bilger, W., and Gauslaa, Y. (2004). The lichens *Xanthoria elegans* and *Cetraria islandica* maintain a high protection against UV-B radiation in Arctic habitats. *Oecologia* 140: 211–216.

10 Ott, S. and Sancho, L.G. (1993). Morphology and anatomy of *Caloplaca coralligera* (*Teloschistaceae*) as adaptation to extreme environmental conditions in the maritime Antarctic. *Plant Systematics and Evolution* 185: 123–132.

11 Kappen, L. (1973). Response to extreme environments. In: *The Lichens* (ed. V. Ahmadjian and M.E. Hale), 311–380. Academic Press.

12 Neupane, B.P., Malla, K.P., Gautam, A. et al. (2017). Elevational trends in usnic acid concentration of lichen *Parmelia flexilis* in relation to temperature and precipitation. *Climate* 5: 40.

13 Meeßen, J., Sánchez, F.J., Brandt, A. et al. (2013). Extremotolerance and resistance of lichens: comparative studies on five species used in astrobiological research I. Morphological and anatomical characteristics. *Origins of Life and Evolution of Biospheres* 43: 283–303.

14 Leinaas, H.P. and Sømme, L. (1984). Adaptations in *Xenylla maritima* and *Anurophorus laricis* (Collembola) to lichen habitats on alpine rocks. *Oikos* 197–206.

15 Pellegrini, E., Bertuzzi, S., Candotto Carniel, F. et al. (2014). Ozone tolerance in lichens: a possible explanation from biochemical to physiological level using *Flavoparmelia caperata* as test organism. *Journal of Plant Physiology* 171: 1514–1523.

16 Sağlam, N.G., Cekic, F., and Dilek, Ü.N.A.L. (2022). The role of gamma-amino butyric acid in short-term high temperature acclimation in lichen *Pseudevernia furfuracea*. *European Journal of Biology* 81: 184–189.

17 Rogers, R.W. (1971). Distribution of the lichen *Chondropsis semiviridis* in relation to its heat and drought resistance. *New Phytologist* 70: 1069–1077.

18 Armstrong, R.A. and Smith, S.N. (2009). Carbohydrates in the hypothallus and areolae of the crustose lichen *Rhizocarpon geographicum* (L.) DC. *Symbiosis* 49: 95.

19 Bjerke, J.W., Lerfall, K., and Elvebakk, A. (2002). Effects of ultraviolet radiation and PAR on the content of usnic and divaricatic acids in two arctic-alpine lichens. *Photochemical & Photobiological Sciences* 1: 678–685.

20 Li, H. and Wei, J.C. (2016). Functional analysis of thioredoxin from the desert lichen-forming fungus, Endocarpon pusillum Hedwig, reveals its role in stress tolerance. *Scientific Reports* 6: 1–10.

21 Strzalka, K., Szymanska, R., and Suwalsky, M. (2011). Prenyllipids and pigments content in selected antarctic lichens and mosses. *Journal of the Chilean Chemical Society* 56: 808–811.

22 Hell, A.F., Gasulla, F., Gonzalez-Houcarde, M. et al. (2021). Polyols-related gene expression is affected by cyclic desiccation in lichen microalgae. *Environmental and Experimental Botany* 185: 104397.

23 Farrar, J.F. (1976). The lichen as an ecosystem: observation and experiment. *Lichenology: Progress and Problems* 385–406.

24 Goga, M., Elečko, J., Marcinčinová, M. et al. (2020). Lichen metabolites: an overview of some secondary metabolites and their biological potential. In: *Co-evolution of Secondary Metabolites* (ed. J.M. Mérillon and K. Ramawat), 175–209. Cham: Springer.

25 Molnár, K. and Farkas, E. (2010). Current results on biological activities of lichen secondary metabolites: a review. *Zeitschriftfür Naturforschung C* 65: 157–173.

26 Kappen, L. (1988). Ecophysiological relationships in different climatic regions. In: *Handbook of Lichenology*, vol. 2 (ed. M. Galun), 37–100. Boca Raton, FL: CRC Press.

27 Kappen, L., Schroeter, B., Scheidegger, C. et al. (1996). Cold resistance and metabolic activity of lichens below 0 °C. *Advances in Space Research* 18: 119–128.

28 Sieglinde, O.T.T. (2004). Early stages of development in Usnea antarctica Du Rietz in the South Shetland Islands, northern maritime Antarctica. *The Lichenologist* 36: 413–423.

29 Henri, D. (2011). The physiological response of sub-Arctic lichens to their abiotic environment. Doctoral dissertation. Durham University.

30 Reiter, R., Höftberger, M., Green, T.A., and Türk, R. (2008). Photosynthesis of lichens from lichen-dominated communities in the alpine/nival belt of the Alps–II: laboratory and field measurements of CO_2 exchange and water relations. *Flora-Morphology, Distribution, Functional Ecology of Plants* 203: 34–46.

31 Robinson, S.A., Wasley, J., and Tobin, A.K. (2003). Living on the edge–plants and global change in continental and maritime Antarctica. *Global Change Biology* 9: 1681–1717.

32 Sancho, L.G., Green, T.A., and Pintado, A. (2007). Slowest to fastest: extreme range in lichen growth rates supports their use as an indicator of climate change in Antarctica. *Flora-Morphology, Distribution, Functional Ecology of Plants* 202: 667–673.

33 Crabtree, D. and Ellis, C.J. (2010). Species interaction and response to wind speed alter the impact of projected temperature change in a montane ecosystem. *Journal of Vegetation Science* 21: 744–760.

34 van Herk, C.V., Aptroot, A., and Van Dobben, H.F. (2002). Long-term monitoring in the Netherlands suggests that lichens respond to global warming. *The Lichenologist* 34: 141–154.

35 Kuldeep, S. and Prodyut, B. (2015). Lichen as a bio-indicator tool for assessment of climate and air pollution vulnerability. *International Research Journal of Environment Sciences* 4: 107–117.

36 Bokhorst, S., Convey, P., Huiskes, A., and Aerts, R. (2016). *Usnea antarctica*, an important Antarctic lichen, is vulnerable to aspects of regional environmental change. *Polar Biology* 39: 511–521.

37 Sancho, L.G., Pintado, A., Navarro, F. et al. (2017). Recent warming and cooling in the Antarctic Peninsula region has rapid and large effects on lichen vegetation. *Scientific Reports* 7: 5689.

38 Insarov, G.E., Semenov, S.M., and Insarova, I.D. (1999). A system to monitor climate change with epilithic lichens. *Environmental Monitoring and Assessment* 55: 279–298.

39 Escolar, C., Martínez, I., Bowker, M.A., and Maestre, F.T. (2012). Warming reduces the growth and diversity of biological soil crusts in a semi-arid environment: implications for ecosystem structure and functioning. *Philosophical Transactions of the Royal Society B: Biological Sciences* 367: 3087–3099.

40 Rai, H., Khare, R., Gupta, R.K., and Upreti, D.K. (2011). Terricolous lichens as indicator of anthropogenic disturbances in a high-altitude grassland in Garhwal (Western Himalaya), India. *Botanica Orientalis: Journal of Plant Science* 8: 16–23.

41 Pisani, T., Paoli, L., Gaggi, C. et al. (2007). Effects of high temperature on epiphytic lichens: issues for consideration in a changing climate scenario. *Plant Biosystems* 141: 164–169.

42 Meyer, A.R., Valentin, M., Liulevicius, L. et al. (2023). Climate warming causes photobiont degradation and carbon starvation in a boreal climate sentinel lichen. *American Journal of Botany* 110: e16114.

43 Aptroot, A. and Van Herk, C.M. (2007). Further evidence of the effects of global warming on lichens, particularly those with *Trentepohlia* phycobionts. *Environmental Pollution* 146: 293–298.

44 Rubio-Salcedo, M., Psomas, A., Prieto, M. et al. (2017). Case study of the implications of climate change for lichen diversity and distributions. *Biodiversity and Conservation* 26: 1121–1141.

45 Aptroot, A. (2009). Lichens as an indicator of climate and global change. In: *Climate Change: Observed Impacts on Planet Earth* (ed. T.M. Letcher), 401–408. Elsevier.

46 Kershaw, K.A. (1985). *Physiological Ecology of Lichens*. Cambridge: Cambridge University Press.

47 MacFarlane, J.D. and Kershaw, K.A. (1978). Thermal sensitivity in lichens. *Science* 201: 739–741.

48 Lang, S.I., Cornelissen, J.H., Shaver, G.R. et al. (2012). Arctic warming on two continents has consistent negative effects on lichen diversity and mixed effects on bryophyte diversity. *Global Change Biology* 18: 1096–1107.

49 Alatalo, J.M., Jägerbrand, A.K., Chen, S., and Molau, U. (2017). Responses of lichen communities to 18 years of natural and experimental warming. *Annals of Botany* 120: 159–170.

50 Allen, J.L. and Lendemer, J.C. (2016). Climate change impacts on endemic, high-elevation lichens in a biodiversity hotspot. *Biodiversity and Conservation* 25: 555–568.

51 Maphangwa, K.W., Musil, C.F., Raitt, L., and Zedda, L. (2012). Experimental climate warming decreases photosynthetic efficiency of lichens in an arid South African ecosystem. *Oecologia* 169: 257–268.

52 Ellis, C.J. (2019). Climate change, bioclimatic models and the risk to lichen diversity. *Diversity* 11: 54.

53 Root, H.T., McCune, B., and Jovan, S. (2014). Lichen communities and species indicate climate thresholds in southeast and south-central Alaska, USA. *The Bryologist* 117: 241–252.

54 Sancho, L.G., Schulz, F., Schroeter, B., and Kappen, L. (1999). Bryophyte and lichen flora of South Bay (Livingston Island: South Shetland Islands, Antarctica). *Nova Hedwigia* 68: 301–338.

55 Nelsen, M.P. and Lumbsch, H.T. (2020). A data-driven evaluation of lichen climate change indicators in Central Europe. *Biodiversity and Conservation* 29: 3959–3971. https://doi.org/10.1007/s10531-020-02057-8.

56 McCune, B. (2000). Lichen communities as indicators of forest health. *The Bryologist* 103: 353–356.

57 Giordani, P., Brunialti, G., and Alleteo, D. (2002). Effects of atmospheric pollution on lichen biodiversity (LB) in a Mediterranean region (Liguria, northwest Italy). *Environmental Pollution* 118: 53–64.

58 IUCN SSC (2013). *Guidelines for Reintroductions and Other Conservation Translocations*, 57. Gland, Switzerland: IUCN SSC, Re-introduction Specialist Group.

59 Mallen-Cooper, M. and Cornwell, W.K. (2020). A systematic review of transplant experiments in lichens and bryophytes. *The Bryologist* 123: 444–454.

60 Gilbert, O.L. (1991). A successful transplant operation involving *Lobaria amplissima*. *The Lichenologist* 23: 73–76.

61 Honegger, R. (1996). Experimental studies of growth and regenerative capacity in the foliose lichen *Xanthoria parietina*. *New Phytologist* 133: 573–581.

62 Muir, P.S., Shirazi, A.M., and Patrie, J. (1997). Seasonal growth dynamics in the lichen *Lobaria pulmonaria*. *Bryologist* 458–464.

63 Piccotto, M., Bidussi, M., and Tretiach, M. (2011). Effects of the urban environmental conditions on the chlorophyll a fluorescence emission in transplants of three ecologically distinct lichens. *Environmental and Experimental Botany* 73: 102–107.

64 Gilbert, O.L. (1977). Lichen conservation in Britain. In: *Lichen Ecology* (ed. M.R.D. Seaward), 415–436. London: Academic Press.

65 Scheidegger, C., Frey, B., and Zoller, S. (1995). Transplantation of symbiotic propagules and thallus fragments: methods for the conservation of threatened epiphytic lichen populations. *Mitteilungen der Eidgenössischen Forschungsanstalt für Wald, Schnee und Landschaft* 70: 41–62.

66 Smith, P.L. (2014). Lichen translocation with reference to species conservation and habitat restoration. *Symbiosis* 62: 17–28.

67 Song, L., Liu, W.Y., and Nadkarni, N.M. (2012). Response of non-vascular epiphytes to simulated climate change in a montane moist evergreen broad-leaved forest in southwest China. *Biological Conservation* 152: 127–135.

68 Allen, J.L. (2017). Testing lichen transplant methods for conservation applications in the southern Appalachian Mountains, North Carolina, USA. *The Bryologist* 120: 311–319.

69 Elbert, W., Weber, B., Burrows, S. et al. (2012). Contribution of cryptogamic covers to the global cycles of carbon and nitrogen. *Nature Geoscience* 5: 459–462.

70 Xu, H., Zhang, Y., Shao, X., and Liu, N. (2022). Soil nitrogen and climate drive the positive effect of biological soil crusts on soil organic carbon sequestration in drylands: a meta-analysis. *Science of the Total Environment* 803: 150030.

71 Porada, P., Weber, B., Elbert, W. et al. (2013). Estimating global carbon uptake by lichens and bryophytes with a process-based model. *Biogeosciences* 10: 6989–7033.

72 Porada, P., Pöschl, U., Kleidon, A. et al. (2017). Estimating global nitrous oxide emissions by lichens and bryophytes with a process-based productivity model. *Biogeosciences* 14: 1593–1602.

73 Asplund, J., van Zuijlen, K., Roos, R.E. et al. (2021). Contrasting responses of plant and lichen carbon-based secondary compounds across an elevational gradient. *Functional Ecology* 35: 330–341.

10

Commercial and Traditional Uses of Lichen

Farak Ali[1], Shahnaz Alom[1], Apurba Gohain[2], Sheikh Rezzak Ali[3], Nilayan Guha[3], and Shuby Kumari[3]

[1] *Girijananda Chowdhury Institute of Pharmaceutical Science, Tezpur campus, Girijananda Chowdhury University, Assam 784501, India*
[2] *Department of Chemistry, Assam University, Silchar, Assam, India*
[3] *Department of Pharmaceutical Sciences, Dibrugarh University, Dibrugarh, Assam, India*

10.1 Introduction

Algae (like *Trebouxia* or *Trentepohlia*) or cyanobacteria (like *Nostoc*) living in a mutualistic relationship with fungal filaments results in the emerging of a composite organism known as lichens [1, 2]. Around 7% of Earth's surface is covered in lichens, and it represent a key part of primary producers in many different substrates and habitats, including some of the most extreme environments on Earth (the deserts, North and South Pole, even glass surfaces, etc.) [3–5]. Lichens have been emerging as the crucial household materials in India since ancient times. Lichens harvested in the temperate Himalayan regions are utilized locally and supplied to various regions. Lichen tends to be prevalent and extensively collected in the Himachal Pradesh and Uttaranchal highlands of India. A small number of ethnic groups from the Madhya Pradesh region of central India as well as from specific Western Ghats locations also gather these lichens. About 2450 different kinds of lichen make up the rich and diversified flora on the Indian subcontinent [6]. Lichens have a very large impact on the environment as they play a crucial role in biomonitoring, bioremediation, bioindicators, environmental indicators, and as soil stabilizers. The oldest fossils of these organisms found in soils, date back 2.2 billion years [7]. Lichens serve as a nutritious food source for both people and animals. A group of Chinese research team found that, lichens constituted 38.4% of the diet of the primate species *Rhinopithecus roxellana* Milne-Edwards [8]. The changes in *Rh. roxellana*'s habitat will also be affected by the lichens that are native to the area. In addition, lichens play an important role in the human diet as food, therapeutic items, and cultural applications [9], particularly in the temperate and polar parts of the world. Although lichens have been put to many different kinds of medicinal uses in different parts of the world, there are a few broad types that tend to crop up repeatedly, like ingestion of a lichen decoction for respiratory and digestive

Chemistry, Biology and Pharmacology of Lichen, First Edition. Edited by Ashoke Kumar Das, Ajay Sharma, Deepika Kathuria, Mohammad Javed Ansari, and Garima Bhardwaj.

disorders [10–12]. This is especially prevalent in the Himalayas and the southeast of China. Lichens have a wide range of additional applications, such as the treatment of gynecological conditions, treatment of eye disorders, wound healing [12], treatment of skin diseases and lesions [9, 13], antibacterial properties [11], treatment of respiratory and digestive disorders, topical applications [11, 12, 14]. Additional lichen carbohydrates that may have a significant medicinal effects include the ubiquitous isolichenins and galactomannans, and the Umbilicariaceae-specific pustulins [9]. The first record of the use of a lichen as medicine is Shipal in Atharvaveda (1500 BC). The vernacular name Charila is extensively utilized in Ayurvedic medicine systems, which is the ancient system of medicine in India. Different communities adopted classical systems of medicine like Ayurveda, Siddha, Unani, Traditional Chinese Medicine (TCM), and Amchi. Some alternative names for lichens in Sanskrit are "Shailaya and ShilaPushp (Shila, rock; Pushp, flower)." Other different kind of species of lichens are burnt in sacred sacrifice fires known commonly as "Hawan or Homa" and are often mixed with a variety of aromatic herbs. Since lichens weigh not much after dried and an enormous amount of raw material is needed in India for the many uses of lichens, a significant amount is also necessary. Every year, 320 tonnes of lichens are used in Nepal and the neighboring parts of India for a variety of applications. The lower-lying branches, trunks, and twigs of fallen trees are where the lichens are picked or occasionally scraped off. The lichens discovered and explored in India possess growth rates of 5 mm every year to about 2 cm every year in case of most quickly grown leafy (foliose) or shrubby (fruticose) lichens. Traditional uses of lichens for food and beverages are still common among the indigenous peoples of the Himalayas (Bhutan, China, India, and Nepal). The Himalayas and the south-western sections of China have become popular destinations for Chinese domestic tourists, the selling of lichens for folk applications, especially for apparently health-promoting teas, has expanded dramatically over the past decade [6, 15]. This chapter will delve into the commercial and traditional uses of lichen and explore why this composite organism is receiving much-needed attention in today's world. We anticipate that the knowledge of lichens' uses in the Himalayas and southern China will be a valuable resource for future lichen studies.

10.2 Historical Background

Although lichens had long been recognized as living things, it was not until Swiss botanist Simon Schwendener put forth his dual theory of lichens in 1867 that the true nature of the lichen association started to come into focus. According to Schwendener's theory, lichens are a combination of fungi with algae or cyanobacteria. As a result of his thorough examination of the anatomy and growth of lichens, algae, and fungi under a light microscope, Schwendener developed his idea, which at the time lacked experimental support. Although it was widely believed that all living things were independent at the time, many of the top lichenologists, including James Crombie and Nylander, rejected Schwendener's claim [11].

Other notable biologists, including Heinrich Anton de Bary, Albert Bernhard Frank, Beatrix Potter, Melchior Treub, and Hermann Hellriegel, were slower to dismiss Schwendener's theories, and the idea quickly permeated other fields of research, including the study of microbial, plant, animal, and human pathogens. Schwendener's concept gained acceptance when the intricate connections between pathogenic bacteria and their hosts were ultimately discovered. When Eugen Thomas published his findings on the first successful re-synthesis experiment in 1939, further experimental evidence of the dual nature of lichens was gained [16].

A novel aspect of the relationship between fungi and algae was identified in the 2010s. Many species of lichen that were previously believed to be ascomycete-algae pairings were actually ascomycete-basidiomycete-algae trios, according to research by Toby Spribille and colleagues [17].

10.3 Lichen as Ethnomedicine

Local and indigenous communities all throughout the world have knowledge, practices, and techniques that are related to traditional knowledge. Indigenous groups acquire traditional knowledge, which is passed down orally from generation to generation. Traditional knowledge is developed from information accumulated through the years and is generally applicable to fields like gardening, forestry, agriculture, and fisheries [18. While the cultural applications of plants from different regions of the world have been acknowledged in a number of literary works, lichens' ethnic uses are less extensively documented. Cryptogams have mostly been disregarded by ethnobotanists historically and today. Nonetheless, lichens have been thoroughly investigated in various temperate nations of Asia, Europe, Africa, and the United States. Ethnic lichen studies from India, Nepal, China, and Tibet are well documented in Asian nations [9]. The first lichen utilized for medicinal purposes was *Pseudevernia furfuracea* (L.) Zopf in the 18th century [19]. Lichens are used in conventional medical practices such as Western medicine, homeopathy, traditional Chinese medicine, and Indian medicine. Crawford identified 60 lichen genera utilized in traditional medicine, stating that the most often treated conditions were wounds, skin, respiratory, gastrointestinal, obstetric, and gynecological. Several disorders are treated with Umbilicaria species and *Parmelia caperata* (L.) Ach. in Chilean traditional medicine [9, 20]. A list of medicinally important lichens is given below with their therapeutic activity (Table 10.1).

10.4 Cultural Aspects of Lichen

While the cultural applications of plants from various regions of the world have been recognized in a number of literary works, lichens' cultural usage is less extensively documented. Even though lichens have been thoroughly investigated in various temperate nations of Asia, Africa, the United States, and Europe. Cultural studies of lichens from India, China, Tibet, and Nepal are well known in Asian nations. Since earlier civilizations, lichens have been employed as traditional recipes, remedies, and sacred sacrifice fires known as "HAVAN" or "HOMA," and they have important effects on both human well-being and ecosystem function. Lichens were a common remedy employed by doctors during the Middle Ages [14]. An Egyptian vase from the 18th Dynasty that contains *Evernia furfuracea* was utilized as a medicine. Due to their ease of accessibility and excellent nutritional content, lichens were primarily consumed as a foodstuff in Europe [48].

Table 10.1 Some medically important lichens along with their distribution and medical applications.

Name	Distribution	Activity	References
Acarospora gobiensis H. Magn.	Himalayas	Antibacterial	[21]
Alectoria ochroleuca (Hoffm.) Massal.	Alpine regions of Nepal and Himalayas	Anticancer	[22]
Arthrorhaphis alpina (Schaer.) R. Sant. in D. Hawksw., P. James and Coppins	Himalayas	Cytotoxicity	[23]
Arthothelium awasthii Patw. and Kulk.	Maharashtra	Antioxidant	[24]
Bulbothrix setschwanensis (Zahlbr.) Hale	South, central India and Himalayas	Inhibition of tyrosinase and xanthine oxidase activity	[25]
Cetraria aculeata (Schreb.) Fr.	Alpine region of Uttarakhand Himalayas	Antibacterial and cytotoxic activity	[26]
Cladonia chlorophaea (Flörke in Sommerf.) Spreng.	Himalayas	Wound healing	[27]
Cladonia crispata (Ach.) Flotow	Eastern Himalayas in Sikkim and Arunachal Pradesh	Antitumor activity	[28]
Cladonia deformis (L.) Hoffm.	Finland	Pulmonary tuberculosis and cough	[29]
Cladonia fimbriata (L.) Fr.	Himalayas and Nilgiri Hills	Cytotoxicity	[23]
Cladonia humilis (With.) J. Laundon	Himalayas	Superoxide dismutase like activity	[30]
Coccocarpia palmicola (Sprengel) Arvid. and Galloway	Himalayas	Cytotoxicity against murine leukemia cells and slow growing BS-C-1 cells	[23]
Collema flaccidum (Ach.) Ach.	Jammu and Kashmir, Manipur and Palni Hills of Tamil Nadu	Antitumor activity	[31]
Dermatocarpon miniatum (L.) Mann	Himachal Pradesh, Uttarakhand, Jammu and Kashmir	Anti-oxidant and antimicrobial activity	[32]
Evernia divaricata (L.) Ach.	Western Himalaya in Jammu and Kashmir	Antimicrobial activity	
Evernia prunastri (L.) Ach.	Himalayas	Antimicrobial activity	
Everniastrum cirrhatum (Fr.) Hale	Tropical and temperate regions of India	Antimicrobial activity	[33]
Everniastrum nepalense (Taylor) Hale	Subtropical and temperate regions of India	Anticancer	[34]

Table 10.1 (Continued)

Name	Distribution	Activity	References
Flavoparmelia caperata (L.) Hale	Himalayas, Nilgiri, Palni Hills and central India	Antimicrobial	[33]
Graphina acharii (Fée) Müll. Arg.	Western Ghats and Eastern Himalayas	Inhibitory properties of tyrosinase and xanthine oxidase and superoxide scavenging capacity	[35]
Graphis scripta (L.) Ach.	Throughout India	Antimicrobial, anticancer and antioxidant	[30]
Heterodermia diademata (Taylor) D.D. Awasthi	Subtropical to temperate parts of India	Wound healing	[36]
Heterodermia microphylla (Kurok.) Skorepa	Eastern Himalayas	Immunomodulator	[37]
Heterodermia podocarpa (Bél.) D.D. Awasthi	Arunachal Pradesh, Maharashtra, Manipur, Nagaland, Sikkim, Uttarakhand, Tamil Nadu and West Bengal	Anti-oxidant	[24]
Heterodermia leucomelos (L.) Poelt	Karnataka, Nagaland, Sikkim, Tamil Nadu, West Bengal	Antibacterial activity	[33]
Hypogymnia enteromorpha (Ach.) Nyl.	Eastern Himalayas	Inhibition of tyrosine and superoxide dismutase like activity	[30]
Hypogymnia tubulosa (Schaer.) Havaas	Himachal Pradesh	Antibacterial activity	[38]
Lecanora muralis (Schreb.) Rabenh. emPoelt	Himachal Pradesh, Uttarakhand, Jammu and Kashmir	Antibacterial activity	[39]
Leptogium cyanescens (Rabenh.) Koerb.	Subtropical and subtemperate regions of India	Cytotoxic	[23]
Mycoblastus sanguinarius (L.) Norman	Himalayas and Nepal	Superoxide dismutase like activity	[31]
Myelochroa entotheiochroa (Hue) Elix and Hale	Himalayas	Inhibition of tyrosine activity	
Myelochroa irrugans (Nyl.) Elix and Hale	Kerala, Nagaland, Sikkim, Tamil Nadu, Uttarakhand and West Bengal	Inhibition of tyrosine activity	
Nephroma expallidum (Nyl.) Nyl.	Himalayas	Cancer chemo preventive and cytotoxic activity	[22]

(*Continued*)

Table 10.1 (Continued)

Name	Distribution	Activity	References
Parmelaria thomsonii (Stirton) D.D. Awasthi	Sikkim	Used as eye pain reliver	[40]
Parmotrema austrosinense (Zahlbr.) Hale	Assam, Himachal Pradesh, Jammu and Kashmir, Kerala, Karnataka, Madhya Pradesh, Tamil Nadu and Uttarakhand	β-glucosidase inhibitor activity	[41]
Parmotrema dilatatum (Vain.) Hale	Kerala, Manipur, Nagaland and Tamil Nadu	Immunomodulating activity	[37]
Parmotrema praesorediosum (Nyl.) Hale	Temperate and tropical regions of India	β-glucosidase inhibitor activity	[41]
Parmotrema reticulata (Taylor) M. Choisy	Tropical and temperate regions of India	Antibacterial activity	[33]
Parmotrema saccatilobum (Taylor) Hale	Andaman, Assam, Goa, Nagaland, Uttarakhand and West Bengal	Superoxide dismutase like activity	[31]
Parmotrema sancti-angelii (Lynge) Hale	Andhra Pradesh, Assam, Kerala, Madhya Pradesh, Meghalaya, Nagaland, Rajasthan and Tamil Nadu	Effective against ring-worm like skin disease	[42]
Parmotrema stuppeum (Taylor) Hale	Nagaland and Nilgiri and Palni Hills of Tamil Nadu	Antioxidant activity	[43]
Peltigera dolichorhiza (Nyl.) Nyl.	Himalayas	Cytotoxicity	[23]
Peltigera polydactylon (Neck.) Hoffm.	Himalayas	Antiseptic	[36]
Ramalina celastri (Sprengel) Krog and Swinsc.	Nilgiri and Palni Hills of Western Ghats	Cytotoxic effect, antitumor, antiviral	[44–46]
Ramalina pacifica Asahina	Karnataka and Kerala	Inhibition of tyrosine activity	[31]
Ramalina pollinaria (Westr.) Ach.	Andhra Pradesh, Himachal Pradesh, Uttarakhand, Jammu and Kashmir	Antimicrobial	[47]
Solorina crocea (L.) Ach.	Himalayas	Cancer chemo preventive and cytotoxic activity	[22]
Stereocaulon foliolosum var. *strictum* (C. Bab.) Lamb.	Eastern and western Himalayas	Used to cure kidney stones	[40]
Thamnolia vermicularis var. *subuliformis* (Ehrh.) W. Culb.	Himalayas	Antibacterial	[22]
Usnea bismolliuscula Zahlbr.	Temperate Himalayas and Western Ghats	Antibacterial	[31]

Table 10.1 (Continued)

Name	Distribution	Activity	References
Usnea subflorida (Zahlbr.) Mot.	Nilgiri Hills of Tamil Nadu	Antibacterial	[47]
Xanthoparmelia coreana (Gyeln.) Kurok.	Uttarakhand Himalayas	Antibacterial	[31]
Xanthoria elegans (Links) Th. Fr.	Western Himalayas	Cancer chemo preventive and cytotoxic activity	[22]

According to Ahmadjian and Nilsson, *Cetraria islandica* is widely advertised in Swedish apothecary stores and touted as having benefits for treating diabetes, lung disease, and catarrh. *Parmelinella salacinifera*, *Heterodermia galactophylla*, and *Parmotrema wrightii* are three lichen species that the Pankararu indigenous community in Brazil used as medicine to cure digestive system problems like vomiting and diarrhea. Moreover, these three species' aqueous extracts are used to treat epilepsy and other illnesses associated with culture by smoking [49]. Many reports on the indigenous medicinal uses of lichens by native peoples for food, fodder, medicine, and beauty have been published in European countries. The following significant lichen species are utilized by the native populations of European nations: *Cladonia pyxidata*, *Cladonia cornuta*, *Cladonia coccifera*, *C. islandica*, *Evernia prunastri*, *Hypogymnia physodes*, *Lobaria pulmonaria*, *Peltigera canina*, *Peltigera aphthosa*, *Usnea plicata*, *Usnea hirta* [9]. *Usnea barbata* was reportedly utilized by African people to cure mammary infections in cattle, according to Afolayan et al. [50].

According to Wang et al., the ethnic groups of Yunnan Province, China, use five species of lichens as food, including *Lobaria kurokawae*, *Lobaria semanderi*, *Lobaria sinensis*, *Thamnolia vermicularis*, *and Thamnolia subuliformis*, as well as five other species as medicinal teas [51]. Song and Gang found that Chinese indigenous societies utilized a wide variety of lichen species. Headaches and vertigo are treated by *Cladonia amaurocraea*. *Cladonia cervicornis* is used for burns, wounds, and bloody coughs. Extract from *Cladonia fruticulosa* is used to control bacterial skin infections. Tuberculosis, intestinal blockage, burns, scalds, skin infections, and pain alleviation are all treated using *Rhizoplaca chrysoleuca*. *Bryoria asiatica* is used to treat heart palpitations, renal deficiencies, dizziness, and trouble urinating. *Usnea ceratina* is used to treat inflammation of the lymphatic system, pulmonary TB, infections, headaches, and coughs. *Usnea pectinata* is used to relieve pain, halt bleeding, and treat bloody feces. *C. islandica* decoction is consumed to strengthen the stomach and promote digestion. Blurred vision, uterine hemorrhage, bleeding from wounds on the outside, ulcers, and swelling are all treated with *Parmotrema tinctorum*. *Punctelia borreri* is used to treat external wounds, uterine bleeding, and hazy eyesight [15].

According to Devkota and coworkers, nine different populations in Nepal use seven different species of lichen, including *Everniastrum cirrhatum, Parmotrema cetratum, and Everniastrum nepalense* for food, *Heterodermia diademata* and Ramalina species for medicinal purposes, *Usnea longissima* for ritual, aesthetic, and bedding purposes, and *Thamnolia vermicularis* for religious and aesthetic purposes. *Heterodermia diademata* and *Eupatorium odoratum* were utilized by the Limbu community in eastern Nepal, according to Limbu and Rai, to treat cuts and wounds [14]. According to Kunwar et al., lichen extract and decoction were used to cure moles in Nepal [52].

Many studies on the use of lichens by various ethnic groups in India are available. Saklani and Upreti claimed that Sikkimese people use *P. canina* to treat liver problems, *H. diademata* to treat cuts and injuries, *Parmelia cirrhata* as a vegetable, *Peltigera polydactyla* to stop bleeding, *Stereocaulon himalayense* to treat urinary diseases and blisters of the tongue [36]. According to Pathak et al. [53], people in Sikkim and Tamil Nadu, India, employ *Hypotrachyna cirrhata* and *Flavoparmelia caperata* to treat wound infections, burns, and bites. The *U. longissima* plant is utilized by the Bhotia tribe and other Garhwal Himalayan dwellers as stuffing for pillows and cushions, and it was utilized by the Madhya Pradesh Baiga tribe to cure bone injuries [54]. According to Upreti et al. [11], 15 kinds of lichen are used by different cultural communities in India and Nepal as food, medicine, or ornaments. *T. vermicularis* is largely utilized by the Bhotia tribes to eradicate worms in milk, according to Upreti and Negi's description in Schaerer ethnobotanical use of the plant [14]. Vinayaka and Krishnamurthy documented the ethnobotanical uses of six species of lichen from distinct tribal communities in southern India for food, medicinal purposes, pillows, and decorative purposes [55]. Lichens were used to make natural dye by the Bhotiya tribal people in Uttaranchal, India's high-altitude Garhwal Himalaya [56]. The Gond and Oraon people of Central India use Parmelia sancti-angeli to heal white patches around the throat that cause a skin condition resembling ringworm. Lai and Upreti recorded three lichens, used as a replacement for "henna" in the Garhwal area [42].

10.5 Commercial Uses of Lichen

Lichens are an important natural resource with a wide range of commercial applications. They consist of a symbiotic relationship between fungi and algae or cyanobacteria and can be found in various habitats worldwide. Here are some of the major commercial uses of lichens:

a) **Dyeing**: Lichens have been used for centuries to produce dyes. Some lichen species produce pigments that can be used to create red, orange, yellow, green, and brown colors. These dyes have been used to color wool, silk, and other textiles.
b) **Perfume Industry**: Lichens are a source of natural fragrances that have been used in perfumes, soaps, and cosmetics. The fragrance industry has been using lichens such as oakmoss and tree moss to create a range of scents, including floral, woody, and musky fragrances. At Kannauj, a town of Uttar Pradesh that has been famous for the past 800 years for its perfume production, lichens are today used in the preparation of an indigenous perfume named "Otto" (the Hina Attar).
c) **Cosmetics**: Lichens are increasingly being used in the cosmetics industry due to their anti-inflammatory and anti-aging properties. Lichen extracts are added to a range of skincare products, including moisturizers and serums.
d) **Textile Production**: Lichen extracts are used in the textile industry to create flame-retardant fabrics. Lichens produce compounds such as atranorin and chloroatranorin, which can be used to make textiles resistant to fire.

e) **Biomonitoring**: Lichens are widely used as bioindicators of environmental pollution. They are highly sensitive to changes in air quality and can accumulate pollutants such as heavy metals and pesticides. By analyzing the concentrations of these pollutants in lichens, scientists can monitor the health of ecosystems and assess the impacts of human activities such as mining and industrial activities.
f) **Bioremediation**: Some species of lichens have been found to have the ability to break down pollutants such as oil and diesel fuel. These lichens can be used in bioremediation projects to clean up contaminated soil and water.
g) **Bioindicators of Climate Change**: Lichens have been used to monitor the effects of climate change on ecosystems. As lichens are sensitive to changes in temperature and moisture, their distribution and abundance can provide valuable information about the impact of climate change on biodiversity.
h) **Environmental Indicators**: Lichens are sensitive to air pollution and changes in the environment. They are used as indicators of air quality and have been used to monitor the effects of acid rain and other forms of pollution. Examples of such lichens are *Usnea* Dill. ex Adans., *Everni* Ach., *Hypogymnia* (Nyl.) Nyl., *Parmelia* Ach., and many more.
i) **Soil Stabilization**: Lichens can be used to stabilize soil and prevent erosion. They can grow on bare rock or soil, and their root-like structures can help bind soil particles together. This makes them useful in areas prone to landslides or erosion, such as steep slopes or construction sites.
j) **Biotechnology**: Lichens have been used in biotechnology research for the manufacturing of novel medications and bioproducts. For example, lichen-derived compounds such as usnic acid have been shown to have anticancer and antiviral properties.
k) **Medicine**: Lichens have been used in medicine preparation commercially apart from its usage in traditional medicine for centuries. A wide range of Ayurvedic and Unani medications, marketed as Charila and Ushna, respectively, are available in Indian marketplaces. These medications are made from several lichen varieties.
l) **Food**: Lichens have been used in the food industry apart from traditional food usage. Lichen-derived powder and the whole portion are major ingredients of the common condiments normally used in Indian food dishes, known as garam, meat, and sambar masala. Lichens are also used in making other different food products available in the market, like cornflakes, etc.

Lichens have a range of commercial uses beyond their traditional uses that are beneficial to various industries. They are a valuable natural resource that can be sustainably harvested and utilized to meet the needs of these industries. However, it is important to ensure that the collection of lichens is done in a responsible and sustainable manner to avoid damaging their habitats and compromising their future availability [14, 15, 42].

10.6 Conclusion

Lichens are one of the algae or cyanobacteria which are highly mutualistic in nature. They used to grow on the woody surface, bark of trees, decaying matters, etc., and a symbiotic relationship can be observed between the lichen and host, which prefer to grow in temperate conditions. From an ancient period, lichen has been an ideal choice for Ayurvedic medicine for their health-beneficial properties. Since earlier civilizations, lichens have been employed as traditional recipes, remedies, and sacred sacrifice fires known as "Havan" or "Homa." Traditionally, lichens have been widely used for ailments such as gynecological infections, eye disorders, wound infections, skin diseases and lesions, respiratory and digestive disorders, etc. Lichen has also been used in the preparation of litmus paper, which is used in chemical labs for testing acidic or basic nature. In the Himalayan region, natives use lichens as a source of food materials. Apart from those traditional uses, extensive research revealed that various species of lichen possess pharmacological properties like antioxidants, antibacterial, anticancer, immunomodulator, anti-tumor, radical scavenging properties, etc. Lichen also has a wide number of commercial applications like dyeing, the perfume industry, cosmetics, textiles, biomonitoring, bioremediation, bioindicator of climate changes, biotechnological applications, etc. As the lichens are enriched with cultural, traditional, and current medicinal properties and maintenance of ecological stability and biodiversity, they have become a target for extinction. People used to destroy the forest where lichen grows, so deforestation is the ultimate reason for the extinction of valuable lichen species. Planting more trees, saving our forest lands, protect and preserve endangered lichen species are the prime concerns for enhancing biological and ecological diversity.

References

1. Spribille, T., Tuovinen, V., Resl, P. et al. (2016). *Science* 353: 488.
2. Brodo, I.M., Sharnoff, S.D., and Sharnoff, S. (2001). *Lichens of North America*. Yale University Press.
3. Rose, C.I. and Hawksworth, D.L. (1981). *Nature* 289 (5795): 289–292.
4. Conti, M.E. and Cecchetti, G. (2001). *Environmental Pollution* 114: 471–492.
5. Andrews, J.H. and Harris, R.F. (2000). *Annual Review of Phytopathology* 38: 145–180.
6. Upreti, D.K., Divakar, P.K., and Nayaka, S. (2005). *Economic Botany* 59: 269–273.
7. Retallack, G.J., Krull, E.S., Thackray, G.D. et al. (2013). *Precambrian Research* 235: 71–87.
8. Liu, X., Stanford, C.B., Yang, J. et al. (2013). *American Journal of Primatology* 75: 860–871.
9. Crawford, S.D. (2019). *Lichen Secondary Metabolites*, 31–97. Switzerland: Springer Cham.
10. Bustinza, F. (1952). *Economic Botany* 6: 402–406.
11. Upreti, D.K., Bajpai, R., and Nayaka, S. (2015). *Plant Biology and Biotechnology*, 1: 263–280. New Delhi: Springer.
12. Prateeksha, Paliya, B.S., Bajpai, R. et al. (2016). *RSC Advances* 6: 21672–21696.
13. Sochting, U. (1999). *Lichen of Bhutan: Biodiversity and Use*. Afdeling for Alger og Svampe, Kobenhavns Universitet.
14. Devkota, S., Chaudhary, R.P., Werth, S., and Scheidegger, C. (2017). *Journal of Ethnobiology and Ethnomedicine* 13. https://doi.org/10.1186/S13002-017-0142-2.
15. Yang, M.X., Devkota, S., Wang, L.S., and Scheidegger, C. (2021). *Diversity* 13: 330.
16. Lokhande, L. (2023). *The Immune Microenvironment in Mantle Cell Lymphoma Targeted Liquid and Spatial Proteomic Analyses*. Department of Immunotechnology, Lund University, Sweden.
17. Toby, S. (2010). *How a Guy from a Montana Trailer Park Overturned 150 Years of Biology*. Washington, DC: The Atlantic.

18 Ludlow, K., Smyth, S.J., and Falck-Zepeda, J. (2014). *Socio-Economic Considerations in Biotechnology Regulation*. New York: Springer.

19 Thiyagarajan, T. (2017). *International Journal of Life Sciences* 5: 387–393.

20 Shahid, M., Rasool, A., Anjum, F., and Rehman, M.T. (2020). *Biomedical Perspective of Lichen-Derived Products*, 263–276. Wiley.

21 Řezanka, T. and Guschina, I.A. (1999). *Journal of Natural Products* 62: 1675–1677.

22 Ingólfsdóttir, K., Kook Lee, S., Bhat, K.P.L. et al. (2000). *Pharmaceutical Biology* 38: 313–317.

23 Perry, N.B., Benn, M.H., Brennan, N.J. et al. (1999). *Lichenology* 31: 627–636.

24 Verma, N., Behera, B.C., Sonone, A., et al. (2008). Lipid peroxidation and tyrosinase inhibition by lichen symbionts grown in vitro. *African Journal of Biochemistry Research* 2 (12): 225–231.

25 Behera, B.C. and Makhija, U. (2002). *Current Science* 82: 61–66.

26 Zeytinoglu, H., Incesu, Z., Tuylu, B.A. et al. (2008). *Phytherapy Research* 22: 118–123.

27 Turner, N.J. (1988). *Economic Botany* 42: 177–194.

28 Nishikawa, Y., Ohki, K., Takahashi, K. et al. (1974). *Chemical & Pharmaceutical Bulletin* 22: 2692–2702.

29 Vartia, K.O. (1973). Antibiotics in lichen. In: *The Lichen* (ed. V. Ahmedjian and M. E. Hale), 547–561. Academic Press. https://doi.org/10.1016/B978-0-12-044950-7.50022-2.

30 Takahagi, T., Endo, T., Yamamoto, Y., and Sato, F. (2008). *Bioscience, Biotechnology, and Biochemistry* 72: 3122–3127.

31 Řezanka, T. and Dembitsky, V.M. (2006). *Natural Product Research* 20: 969–980.

32 Aslan, A., Güllüce, M., Sökmen, M. et al. (2006). *Pharmaceutical Biology* 44: 247–252.

33 Gupta, V.K., Darokar, M.P., Saikia, D. et al. (2007). *Pharmaceutical Biology* 45: 200–204.

34 Kumar, K.C. and Müller, K. (1999). *Journal of Natural Products* 62: 821–823.

35 Behera, B.C., Adawadkar, B., and Makhija, U. (2003). *Phytomedicine* 10: 536–543.

36 Saklani, A. and Upreti, D.K. (1992). *Journal of Ethnopharmacology* 37: 229–233.

37 Santos, L., Honda, N., Carlos, I., and Vilegas, W. (2004). *Fitoterapia* 75: 473–479.

38 Yılmaz, M., Tay, T., Kıvanç, M. et al. (2005). *Zeitschrift für Naturforsch. C* 60: 35–38.

39 Saenz, M.T., Garcia, M.D., and Rowe, J.G. (2006). *Fitoterapia* 77: 156–159.

40 Sinha, G. and Singh, K. (2005). *Macroliches of Sikkim*. Botanical Survey Of India.

41 Le, K.A. and Kim, M.S. (2000). *Canadian Journal of Microbiology* 46: 1077–1081.

42 Lai, B. and Upreti, D.K. (1995). *Lichenologist* 27: 77–79.

43 Jayaprakasha, G.K. and Rao, L.J. (2000). *Zeitschrift für Naturforsch. C* 55: 1018–1022.

44 Leão, A.M., Buchi, D.F., Iacomini, M. et al. (1997). *Journal of Submicroscopic Cytology and Pathology* 29: 503–509.

45 Stuelp-Campelo, P.M., de Oliveira, M.B.M., Leão, A.M.A. et al. (2002). *International Immunopharmacology* 2: 691–698.

46 Fazio, A.T., Adler, M.T., Bertoni, M.D. et al. (2007). *Zeitschrift fur Naturforsch. - Section C Journal of Bioscience* 62: 543–549.

47 Cansaran, D., Kahya, D., Yurdakulol, E., and Atakol, O. (2006). *Zeitschrift für Naturforsch. C* 61: 773–776.

48 Llano, G.A. (1948). *Economic Botany* 2: 15–45.

49 Londoño-Castañeda, P.A., Buril, M.L.L., Rego-Cunha, I.P. et al. (2017). *Global Journal of Science Frontiers Research C Biological Science* 17: 15–22.

50 Afolayan, A.J., Grierson, D.S., Kambizi, L. et al. (2002). *South African Journal of Botany* 68: 72–76.

51 Wang, L.S., Narui, T., Harada, H. et al. (2001). *Bryologist* 104: 345–349.
52 Kunwar, R.M., Shrestha, K.P., and Bussmann, R.W. (2010). *Journal of Ethnobiology and Ethnomedicine* 6: 35.
53 Pathak, A., Shukla, S.K., Pandey, A. et al. (2016). *Proceedings of the National Academy of Sciences India Section B: Biological Sciences* 86: 863–868.
54 Gautam, A.K., Yadav, D., Bhagyawant, S.S. et al. (2021). *Progress in Nutrition* 23. https://doi.org/10.23751/pn.v23i3.9833.
55 Kekuda, T.R.P., Vinayaka, K.S., and Sachin, M.B. (2018). *Journal of Applied Pharmaceutical Science* 8: 148–155.
56 Kala, C.P. (2002). *Current Science* 83: 814–817.

11

Bioactive Compounds in Lichens and Their Therapeutic Potential

Farhan Saeed, Muhammad Afzaal, Muhammad Ahtisham Raza, Habiba Arooj, Mariam Islam, and Rimsha Gulzar

Department of Food Science, Government College University, Faisalabad, Pakistan

11.1 Introduction

Herbal therapy is a traditional method of treating illnesses that has been used for five decades in many cultures. Plant-based materials play an important part in basic medical care as therapeutic agents in several underdeveloped nations even now. Lichens and Bryophytes, which grow outside angiosperms, have the capacity to treat or heal.

The pharmaceutical sector is constantly on the lookout for and developing novel pharmacologically active compounds. Lichens, like higher plants, are thought to be a source of unique physiologically active chemicals. The lichens involve complicated mutually beneficial interactions between algae (photobiont) and fungi (mycobiont) each of which exhibits characteristics unique to the plant kingdom. Lichens are thought to number around 25,000 different species [1].

Lichens are plant-like organisms that live in symbiosis with mycobiont (cyanobacterium) or green alga, a fungal partner, including a couple of other organisms that are involved in photosynthesis (photobionts) [2]. Lichen species account for more than 20% of worldwide fungal biodiversity and are distinctive mutual creatures found in severe settings on the planet, including hot deserts, polar tundra, rocky beaches, poisonous slag heaps, and so on. Lichens manufacture unique but little-understood chemicals to thrive in these harsh settings. More focus is being given to lichens as a potential source for medicinal product development as we learn more about the bioregulatory function of numerous indigenous macromolecules and their modes of action [3]. Currently, there are around 20,000 different species of lichens scattered worldwide, and their medical potential has been discovered [4]. Through fungal metabolism, there are numerous secondary metabolites produced by lichens [5].

Bioactive compounds of lichens have a wide range of biological activities, including antibacterial, antiviral, antimycotic, analgesic, anti-inflammatory, antipyretic, and cytotoxic activities [6]. No instances of lethal infections caused by bacteria in lichens have

Chemistry, Biology and Pharmacology of Lichen, First Edition. Edited by Ashoke Kumar Das, Ajay Sharma, Deepika Kathuria, Mohammad Javed Ansari, and Garima Bhardwaj.

been discovered, in spite of the predicted durations of each person ranging from centuries to millennia. Lichens' ability to form a diverse variety of intriguing biochemical composites helps to describe why they have historically been employed as sources of colors, fragrances, human medications, and so on [7]. Numerous chemical substances have been frequently connected to lichens, and the fact that they are present in axenic cultures or other fungi that do not produce lichens shows that the mycobiont is the source of these substances. It is believed that non-reducing fungus polyketide synthases produce repetitions of aromatic acids like orsellinic acid (usnic acids, depsidones, and depsides) during their synthesis and have been recovered directly from lichens [8].

Lichen mycobiont culture is an important technique for studying the metabolic routes of lichen compounds as well as the circumstances required for their formation. A large range of lichens produce compounds with cytotoxic activity against cancer cells, such as extracts from Xanthoria elegans [9]. Some bioactive substances extracted from lichens include gyrophoric acid, usnic acid, polysaccharides, diffractaic acid, terpene, anthraquinones, and some of these molecules have been used in clinical therapies [10]. Slow-moving creatures that flourish in low-resource habitats create a huge number of defensive metabolites to protect themselves from several slayers. Lichens and their symbiotic creatures, particularly actinomycetes, grow at a slow rate. They are natural environments where helpful bioactive chemicals or metabolites are produced.

The diverse range of lichens can give new bases of actinomycetes for application in drug research [11]. Actinomycete lichen resources have only been studied by some international and national research groups. Furthermore, only a few studies have shown active metabolites produced from actinomycetes. Actinomycetota produced from lichens are prolific makers of bioactive compounds.

For a very long time, countries including Japan, China, India, Iceland, Europe, and others have employed lichens in traditional meals. For instance, following simple processing, edible lichens from China, including *Lobaria pulmonaria, Dermatocarpon miniatum, Umbilicaria esculenta*, as well as *Alectoria asiatica* have been utilized in recipes [12]. Studies on the edibility of lichens and comprehensive analyses of lichen intake are still lacking. This study explored the nutritional or eatable characteristics of lichen along with the applications of lichens in the food sector. Additionally, a study of the bioactive chemicals found in lichen extracts and their potential health benefits provided new properties for the creation of useful meals.

11.2 Diversity in Lichens

Lichens are a group of organisms with a mutualistic partnership between a mycobiont (fungi) and a photobiont (algae) [13, 14]. This symbiotic relationship enables the lichen to grow in a large number of areas from deserts to polar regions. Due to this diversity, there are about 15,000–20,000 known species of lichens that have been discovered now [15], and they are present in different colors, shapes, and ecological niches [16]. In the light of recent studies conducted in the exploration of lichens, researchers reported the presence of lichens on a great number of natural resources and plant fractions, including leaves (foliicolous), bark (corticolous), soil (terricolous), rocks (saxicolous), wood (lignicolous) [17]. Their presence in different parts of natural ecosystems determines their activity and health benefits based on their origin and nutrition. However, their secondary metabolites are still a great addition to the treatment of various diseases and are currently being used as food components for nutrition purposes [18, 19]. Different types of lichens and their characteristics are discussed below.

11.2.1 Reindeer Lichens (*Cladonia* spp)

Reindeer lichens are well known and the largest group with the genera Cladonia. They contain almost 288 species, which have been discovered till now [20]. It forms tufted structures resembling antlers and is commonly found in arctic and subarctic regions. Reindeer lichens have a branching structure like reindeer, and this is the way their name originated. Moreover, reindeer rely on them in harsh environmental conditions for their survival [21, 22]. Reindeer lichens are well-adjusted to the cold atmosphere and are generally found in open tundra, heathland, and boreal forests. With this, they can grow on many substrates, including rock, leaves, mosses, soil, and bark of the plants. Like all lichens, reindeer lichens also form a symbiotic relationship between mycobiont and photobiont; however, the photobiont for reindeer lichens is generally green algae.

11.2.2 British Soldier Lichen (*Cladonia* genera)

British soldier lichens took their name due to their red apothecia (a cup-like structure), which closely resembles the color of the uniform of the British Army [23]. This cup (apothecium) contains reproductive structures called ascospores [24]. *C. cristatella* was first described scientifically by an American Botanist, Edward Tuckerman, in 1858. It is often found on soil, red mosses, and decaying wood. *Cladonia* lichens contribute greatly to the regulation of the ecosystem, including soil formation, nitrogen fixation, and provide habitat to small organisms [25, 26]. Like other lichens, *Cladonia* can produce both sexually and asexually by producing spores and forming new colonies. However, in some cultures, *Cladonia* is used for commercial purposes, including dyeing, herbal medicine, and reindeer feed [27]. Various species are present in this genera with distinct characteristics and attributes including *Cladonia cristatella* (pixie cup lichen), *Cladonia rangiferina* (reindeer lichen), *Cladonia chlorophaea* (green cup lichen), *Cladonia pyxidata* (cup-like structure with thallus), *Cladonia fimbriata* (intricate structure), *Cladonia stellaris* (star-tripped reindeer and radiating branches), *Cladonia squamosa* (crustose base with squamules), *Cladonia portentosa* (devil's matchstick), and *Cladonia pocillum* (cup-like structure with a long stalk).

11.2.3 Crustose Lichen

Crustose lichens are named due to their property of tight adherence with their substrate like a crust or layer [16], form thin, flat, and inconspicuous thallus that adhere with the surface on which they grow, such as bark, small twigs, rock, soil, and sometimes with other lichens [28, 29]. Crustose lichens are considered pioneers in the regulation of the ecosystem at the very initial stage through a process of weathering [30], by growing on rocks and producing the organic and inorganic acids that gradually degrade the rock surface and contribute to the soil formation [31]. However, crustose lichens are particularly sensitive to the composition of the air and pollutants, and changes in crustose lichens' population and diversity can be considered as an indicator of air quality or other ecological changes [32, 33].

Crustose lichens reproduce through the release of spores in a reproductive structure called apothecia [34], and also with the dispersion of spores in the environment through the help of wind, water, and other pollinators and form new colonies [35]. As crustose lichens often grow in exposed environments, they have developed the adaptation to grow in harsh conditions and can tolerate extreme temperature fluctuations and sunlight [36, 37]. Further, some species of lichens have pigments that protect them from harmful UV light. Some common examples of crustose lichens include Rhizocarpon, Lecanora, and Buellia.

11.3 Bioactive Compounds in Lichens

Lichens are the repository of bioactive compounds that impart positive health impacts and prevent the body from various chronic complications, including tumor progression, cell cycle arrest, and insulin resistance. The typical way of cancer prevention of lichens involves induction of cell cycle arrest, suppression of chromosome dysfunction, inhibition of tumor metastasis, tumor-friendly inflammation as well as antitumor immunity [38]. The presence of bioactive compounds in lichens is typically responsible for the treatment of these abnormalities in human bodies. These phenols and flavonoids have antimicrobial and antioxidant properties of extracts. However, the mode of action and complete pharmacokinetics of these compounds are still a mystery, and various research is being conducted to fully understand the mode of action of these drugs, which can be implied later for the treatment of rapidly progressed diseases in humans. Various *in vivo* and *in vitro* studies suggested that these bioactive compounds are potent and can be used as consolidated antimicrobial, anti-inflammatory, and antioxidant compounds. Dar et al. [38] designed an *in vitro* study and concluded that *Parmelia sulcata* contains 7-hydroxy-3-(2-methylbut-3-en-2-yl)-chromen-2-one with prominent antibacterial, antifungal, and antioxidant properties. Keeping in mind the diversity of bioactive compounds in lichens, we will explore some deep insights into these bioactive compounds and their potential health benefits for the human body. Some major bioactive compounds found in lichens include lichen acids, lichen polyphenols, lichen phenolics, lichen depsides and dapsone derivatives, lichen alkaloids, lichen pseudomolecules, and lichen terpenes.

11.3.1 Lichen Acids

Lichen acids are chemically diverse and broadly classified into two distinct groups, i.e., depsides and depsidones [39]. These chemical compounds are characterized by their structure, typically consisting of one or more phenolic units connected by ester linkages. Some common lichen acids include usnic acid, atranorin, fumarprotoceteraric acid, lecanoric acid, and stictic acid. The lichen acids are important not only for their pharmacological properties but also for their ecological importance due to their potential to protect the lichens from environmental factors including pathogens, and herbivores, and shield them from UV light [25]. However, various studies have reported the toxic effects of some lichen acids, such as usnic acid (a photoprotective compound). Further studies are being conducted on its structural changes to make it suitable for its applicability in daily life applications. Usnic acid is an extensively studied metabolite present in lichens, and the historical evidence in traditional medicine endorsed that lichen extract enriched with usnic acid was employed to treat malaria, wound healing, snakebites, and cough at the dosage of 6–9 gm of dehydrated lichens in the form of decoction or tea, which is approximately equivalent to the 60–120 mg of usnic acid per day [40]. Usnic acid gained significant research attention after the Second World War for the treatment of a large number of diseases; however, in the 1960s, the development of chemical antibiotics and its poor solubility limited its use. Usnic acid again became the top research subject in the 1980s after the detection of antibiotic-induced nosocomial infection. However, its uncontrolled use could provoke severe medical complications, including hepatotoxicity [41].

Atranorin is a lichen depside with two phenolic rings connected by ester linkages, produced by the symbiotic association of fungi and cyanobacteria. Its chemical structure consists of dibenzofuran and phenolic rings, which impart various properties in atranorin. Atranorin is widely spread among lichen species and is often detected in thalli (lichen bodies) of both crustose and foliose lichens. Among all other secondary metabolites of lichens, atranorin also exhibits antioxidant, photoprotective, antinociceptive, antitumoral, wound healing, and anti-inflammatory characteristics, which make it significant for utilization for medicinal purposes. However, the mode of action for this depsides is poorly understood, which limits its utilization despite its excellent therapeutic potential. To the best of our knowledge, very limited efficacies of atranorin have been reported till now. Harikrishnan et al. [42] conducted research and reported that atranorin is a compound with molecular target potential for the treatment of breast cancer.

11.3.2 Fumarprotoceteraric Acid

Fumarprotoceteraric acid is a lichenized depside found abundantly in lichens and exhibits excellent pharmacological properties in the *in vivo* system. Its chemical structure consists of two phenolic rings linked together with ester linkage. These two phenolic units are derived from the aromatic compounds resorcinol and orcinol. Further, the structure of fumarprotoceteraric acid includes a dibenzofuran ring and a phenolic ring, which give this compound the specific characteristics that make it distinct in its pharmacological applications. This compound is least explored in order to get knowledge. Hence, few studies have been conducted on the health benefits of this compound. However, the proposed chemical structure from the spectroscopic studies endorsed the view that these lichenized depsides could impart a valuable contribution to the treatment of various medical conditions.

11.3.3 Lecanoric Acid

Lecanoric acid is also known as 2′,3,3′,4′-tetrahydroxy-6′-methoxy-5,7-dioxo-[1,1′:3′,1″] terphenyl-8-carboxylic acid, has a unique and complex chemical structure, with multiple phenolic rings linked together through the ester linkage. Shcherbakova et al. [43] reported that *Evernia prunastri* extract is an antimicrobial agent due to its inhibition potential against *Staphylococcus aureus* and *C. albicans,* and this species of lichens contains a significant amount of lecanoric acid. Furthermore, lecanoric acid is reported as antifungal due to its inhibition potential against different fungi. Rajendran et al. [44] conducted a study, through the extraction and purification of lecanoric acid from the thallus of *Parmotrema austrosinense* lichen through the microcrystallization method and assessed their activity against tea pathogens. The authors reported that lecanoric acid's antifungal activities may have contributed to its considerable inhibiting impact on tea microbes. Furthermore, Kocovic et al. [45] reported the extraction and characterization of lecanoric acid through HPLC. The authors reported strong antimicrobial activities against *Bacillus subtilis*, *Bacillus cereus*, and *S. aureus*; further, lecanoric acid showed a strong cytotoxic impact against HeLa and HCT-116 cell lines.

11.3.4 Therapeutic Potential of Lichens

The Parmelia genus of lichens has not been frequently used in traditional medicine as other species (such as Cetraria, Usnea, and Cladonia), there are several links to their use of ethnopharmacology in literature [46]. The crottle is also known as "skull lichen," or *Parmelia saxatilis* (L.). Ach was long thought to be a treatment for epilepsy and the plague; in addition, it was used as a reddish-brown fabric dye. Wart removal using *P. saxatilis* has also been practiced in Sweden [47]. The medicinal value of Parmelia species' extract and extracted compounds has only lately begun to be thoroughly researched. Over the past two to three decades, they have been moderately examined, particularly for their antibacterial, cytotoxic, and antioxidant characteristics. This information about the lichen varieties of the Parmeliaceae family is deeper than that of other genera [48]. Antioxidants, both natural and synthetic, may benefit human bodies by lowering the abundance of reactive species formation. Butylated hydroxytoluene (BHT), Butylated hydroxyanisole (BHA), propyl gallate (PG), and tert-butyl hydroquinone (TBHQ) are examples of synthetic antioxidants that are currently supposed to have hazardous and carcinogenic effects [49]. For external protection, antioxidants are required in order to lessen the harmful impacts of oxidative stress. Since lichens are a rich source of bioactive chemicals, lichen extracts or their active ingredients have been studied and evaluated for their antioxidant effects. However, most of these are assessments of *in vitro* studies (Table 11.1). Lichens can produce bioactive substances, primarily phenolic compounds containing reactive hydroxyl groups, which have antioxidant properties [61].

11.3.5 Antimicrobial Activity

Parmelia caperata and *Parmelia perlata* extracts were tested for their antibacterial effects against *B. cereus, B. subtilis, S. aureus, Escherichia coli, Bacillus megaterium, Proteus vulgaris,* and *Serratia marcescens.* Both Gramme-positive and Gramme-negative bacteria were susceptible to the effects of *P. caperata* extracts. The strongest antiseptic effect against *S. aureus* was specifically found in *P. caperata* extracts, which produced an inhibitory halo of about 19 mm. The ether extract against *S. aureus* was the only antibacterial action that *P. perlata* showed, though [62]. In this section, the antibacterial qualities of a methanol extract of *P. saxatilis* were assessed alongside those of other lichen species. The proliferation of 9 out of 35 examined bacterial microbes and 2 out of 18 tested that fungus was suppressed by this lichen species, although it did not show anti-Candida efficacy [63]. Methanol extracts of lichen species suppressed the

Table 11.1 Bioactive compounds in lichens and their therapeutic potential.

Bioactive compound	Structure	Therapeutic activity	Experimental details	References
Usnic acid	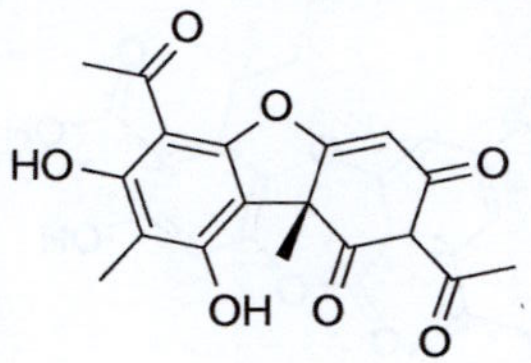	Antimicrobial activity	*In vitro* study Effective against all tested Pathogens including gram negative bacteria (*Escherichia coli*, *Mycobacterium* sp.), gram positive bacteria (*Bacillus cereus*, *Staphylococcus aureus*), fungi (*Aspergillus* sp.) and yeast (*Candida* sp., *Saccharomycescerevisiae*)	[50]
		Antitumor activity	*In vivo study* Inoculated usnic acid into the rim of nude female rats reduced the weight and tumor volume.	
Atranorin	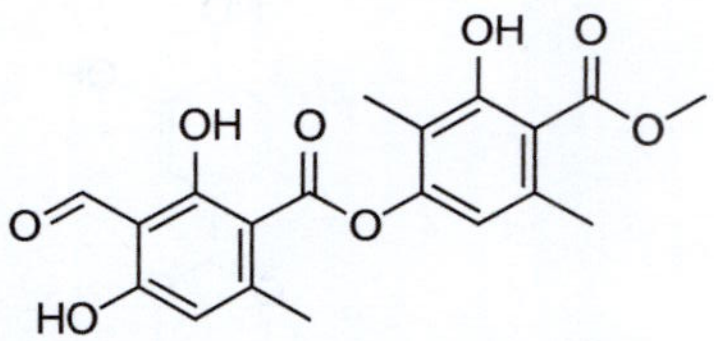	Antitumor activity	*In vivo* study: Antitumor potential against human melanoma (Fem X) and human colon carcinoma (LS-174) cell lines, and also examined that atranorin reduced the tumor volume and augmented apoptosis of tumor-bearing mice.	[38]

(*Continued*)

Table 11.1 (Continued)

Bioactive compound	Structure	Therapeutic activity	Experimental details	References
Depside	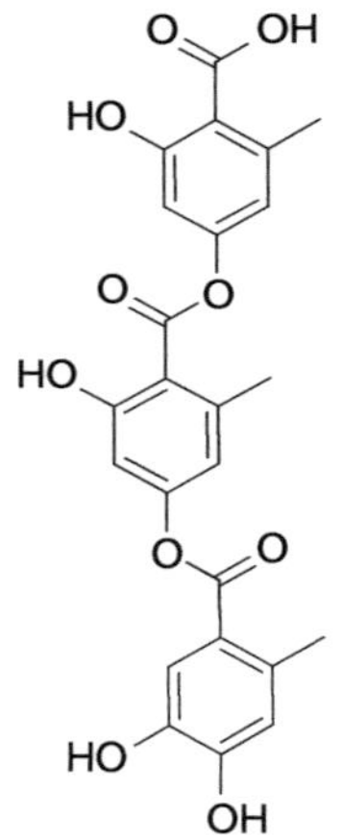	Antiangiogenic activity	*In vitro* study: Depside showed antiangiogenic effect on (RATECs) rat adipose tissue endothelial cells and interruption of endothelial spout development.	[51]
Lobaric acid	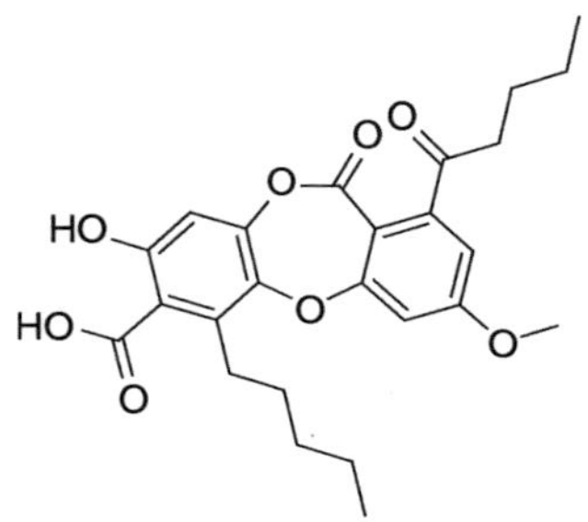	Antitumor activity	*In vitro* study: Lobaric acid cause anticancer effect on colon carcinoma cells (HCT116) and human cervix adenocarcinoma cells (HeLa). Lobaric acid treated with these cells cause apoptosis in both cell lines.	[52]
Depsidones	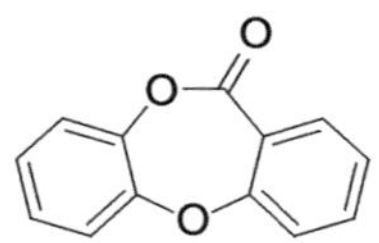	Cytotoxic activity Antioxidant activity	*In vitro* and *in vivo* study: Cytotoxic affect against cell lines of animal and human origin and also showed antioxidant properties by reducing in oxidative stress in both *in vitro* and *in vivo* experiments.	[53]

Dibenzofuran	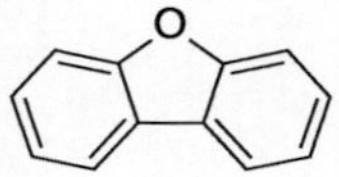	Anti-inflammatory activity	*In vivo* study: Dibenzofuran hindering the activities of lipoxygenase, cyclooxygenase and prostaglandin synthase by exerting anti-inflammatory effects and lowering the cytokine levels and controlling the (NF-KB) nuclear factor kappa-B channel.	[54, 55]
Lecanoric acid	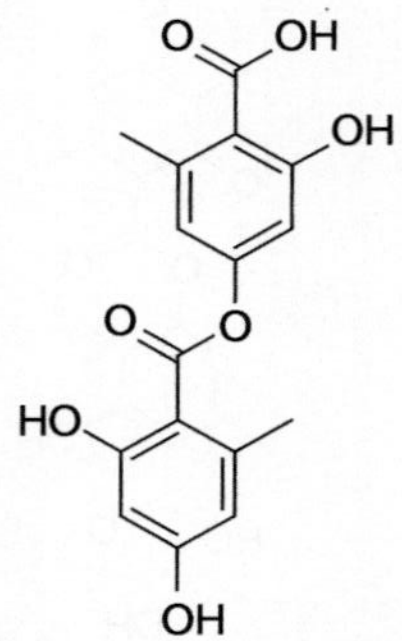	Anticancer activity	*In vitro* study: Lecanoric acid reduced the feasibility of the colon cancer epithelial cell line (HCT-116) and by arresting the cells in Mphaselecanoric acid block the G2 cell cycle in numerous cell lines.	[56]
Vulpinic acid	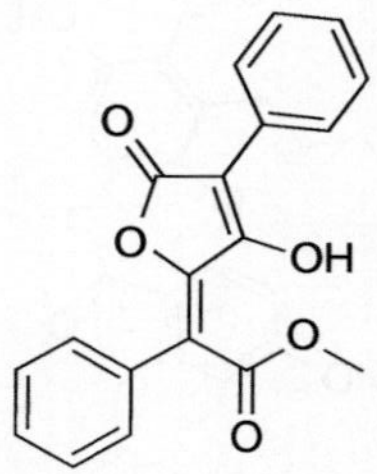	Antiproliferative activity	*In vitro* study:	[57, 58]

(*Continued*)

Table 11.1 (Continued)

Bioactive compound	Structure	Therapeutic activity	Experimental details	References
Physodic acid	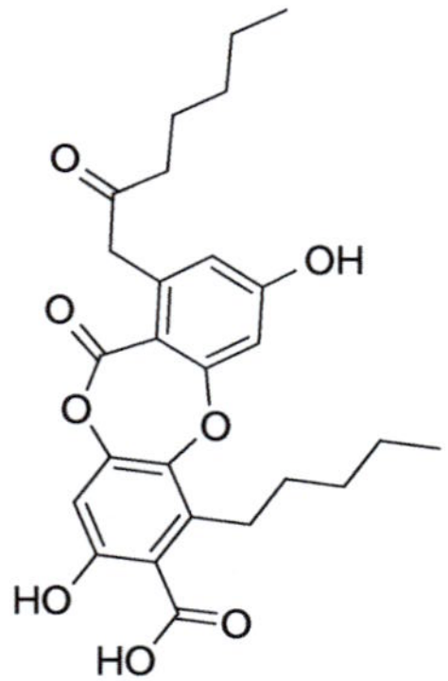	Cytotoxic activity	*In vitro* study: Tyrosinase activity, an enzyme connected to neurological illnesses, was suppressed by the cytotoxic activity of physodic acid in the glioblastoma cell lines (A-172, T98G, U-138 MG) and decreased cyclooxygenase-2 activity enzymes shown in individuals with malignant glioma.	[59]
Diffractaic acid	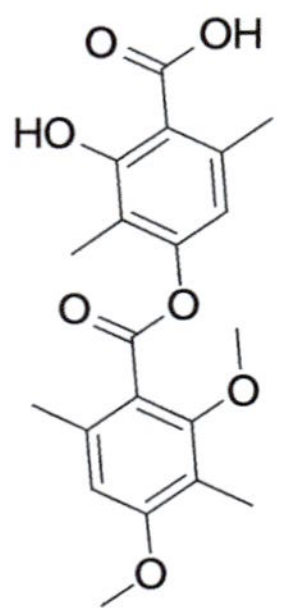	Antiviral activity	Cell based study: A mosquito-borne antiviral is produced by diffractaic acid because of its excellent the selectivity indices and hypothesized target of viral propagation.	[60]

proliferation of the 6 bacteria examined in another study having the zones of inhibition of 15 and 28 mm and MIC levels of 0.94 and 3.75 mg/mL, against *E. coli* and Micrococcus lysodeikticus correspondingly. The antimicrobial activities of *Parmelia crinita* are especially emphasized. Additionally, all investigated species of fungus were inhibited by the methanol extracts of *P. crinita*, apart from *Penicillium verrucosum* and *Aspergillus niger*, with disc diffusion diameters of 11–18 mm and MIC values ranging from 3.75 to 30 mg/mL [64].

The antibacterial activity of usnic acid and lecanoric acid (derived from *Parmelia subrudecta*) against gram-negative and gram-positive infections was further demonstrated by Ivanova et al. [65]. Atranorin and lecanoric acid are both powerful suppressors of *S. aureus* germs; it is crucial to highlight. Additionally, lecanoric acid and atranorin were effective against mycobacteria and *B. subtilis*, respectively.

11.3.6 Antioxidant Activity

The most effective technique to get rid of reactive species, which cause oxidative stress, is with antioxidants. The high activity of lichen species allowed them to be considered natural antioxidant resources. Under typical physiological circumstances, the body continuously produces free radicals, which have a substantial impact on several chemical processes [66]. The investigation of the antioxidant capacities of Parmelia species has been covered in several published papers. In fact, several *in vitro* techniques like DPPH, FRAP, linoleic acid peroxidation, and superoxide anion assay methods have been used to investigate the antioxidant potential of extracts obtained from *P. caperata (L.)* Ach, Ach., *P. pert*, *P. perlata (Huds)*, *Parmelia arseneana Gyeln* and *P. crinita Ach.*, [61]. Last but not least, the antioxidant properties of methanol extract from *P. saxatilis* were evaluated in comparison to those of other lichen species; nevertheless, as indicated by linoleic acid peroxidation and DPPH methods, its methanol extract was ineffective as an antioxidant [63].

11.4 Conclusion

Lichen a symbiotic relation between photobionts and mycobionts are the agents responsible for the maintenance of nature in multiple aspects. They can produce strong secondary metabolites, which could be used for the treatment of various medical conditions, but their limited clinical trials limited the physicians to prescribe or commercialize these compounds due to their unknown dosages and safety limits. Furthermore, after a thorough review, we can conclude that lichens and lichens-based extracts can be used as food as well as medicines due to their medicinal effect against various physiological complications.

References

1 Chapman, A.D. (2009). *Numbers of living species in Australia and The World*. Canberra: Australian Biological Resources Study (ABRS).

2 Sre-Indrasutdhi, V. (2005). Isolation and optimization of lichenized fungi for bioactive compound screening. Doctoral dissertation, Master Thesis. Mahidol University.

3 Karthikaidevi, G., Thirumaran, G., Manivannan, K. et al. (2009). Screening of the antibacterial properties of lichen *Roccella belangeriana* (Awasthi) from Pichavaram mangrove (Rhizophora sp.). *Advances in Biological Research* 3 (3-4): 127–131.

4 Ramya, K. and Thirunalasundari, T. (2017). Lichens: a myriad hue of bioresources with medicinal properties. *International Journal of Life Sciences* 5 (3): 387–393.

5 Gandhi, A.D., Umamahesh, K., Sathiyaraj, S. et al. (2022). Isolation of bioactive compounds from lichen *Parmelia sulcata* and evaluation of antimicrobial property. *Journal of Infection and Public Health* 15 (4): 491–497.

6 Manojlović, N.T., Vasiljević, P., Jusković, M. et al. (2010). HPLC analysis and cytotoxic potential of extracts from the lichen, *Thamnolia vermicularis var. subuliformis*. *Journal of Medicinal Plants Research* 4 (9): 817–823.

7 Suzuki, M.T., Parrot, D., Berg, G. et al. (2016). Lichens as natural sources of biotechnologically relevant bacteria. *Applied Microbiology and Biotechnology* 100: 583–595.

8 Robinson, C. (1999). The genetics of industrial microorganisms: the first half century. *Trends in Biotechnology* 17 (5): 178–181.

9 Ingólfsdóttir, K., Kook Lee, S., Bhat, K.P. et al. (2000). Evaluation of selected lichens from Iceland for cancer chemopreventive and cytotoxic activity. *Pharmaceutical biology* 38 (4): 313–317.

10 Yang, Q., Song, Z., Li, X. et al. (2023). Lichen-derived Actinomycetota: Novel Taxa and bioactive metabolites. *International Journal of Molecular Sciences* 24 (8): 7341.

11 Ding, T., Yang, L.J., Zhang, W.D., and Shen, Y.H. (2019). The secondary metabolites of rare actinomycetes: chemistry and bioactivity. *RSC Advances* 9 (38): 21964–21988.

12 Huang, X., Ma, J., Wei, L. et al. (2018). An antioxidant α-glucan from *Cladina rangiferina* (L.) Nyl. and its protective effect on alveolar epithelial cells from Pb^{2+}-induced oxidative damage. *International Journal of Biological Macromolecules* 112: 101–109.

13 Schweiger, A.H., Ullmann, G.M., Nürk, N.M. et al. (2022). Chemical properties of key metabolites determine the global distribution of lichens. *Ecology Letters* 25 (2): 416–426.

14 Mansournia, M.R., Bingyun, W.U., Matsushita, N., and Hogetsu, T. (2012). Genotypic analysis of the foliose lichen *Parmotrema tinctorum* using microsatellite markers: association of mycobiont and photobiont, and their reproductive modes. *The Lichenologist* 44 (3): 419–440.

15 Khare, R., Upreti, D.K., Haq, M.U., and Behera, B.C. (2020). *Diversity of Lichens in Jammu and Kashmir State*, 343–377. Jammu and Kashmir: Biodiversity of the Himalaya.

16 Tina-Picaza, P.L.D. and Picaza, R.R. (2023). Identification and diversity of lichen species in Lake Leonard, New Leyte, Maco, Davao De Oro Province. *East Asian Journal of Multidisciplinary Research* 2 (7): 2967–2982.

17 Wang, X., Yuan, W., Lin, C.J. et al. (2020). Underestimated sink of atmospheric mercury in a deglaciated forest chronosequence. *Environmental Science & Technology* 54 (13): 8083–8093.

18 Elkhateeb, W.A. (2023). Mushrooms and lichens the factory of important secondary metabolites. *Journal ISSN*. https://doi.org/10.37871/jbres1770.

19 Thakur, M., Kasi, I.K., Islary, P., and Bhatti, S.K. (2023). Nutritional and health-promoting effects of lichens used in food applications. *Current Nutrition Reports* 1–12. https://doi.org/10.1007/s13668-023-00489-6.

20 Kono, M. and Terai, Y. (2023). The third symbiotic partner of the volcano lichen Cladonia vulcani Savicz drove adaptation to an extreme environment. *bioRxiv*. https://doi.org/10.1101/2023.04.05.535799.

21 Kater, I. and Baxter, R. (2022). Abundance and accessibility of forage for reindeer in forests of Northern Sweden: impacts of landscape and winter climate regime. *Ecology and Evolution* 12 (4): e8820.

22 Axelsson-Linkowski, W., Fjellström, A.M., Sandström, C. et al. (2020). Shifting strategies between generations in Sami reindeer husbandry: the challenges of maintaining traditions while adapting to a changing context. *Human Ecology* 48: 481–490.

23 Bunyard, B. (2022). *The Lives of Fungi: A Natural History of Our Planet's Decomposers.* Princeton University Press.

24 Dongre, M.A., Ashraf, M.S., Koshariya, A.K., and Reddypriya, P. (2023). *Plant Pathology and Plant Diseases.* AG publishing house (AGPH Books).

25 Grimm, M., Grube, M., Schiefelbein, U. et al. (2021). The lichens' microbiota, still a mystery? *Frontiers in Microbiology* 12: 714.

26 Ellis, C.J., Asplund, J., Benesperi, R. et al. (2021). Functional traits in lichen ecology: a review of challenge and opportunity. *Microorganisms* 9 (4): 766.

27 Richardson, D.H.S. (2019). Medicinal and other economic aspects of lichens. In: *Handbook of Lichenology* (ed. D.H.S. Richardson), 93–108. CRC press.

28 Ekman, S., Svensson, M., Westberg, M., and Zamora, J.C. (2019). Additions to the lichen flora of Fennoscandia III. *Graphis Scripta* 31 (5): 34–46.

29 Tripp, E.A. and Lendemer, J.C. (2019). Highlights from 10+ years of lichenological research in Great Smoky Mountains National Park: celebrating the United States National Park Service Centennial. *Systematic Botany* 44 (4): 943–980.

30 Armstrong, R.A. (2021). The potential of pioneer lichens in terraforming Mars. In: *Terraforming Mars* (ed. M. Beech, J. Seckbach, and R. Gordon), 533–553. Wiley.

31 Wu, Q., Hu, H., and He, L. (2019). Preliminary study on the rock weathering effect of *Bacillus mucilaginosus. Applied Ecology & Environmental Research* 17 (6).

32 Frati, L. and Brunialti, G. (2023). Recent trends and future challenges for lichen biomonitoring in forests. *Forests* 14 (3): 647.

33 Abas, A., Asnawi, N.H., Aiyub, K. et al. (2022). Lichen biodiversity index (LBI) for the assessment of air quality in an industrial city in Pahang, Malaysia. *Atmosphere* 13 (11): 1905.

34 Öztürk, Ş. (2021). Lichens: characteristics, importance, uses, and distribution in Turkey. In: *Biodiversity, Conservation and Sustainability in Asia: Volume 1: Prospects and Challenges in West Asia and Caucasus* (ed. M. Öztürk, V. Altay, and R. Efe), 39–59. Cham: Springer International Publishing.

35 McGrath-Blaser, S.E. and Longo, A.V. (2022). Fungal invasions and potential spread through tourism and recreation. In: *Tourism, Recreation and Biological Invasions* (ed. A. Barros, R. Shackleton, L. Rew, et al.), 27–35. CABI.

36 Beckett, R.P., Minibayeva, F., Solhaug, K.A., and Roach, T. (2021). Photoprotection in lichens: adaptations of photobionts to high light. *The Lichenologist* 53 (1): 21–33.

37 Bajpai, R., Singh, C.P., and Upreti, D.K. (2023). Community-level lichen diversity assessment in Alpine Zone of Indian Himalaya: climate change implications. In: *Ecology of Himalayan Treeline Ecotone* (ed. S.P. Singh, Z.A. Reshi, and R. Joshi), 339–359. Springer Nature Singapore: Singapore.

38 Dar, T.U.H., Dar, S.A., Islam, S.U. et al. (2022). Lichens as a repository of bioactive compounds: an open window for green therapy against diverse cancers. *Seminars in Cancer Biology* 86: 1120–1137.

39 Asplund, J., van Zuijlen, K., Roos, R.E. et al. (2021). Contrasting responses of plant and lichen carbon-based secondary compounds across an elevational gradient. *Functional Ecology* 35 (2): 330–341.

40 Croce, N., Pitaro, M., Gallo, V., and Antonini, G. (2022). Toxicity of usnic acid: a narrative review. *Journal of Toxicology* 2022: 8244340.

41 Kwong, S.P. and Wang, C. (2020). Usnic acid-induced hepatotoxicity and cell death. *Environmental Toxicology and Pharmacology* 80: 103493.

42 Harikrishnan, A., Veena, V., Lakshmi, B. et al. (2021). Atranorin, an antimicrobial metabolite from lichen *Parmotrema rampoddense* exhibited in vitro anti-breast cancer activity through interaction with Akt activity. *Journal of Biomolecular Structure and Dynamics* 39 (4): 1248–1258.

43 Shcherbakova, A., Strömstedt, A.A., Göransson, U. et al. (2021). Antimicrobial and antioxidant activity of *Evernia prunastri* extracts and their isolates. *World Journal of Microbiology and Biotechnology* 37: 1–14.

44 Rajendran, K., Ponmurugan, P., Gnanamangai, B.M. et al. (2023). Bioefficacy of lecanoric acid produced by *Parmotrema austrosinense* (Zahlbr.) hale against tea fungal pathogens. *Horticulturae* 9 (6): 705.

45 Kocovic, A., Jeremic, J., Bradic, J. et al. (2022). Phytochemical analysis, antioxidant, antimicrobial, and cytotoxic activity of different extracts of *Xanthoparmelia stenophylla* lichen from Stara Planina, Serbia. *Plants* 11 (13): 1624.

46 Illana-Esteban, C. (2012). Lichens used in traditional medicine. *Boletin de La Sociedad Micologica de Madrid* 36: 163–174.

47 Crawford, S.D. (2019). Lichens used in traditional medicine. In: *Lichen Secondary Metabolites: Bioactive Properties and Pharmaceutical Potential* (ed. B. Ranković), 31–97. Springer Cham.

48 Gómez-Serranillos, M.P., Fernández-Moriano, C., González-Burgos, E. et al. (2014). Parmeliaceae family: phytochemistry, pharmacological potential and phylogenetic features. *RSC Advances* 4 (103): 59017–59047.

49 Zhang, W., Li, B., Han, L., and Zhang, H. (2009). Antioxidant activities of extracts from areca (*Areca catectu* L.) flower, husk and seed. *African Journal of Biotechnology* 8 (16).

50 Sepahvand, A., Studzińska-Sroka, E., Ramak, P., and Karimian, V. (2021). Usnea sp.: Antimicrobial potential, bioactive compounds, ethnopharmacological uses and other pharmacological properties; a review article. *Journal of Ethnopharmacology* 268: 113656.

51 Ureña-Vacas, I., González-Burgos, E., Divakar, P.K., and Gómez-Serranillos, M.P. (2023). Lichen depsides and tridepsides: progress in pharmacological approaches. *Journal of Fungi* 9 (1): 116.

52 Hong, J.M., Suh, S.S., Kim, T.K. et al. (2018). Anti-cancer activity of lobaric acid and lobarstin extracted from the antarctic lichen Stereocaulon alpnum. *Molecules* 23 (3): 658.

53 Ureña-Vacas, I., González-Burgos, E., Divakar, P.K., and Gómez-Serranillos, M.P. (2022). Lichen depsidones with biological interest. *Planta Medica* 88 (11): 855–880.

54 Zhao, Y., Wang, M., and Xu, B. (2021). A comprehensive review on secondary metabolites and health-promoting effects of edible lichen. *Journal of Functional Foods* 80: 104283.

55 Studzinska-Sroka, E., & Dubino, A. (2018). Lichens as a source of chemical compounds with anti-inflammatory activity. *Herba Polonica* 64 (1): 56–64. https://doi.org/ 10.2478/hepo-2018-0005.

56 Roser, L.A., Erkoc, P., Ingelfinger, R. et al. (2022). Lecanoric acid mediates anti-proliferative effects by an M phase arrest in colon cancer cells. *Biomedicine & Pharmacotherapy* 148: 112734.

57 Solárová, Z., Liskova, A., Samec, M. et al. (2020). Anticancer potential of lichens' secondary metabolites. *Biomolecules* 10 (1): 87.

58 Jeon, Y.J., Kim, S., Kim, J.H. et al. (2019). The comprehensive roles of ATRANORIN, a secondary metabolite from the antarctic lichen Stereocaulon caespitosum, in HCC tumorigenesis. *Molecules* 24 (7): 1414.

59 Studzińska-Sroka, E., Piotrowska, H., Kucińska, M. et al. (2016). Cytotoxic activity of physodic acid and acetone extract from Hypogymnia physodes against breast cancer cell lines. *Pharmaceutical biology* 54 (11): 2480–2485.

60 Loeanurit, N., Tuong, T.L., Nguyen, V.K. et al. (2023). Lichen-derived diffractaic acid inhibited dengue virus replication in a cell-based system. *Molecules* 28 (3): 974.

61 Fernández-Moriano, C., Gómez-Serranillos, M.P., and Crespo, A. (2016). Antioxidant potential of lichen species and their secondary metabolites. A systematic review. *Pharmaceutical Biology* 54 (1): 1–17.

62 Rowe, J.G., Saenz, M.T., and Garcia, M.D. (1989). Antibacterial activity of some lichens from Southern Spain. *Annales Pharmaceutiques Françaises* 47 (2): 89–94.

63 Gulluce, M., Aslan, A., Sokmen, M. et al. (2006). Screening the antioxidant and antimicrobial properties of the lichens *Parmelia saxatilis, Platismatia glauca, Ramalina pollinaria, Ramalina polymorpha* and *Umbilicaria nylanderiana*. *Phytomedicine* 13 (7): 515–521.

64 Rankovič, B., Rankovic, D., and Maric, D. (2010). Antioxidant and antimicrobial activity of some lichen species. *Microbiology* 79: 809–815.

65 Ivanova, V., Bačkor, M., Dahse, H.M., and Graefe, U. (2010). Molecular structural studies of lichen substances with antimicrobial, antiproliferative, and cytotoxic effects from *Parmelia subrudecta*. *Preparative Biochemistry & Biotechnology* 40 (4): 377–388.

66 Ganesan, K., Sukalingam, K., and Xu, B. (2017). *Solanum trilobatum* L. ameliorate thioacetamide-induced oxidative stress and hepatic damage in albino rats. *Antioxidants* 6 (3): 68.

12

Antioxidant Properties of Lichen

Hadiqa Faiz ul Rasul[1], *Fareed Afzal*[2], *Waseem Khalid*[3], *Mateen Ahmad*[4], *Shirin Gull*[5], *Izza Faiz ul Rasool*[2], *Maryam Ilyas*[6], *and Miral Javed*[7]

[1] *Center of Agricultural Biochemistry and Biotechnology (CABB), University of Agriculture, Faisalabad, Pakistan*
[2] *Department of Food Science, Faculty of Life Sciences, Government College University, Faisalabad, Pakistan*
[3] *University Institute of Food Science and Technology, The University of Lahore, Lahore, Pakistan*
[4] *Department of Food Sciences, Faculty of Agricultural Sciences, University of the Punjab, Lahore, Pakistan*
[5] *Department of Nutrition Science, Faculty of Life Sciences, Government College University, Faisalabad, Pakistan*
[6] *Department of Human Nutrition & Dietetics, Faculty of Life Sciences, Minhaj University, Lahore, Pakistan*
[7] *College of Biosystem Engineering and Food Science, Zhejiang University, Hangzhou, P.R. China*

12.1 Introduction

Lichens are living organisms made up of a symbiotic interaction between a cyanobacterium or alga and a fungus. According to Calcott et al. [1] and Choi et al. [2], it is a synergetic organism with a stable and unique structure made up of an algal partner (photobiont) and a fungal partner (mycobiont). Most of the mycobionts are Ascomycetes, whereas Basidiomycetes, Deuteromycetes, and Phycomycetes make up the photobionts [3]. These types of lichens have mutualistic associations. Chlorophyll is a component of the photobiont and serves as a site for photosynthesis. Apart from the development and growth of the photobiont, utmost of the organic nutrients are delivered from the mycobiont. Moist and inorganic salts that the mycobiont's hyphae have taken in from the environment are supplied to the photobiont. The algae cells wind and encircle the fungal hyphae, which also give the lichen its structure and shield the photobiont from environmental harm. Lichens' secondary metabolites are a distinctive collection of naturally occurring organic chemicals that have significant biological importance because of the specifics of the lichen symbiosis [1].

The overall biodiversity of Lichen was predicted to be almost 13,500–30,000 by the year 2000, and this figure may logically rise with the broadening of research areas and advancements in sequencing techniques [1]. Lichens can be broadly categorized into three categories. *Usnea*, *Umbilicaria*, and *Lobaria* are foliose lichens, and *Rhizocarpon* is an example of crustose lichen. In addition, some lichens, including *Cladonia* are intermediate between the three categories above. According to Molnar and Farkas [4], lichens can thrive in various ecological environments, including the polar area, grassland, desert, plateau, and seaside.

Chemistry, Biology and Pharmacology of Lichen, First Edition. Edited by Ashoke Kumar Das, Ajay Sharma, Deepika Kathuria, Mohammad Javed Ansari, and Garima Bhardwaj.

Lichens also possess the dogged life because they frequently constitute the dominant flora in conditions that are unfavorable to life, such as the Alpine zone, the tundra, the polar area, and numerous drought regions [4]. They therefore play a significant role in the development of our ecosystems. Lichens are quite delicate to their surroundings. Lichen diversity will suffer if the air's sulfur dioxide (SO_2) concentration exceeds 30 g/m^3. In addition, lichens are more prone to accumulating heavy metals and are susceptible to fluoride, automobile smoke, hydrocarbons, and dust. Lichens are now a crucial indication for tracking air pollution and assessing the state of the ecosystem. Lichens have an extended tradition of being utilized as traditional foods and medicines in locations like Japan, China, and the Nordic nations of Iceland, India, and Europe. Different types of usnic acid salt have been used to cure the burns, ulcers, and wounds in Switzerland, Finland, Austria, and other states [3, 5]. For instance, simple processing enables China to employ edible lichens such as *Dermatocarpon miniatum*, *Lobaria pulmonaria*, *Umbilicaria esculenta*, and *Alectoria asiatica* as ingredients in recipes. In addition, lichens are used extensively as a valuable resource in various sectors, including geological prospecting, estimating the glacial age, reagents, spices, dyes, and feed. The medicinal value of lichens has drawn the attention of many researchers because of the special chemicals found in lichens. There is still a shortage of publications on the edible qualities of lichens as well as thorough analyses of both their culinary and therapeutic uses [6]. The historic uses of lichens in our food, and their antioxidant properties are summarized in this chapter, which also examines the lichen's edible and nutritional qualities. Lichen extracts and their bioactive ingredients potential health advantages are also highlighted, opening up new possibilities for creating pharmaceuticals and functional foods [7].

12.2 Botanical History of Lichens

Lichens appear to be one organism, but they result from symbiotic relationships between several organisms. On the one hand, it appears that the fungus and the photobiont have been in a mutualistic association since very long. They can handle ecological situations that neither of them could handle independently. Only mosses are considered plants, even if lichens are also referred to as nonvascular plants. Mosses belong to the bryophytes, a class of nonvascular plants. The plants we have and observe today, such as trees, flowers, and ferns, are thought to have their origins in mosses. On the other hand, lichens bear no resemblance to mosses or the rest of the plant family. Lichens are entirely distinct because they don't have roots, branches, or leaves; instead, their chloroplasts are only present in the algae on the top surface of the lichen [5].

12.3 Classification of Lichen

The classification based on the special features is given below.

12.3.1 Crustose Lichens

Lichens that have the name crustose are "crust-like." Crustose lichens are firmly embedded in or connected to their substrate and lack a lower cortex. They make up approximately 75% of all lichens on Earth. They form a crust over surfaces like a boulder, the ground, a car, or roof shingles. Greys, greens, and dazzling hues like yellow, orange, and red are just a few of the rich colors lichens have. Crustose lichens push up against the substrate. A lower cortex is never present in crustose lichens. The hyphae of the medulla attach them to dirt, rocks, or tree bark, and because of the proximity to the substrate, they are essentially indistinguishable. Even a patch of crustaceous lichen may be made up of numerous individuals that are fused in form and still a member of the same species. Homoiomerous lichens are basic crustose lichens. They are granular in structure because they lack a cortex. A thin filamentous network of mycelium surrounds the algae as it grows across the substrate [8].

Most crustaceous lichens have tiny scales called areolas that make up the thallus. Typically, the lower hyphae of the areolas grow more quickly than the main body of the thallus and surround it with a thin layer. A prothallus is a mat of hyphae that is typically black. The thin filaments connecting the areoles of the inner part of the thallus are more appropriately referred to as the hypothallus. This word has also previously been used for this structure. Lichens may produce a fractured surface in one of two ways. In many species, the thallus develops uniformly and later cracks, frequently inadvertently. In other lichen groupings, the thallus develops on an advancing prothallus in small, defined patches that initially seem like distinct entities before gradually becoming more closely compacted towards older parts of the thallus [9].

12.3.2 Fruticose Lichens

The cross-section of fruticose lichens is usually round and lacks a distinct top and bottom. Their thalli may be shrubby, upright, or drooping. While some fruticose lichens have circular branches with a core at the center, others have hollow cores in the middle of their branches. Other fruticose lichens' flat branches tangle with one another. They are three-dimensional and frequently take the form of hanging or branching formations. The photobiont, an algal or cyanobacterial companion, and the mycobiont, a fungal partner, make up these lichens. The photobiont performs photosynthesis to create energy for both partners, while the fungal partner gives the lichen structural support and protection. Fruticose lichens are often found in various environments, including rocky surfaces, tundras, and woodlands. They serve crucial ecological functions such as supplying habitat and food for creatures, aiding soil creation, and detecting air quality [10].

12.3.3 Foliose Lichens

Leafy lichens foliose lichens are lichens that have a thallus that resembles a leaf and is made up of flattened lobes. They are loosely bound to the surface on which they grow. They have two sides that are simple to separate; one is a top, and the other one is a bottom side. They might be rigid, bumpy, convoluted, and have very flat leaves like lettuce. Because lichens are capable of drying, they may endure extreme conditions. If there is not enough water available, photosynthesis is halted. Some lichens can withstand significant temperature changes while floating in the air. Lichens' slow growth may be due to their low photosynthetic rate. Laciniate and umbilicate are the most prevalent growth forms for foliose lichens, which are generated in various thallus morphologies and sizes [11]. The lower surface of laciniate lichens has lobes and is either entirely intact for the substrate, or the edge is free or bends upwards. The architecture of the laciniate lichen is the most intricate of all the lichens; some resemble enormous plants, while others are only a few millimeters in size. According to Rosentreter et al. [11], the lower cortex of the majority of laciniate lichens is coated in rhizinae, which frequently adheres to the stratum's surface [12].

On the other hand, umbilicate lichens have a disc-like shape and can have a single, unbranched lobe or several lobes. According to Rosentreter et al. [11], umbilicate lichens are observed to be anchored to the surface by a central holdfast.

12.4 Source and Formation of Lichen

The fungal threads, which enclose and extend into the algal cells, are primarily responsible for the physiological bulk overall shape of the lichen. A fungal reproductive structure called an apothecium is where the fungus creates spores to carry out its self-reproduction. These spores spread and develop into new fungi, but they are unable to produce the emergence of fresh lichens [13]. The fungus and the alga must disseminate together for Lichen to proliferate. Lichens can procreate in one of two ways. First, soredia, or lichen can create a clump of algae cells enveloped in fungus filaments. These might grow and produce more lichens. Isidia serves as the lichen's second means of self-reproduction; they resemble soredia but are encased in a layer of shielding cortical tissue [14].

Lichens can flourish practically everywhere there is a sturdy surface with adequate lighting. They may be present on the sides of trees, the soil, or even the rock. Any of these substrates that lichen grows on may allow it to absorb specific mineral nutrients, but in general, lichen feeds itself by photosynthesis in the algal cells; as a result, tree-growing lichens are not parasitic and do not eat the trees. But lichens that develop on rocks may secrete substances that speed up the transformation of the rock into soil and promote the emergence of new soils [15]. Although numerous species of tropical and desert lichens exist, most are temperate or arctic. Lichens appear to thrive in drier conditions where they are less likely to be left in pools of water. However, what humans define as dry might not be what the lichen does [16]. Large lichens known as "old man's beard" are frequently found hanging from tree branches in bayous and cold rainforests. Even though these habitats contain a lot of water, the air is not completely saturated, and drying breezes may help to desiccate arboreal creatures. Lichens are most evident on the tundra, where they make up the majority of the ground cover together with mosses and liverworts. Additionally, to provide fodder for grazing animals, this cover aids with ground insulation [17].

Lichens may thrive in freezing tundra and hot deserts, making them challenging organisms. However, the reasons behind their effectiveness are not fully known. They are said to be able to withstand drying and have complicated chemistry, two major characteristics considered to play essential roles. Lichens may entirely dehydrate if moisture content is not present. This is a total loss of bodily water, not just dehydration as we know it from plants and animals, which makes the lichen exceedingly fragile. When moisture is once more present, they swiftly absorb it and revert to soft and fleshy. Lichens can dry up, and when they are brittle and dry, portions may flake off and eventually develop into new lichens.

Lichens have a somewhat complicated yet well-researched chemistry. Numerous compounds that lichens produce are thought to deter predatory attacks. Lichens are only eaten by a few numbers of insects, including moths and beetles. So, predation is not the most significant danger to the survival of lichens; relatively, it is the century's rising pollution. Numerous studies have shown how manufacturing and urban air pollution negatively impact on lichen growth and health. As some lichens are so delicate, they are now being utilized in Europe and North America to swiftly and affordably measure the concentrations of air pollutants (Kosani et al. [18]).

12.5 Antioxidant Property

The unchecked and ongoing use of synthetic medications has created a demand for novel natural-source preparations for treating and preventing numerous human, animal, and plant ailments. Long-term use of synthetic medications is known to frequently result in various negative effects, including resistance. Unlike synthetic drugs, bioactive natural materials don't have any unfavorable side effects and instead benefit organisms [19].

Lichens are crucial components of many ecosystems. They are symbiotic organisms comprising fungi and algae. According to Fabian et al. [20], typically, they develop as epiphytes upon trees and leaves, on rocks, and in infertile soil. Lichens are utilized to produce alcohol, colors, perfumes, and food for both humans and animals. Additionally, lichens have long been employed as a human disease treatment in many nations. For treating pulmonary and cranial conditions, respectively, *L. pulmonaria* and *Parmelia sulcata* have been successfully employed. Similar remedies for jaundice include *Xanthoria parietina* and *Letharia vulpina* [21]. Numerous studies that supported the various biological actions of particular lichens, which have been utilized for several years in traditional medicine, later, justified [22].

Free radicals are continuously created in the body under normal physiological circumstances and are involved in various chemical reactions, but they can also harm cells [23]. Free radicals can be eliminated by the body's network of antioxidant enzymes, which also helps to support them at a medically safe level. When the regular state is broken, oxidative stress results [4, 24]. Throughout history, lichens have been utilized for medical purposes. Certain lichens, like *Cladonia speres*, *L. pulmonaria*, and *Cetraria islandica*, have even been known to be successful in treating pulmonary tuberculosis. In Turkey, lichen species are widely distributed. In particular, one of the most prevalent lichen species found in Turkey's western regions is *C. islandica*. In Turkish traditional medicine, some types of lichen are used as antidiabetic and stomachic drugs. *C. islandica* is widely used in Turkish traditional medicine to treat many conditions like diarrhea, pneumonia, hemorrhoids, and tuberculosis. Furthermore, this kind of lichen has been employed as a hemostatic agent [25]. Antioxidants are required for exogenic modulation in order to lessen the detrimental effects of free radicals. Since lichens are a rich source of bioactive chemicals, lichen extract and secondary metabolites have been investigated and studied for their antioxidant properties. However, the vast majority of them include *in vitro* chemical tests. Pharmacological studies on lichens concentrate on the synthesis of secondary metabolites, which are primarily phenolic substances containing reactive hydroxyl groups that may have antioxidant properties (Fernandez-Moriano et al. [26, 27]). Antioxidant qualities of lichens and their component parts have been studied in the past using both *in vitro* and *in vivo* methods. Depsides, depsidones, dibenzofurans, and pulvinone derivatives are the main classes of the significant secondary metabolites known as phenolic compounds in lichens (Fernandez-Moriano et al. [26, 27]).

Strong antioxidant activity was shown by the acetone extracts of the lichens *Evernia prunastri* and *Pseudevernia furfuracea*, as well as their primary metabolites, physodic acid and evernic acid, in *in vitro* tests using superoxide anion free radical scavenging, reducing power, and 1,1-diphenyl-2-picrylhydrazyl (DPPH) radical scavenging.

Additionally, they discovered that higher total phenol content corresponded to a stronger capacity for free radical scavenging. It is well known that phenolic chemicals help increase oxidation resistance. However, some researchers have hypothesized that lichens' antioxidant activities may not be exclusively related to phenolic chemicals but involve additional potentially active lichen components [28].

Following the research evaluation of the polysaccharide, utter polyphenol, and antioxidant activity of the components from the lichens *Parmelia reticulate* and *Everniastrum cirrhatum*, Pol et al. [29] hypothesized that higher polysaccharide content was associated with the lichen compounds antioxidant potential. According to Wang et al. [30], *U. esculenta* polysaccharide extracts have high antioxidant activity, concentration-dependent ferrous ion chelating, and free radical scavenging properties. Odabasoglu et al. [28] study also revealed that none of the polyphenols present in lichens exhibit antioxidant effects. Researchers can better understand how lichen antioxidants function by researching the cell forge and using animal models. The study by Kotan et al. [31] concluded that by increasing the activities of the enzymes glutathione peroxidase (GPx) and Cu, Zn-superoxide dismutase (SOD), methanol extracts of *C. islandica* at concentrations of 5 and 10 g/mL could significantly lessen the oxidative stress induced by aflatoxin B1 in human lymphocytes. Çolak et al. [32] found relevant results in their experiment. The similar doses of *C. islandica* water extract raised the number of antioxidant enzymes (catalase, GPx, and SOD) in the blood cells of children with type 1 diabetes and nearly reached the control level. In addition to enhancing the antioxidant apparatus (GPx, catalase, SOD, reduced glutathione) of Pb2+ damaged alveolar epithelial cells A549, the water-soluble polysaccharides of *Cladina rangiferina* also, in a dose-dependent manner, lessen lipid peroxidation and reactive oxygen species (ROS) aggregation in the cells. Diffractaic acid from *Usnea longissima* lichen and five doses of *L. pulmonaria* methanol extract reduced lipid peroxidation in rats with stomach injury caused by indomethacin. Additionally, it strengthened the tissues' antioxidant defense mechanism [33]. They include a considerable quantity of phenolic hydroxyl groups, which are known to have the potential to scavenge free radicals (Fernandez-Moriano et al. [4, 26, 27]); polyphenols are the chemicals that are most intriguing in terms of antioxidation. Phenolic substances typically initiate the lipid oxidation cascade chain reaction, add hydrogen to the reactive radicals, and stop the disease process ([34]). Several studies have shown that phenolic chemicals and antioxidant activity are strongly correlated [17, 30]. This study indicated that lichens have high total polyphenol contents, particularly depsides, depsidones, and dibenzofurans.

12.6 Constituents Responsible for Antioxidant Activity in Lichens

12.6.1 Phenolic Compounds

The presence of phenolic chemicals, which are widely distributed by products of lichen secondary metabolism, is most commonly linked to antioxidant action. While tannins, lignins, and flavonoids are among the phenolics found in vascular plants, lichen phenolics mostly consist of depsides, depsidones, and dibenzofurans. The building block of lichen phenolic biosynthesis is orsellinic acid. The fungal partner typically secretes lichen phenolics, which are then left behind as crystals on the fungal hyphae's cell wall. With the exception of pulvinic acid derivatives, which are produced via the shikimic acid route, lichen phenolics are predominantly obtained from acetate-polymalonate. Ross Watson [35] demonstrated that lichen phenolics consist of two monocyclic phenols joined by an ester link (depsides), ether and ester bond (depsidones), or a furan heterocyclic bond (dibenzofurans, such as usnic acid). Phenolic compounds have an aromatic ring made up of one or more hydroxyl groups in their structure. According to the studies, the primary mechanism by which phenols exhibit their antioxidant activity is thought to be their capacity to act as hydrogen atom donors, eliminating free radicals and forming less reactive phenoxyl radicals.

It is commonly known that an antioxidant's activity rises as its side chain lengthens and as the number of hydroxyl groups in the molecule increases. This is most likely because the antioxidant compounds may stabilize free radicals and form side-chain conjugations. The phenol molecule has the capacity to react simultaneously with the radicals due to its numerous potential reactive centers. Vitamins E and C have been shown to be less effective *in vitro* than phenolics on a molar basis and have the perfect structural chemistry for antioxidant activity [36].

It is possible to conclude that the phenolic components of lichens influence their antioxidant capacities when compared with other constituents. Antioxidant activity and total phenolic content of lichens have been linked in numerous studies. For example, in the methanol extract of *L. pulmonaria*, Odabasoglu et al. [28] discovered a substantial association between antioxidant activity and total phenolic content.

12.6.2 Usnic Acid (UA)

Among the most prevalent and plentiful metabolites found in lichens is usnic acid (UA). It can be separated from a variety of species, including *U. longissima*, *Usnea barbata*, and *Cladonia arbuscula*. It is a member of the dibenzofuran family. This compound has been shown to have a number of biological properties, including antimicrobial, anti-inflammatory, cytoprotective, immunostimulatory, gastroprotective, cardiovascular, and anticarcinogenic activities. These properties are primarily attributed to its antioxidant action in minimizing the effects of oxidative stress [37].

Jin and coworkers explored the molecular processes that underlie UA's anti-inflammatory benefits in 2008. They discovered that when lipopolysaccharide (LPS) was exposed to macrophages (RAW 264.7), this chemical suppressed the generation of nitric oxide (NO) and tumor necrosis factor-α (TNF-α) in a dose-dependent manner [38]. This effect might be related to decreased synthesis of inducible nitric oxide synthase (iNOS) protein and TNF-α mRNA [38].

12.6.3 Atranorin

Atranorin, a significant component of the depside group, is present in several lichen species, including *Usnea articulata*, *Lethariella canariensis*, *Cladina kalbii*, *Caulastraea furcata*, and *Hypotrachyna revoluta*. Additionally, it has a number of biological qualities, including pro-oxidant, cytoprotective, antimicrobial, and anticarcinogenic activity [39].

In their (minimal inhibitory concentration) MIC assay, Kosanic et al. [3] discovered that atranorin exhibited potent antimicrobial activity against a variety of bacteria and fungi, including *Bacillus Mycoides*, *Bacillus subtilis*, *Escherichia coli*, *Klebsiella pneumoniae*, *Staphylococcus aureus*, *Aspergillus flavus*, *Aspergillus fumigatus*, *Candida albicans*, *Penelope purpurascens*, and *Penicillium verrucosum*. Marante et al. [40] found moderate antibacterial activity but only against *S. aureus*, which is consistent with this data. Melo et al. [39], Marante et al. [40], and Kosanic et al. [3] all reported antioxidant effects. However, Dévéhat et al. [41] and Thadhani et al. [42] did not find any antioxidant activity of atranorin on stimulus organism response (SOR) or statistical analysis system (SAS), respectively. The SOR activity requires at least one free OH group to be attached to ring A or B; atranorin has a deactivating aldehyde group at ring A's C-3. The superoxide scavenging effect may also be influenced by the amount of free hydroxide (OH). To find out if these substituents are connected to the activity, other compounds that are similarly substituted should be examined [43].

12.6.4 Lecanoric Acid

A number of lichens, including *Usnea subvacata Motyka*, *Parmotrema stuppeum*, *Parmotrema tinctorum*, and *Parmotrema grayana*, can yield lecanoric acid, which is a member of the depsidone family.

Thadhani et al. [42] showed that lecanoric acid had a high SOR activity that was on par with standards like propyl gallate (PG) and butylated hydroxyanisole (BHA). In NOR and DPPH tests, they also confirmed a moderate activity, supporting the findings of Lopes et al. [44]. Jayaprakasha and Rao [45] also noted that the β-carotene-linoleate model system exhibited moderate antioxidant potential.

The two hydrogen bonds between the 2′–OH and 1′–COOCH3/COOH groups and the 2–OH and 1–COO– groups, as well as the presence of the COO– group, coupled with an aromatic ring, which is also electron-attracting, could be used to justify lecanoric acid's moderate antioxidant activity (Jayaprakasha and Rao [45]). In any instance, these investigations imply that lecanoric acid may be able to communicate with superoxide radicals.

12.6.5 Fumarprotocetraric Acid

According to Kosanić et al. [46], fumarate belongs to the depsidone family member and can be separated from *Cladonia verticillaris Raddi*, *C. rangiferina*, and *U. articulata*. Stronger antioxidant (DPPH, SAS, and reducing power assays), antimicrobial (MIC assay), and anticarcinogenic (MTT assay) activity against the FemX and LS174 cell lines were found during research on its properties; these results were consistent with the stimulation of apoptosis in a cell cycle-dependent manner. Another study by Santos et al. [47] found that fumarprotocetraric acid increased the release of NO from macrophage cells. It is well-recognized that macrophages are crucial to host defense mechanisms. Reactive oxygen intermediates (ROI) and nitric oxide (NO) are frequently used in conjunction by the immune system to produce immunostimulatory effects. One such example is the killing of bacteria and tumor cells by macrophages [47].

12.6.6 Orcinol, Orsellinic Acid and Orsellinates

Jayaprakasha and Rao [45] and Thadhani et al. [42] found orsellinic acid and methyl orsellinate, as well as methyl-β-orcinolcarboxylate and orcinol (benzoic acid derivatives and a benzenoid, respectively), from *Parmotrema stuppeum*. In the NOR assay and the β-carotene-linoleate model system, the compound's antioxidant activity was noted in both investigations [42].

The MIC assay showed antibacterial activity against *S. aureus* for orsellinic methyl-β-orsellinic or methyl-β-orcinolcarboxylate, while the NOR assay showed high activity [42]. On the other hand, the SOR assay revealed reduced activity, and the DPPH assay confirmed little to no effect [48].

The structures of the standard butylated hydroxyanisole (BHA) and propyl gallate (PG), each of which has a single aromatic ring, can be used to explain the SOR assay's relative inactivity. Despite being mononuclear aromatic compounds, the latter two exhibit higher SOR activity because unlike the orsellinates, they have ortho and para OH/OR groups instead of a Meta link between OH/OR. Conversely, simple aromatics' redox potentials (LUMO) are more compatible with the NO radical HOMO, as indicated by the greater activity of mononuclear aromatic compounds in the NOR assay [42].

12.7 Antioxidant Activity in *Parmelia sulcata, Lasallia pustulata, Hypogymnia physodes*

Kosanić et al. [18] reported the antioxidant properties of the acetone, methanol, and aqueous extracts of the lichens *P. sulcata*, *Lasallia pustulata*, *Hypogymnia physodes*, DPPH radical scavenging, superoxide anion radical scavenging, minimizing power, and determination of total phenolic compounds as well as total flavonoid content have all been evaluated *in vitro* using various techniques. *L. pustulata* had the strongest antioxidant activity of all the studied lichens. This lichen exhibited DPPH radical scavenging capabilities in acetone, methanol, and aqueous extracts. The examined extracts also demonstrated strong reducing ability and superoxide anion radical quenching. The typical antioxidants, including ascorbic acid, butylated hydroxyanisole (BHA), and α-tocopherol, were compared to these different antioxidant activities. Furthermore, the total flavonoid and phenol contents in the extracts were calculated as the equivalent of rutin and pyrocatechol, respectively. The studied extracts of antioxidative properties and their total phenolic and flavonoid contents have substantial correlations, indicating that these chemicals are important contributors to antioxidant activity. According to the current study, the investigated lichen species showed significant antioxidant activity and are therefore valuable natural antioxidant suppliers (Kosani et al. [18]).

12.8 Constituents Responsible for Antioxidant Behavior in *Cetraria islandica*

From the 19th century until the present, numerous scientists have studied the chemical makeup of the lichen *C. islandica*. However, the precise nature of the lichen has not yet been determined. Furthermore, pharmacological research has been conducted regarding the makeup of this particular species of lichen. Protolichesterinic acid, which was separated from *C. islandica*, inhibits arachidonate 5-lipoxygenase *in vitro*. Fumar Proto Citric acid, β-orcinol depsidone, α-methylene-γ-lactone, and protolichesterinic acid are thought to be the main physiologically active secondary metabolites in the lichen *C. islandica*. *C. islandica's* lichen metabolites showed the strongest antimicrobial activity. It was discovered that the lichen *C. islandica* contained aliphatic α-methylene-γ-lactone, which exhibited strong inhibitory effects on HIV-1 RT's DNA polymerase activity. On the other hand, nothing is known regarding the antioxidant activity of the aqueous extract of lichen *C. islandic*. Furthermore, the exact components that give *C. islandica* its antioxidative capacity are still unknown. Therefore, it is recommended that more research be done on the separation and identification of *C. islandica*'s antioxidant components. According to studies, *C. islandica* may be a natural antioxidant source [49].

12.8.1 *Antarctica* Lichens Extract

The extracts of the lichenized fungal species composeda variety of bioactive chemicals of the aromatic, carbohydrate, acid, lipid, and depside categories. These substances assess the potential for beneficial impacts on a range of biological processes both *in vivo* and *in vitro* [50]. The antioxidant activity of all three species was found to be moderately significant and correlated with the concentration of total phenols, with *O. frigida* lichen exhibiting the highest antioxidant activity. The extracts from all three species demonstrated significantly superior values for the α-glucosidase enzyme than the standard and less effective results for the α-amylase and pancreatic lipase enzymes in terms of the *in vitro* inhibitory activity of the enzymes. On the other hand, the catalytic sites of the enzymes α-glucosidase, α-amylase, and pancreatic lipase were detected with the in silico assessment of a strong intermolecular interaction of compounds 2,5-dihydroxyterephthalic acid, cyperine, diospyros, hypoxyphenone, lecanoric acid, orsellinic acid, prephenic acid, succinyl salicylic acid, and o-(4-biphenylcarboxylic) benzoic acid (Fernandez-Moriano et al. [26, 27]). The potential for these species' extracts and compounds are treatable metabolic disorders in a promising way. In addition, the biodirected study of ethanolic extract fractions, the isolation, clarification, and biological test of major compounds, and the testing of extracts, active fractions, and compounds in murine models are the prospects for this kind of research in the near future. These efforts are intended to reinforce the pharmacological, nutritional, and healthcare potential of the chemical world of lichenized mushrooms [51].

12.9 Techniques Used to Determine the Antioxidant Activities in Lichens

Antioxidant assays are based on Total antioxidant capacity (TAC). The quantity of free radicals quenched by a test solution used to gauge a biological sample's antioxidant (AO) ability is known as its total antioxidant capacity. Following are different techniques by which one can determine the antioxidant activity among lichens.

12.9.1 Cyclic Voltammetric Technique

The potentiodynamic electrochemical evaluation forms the basis of this technique. The working electrode in this process ramps up linearly with time. To determine the current intensity at each electrode, the functional electrode potential is measured at various intervals. The working electrode's potential ramp is reversed upon reaching the established potential. This can happen multiple times in a single experiment. The working electrode's current is then plated to the applied voltage to create the cyclic voltage photometer. Relevant parameters such as cathodic oxidation potential, anodic oxidation potential, anodic peaks, and cathodic peaks are obtained from a cyclic voltammetry graph [52, 53].

12.9.2 Amperometric Technique

The core of it is the idea that the current intensity between a working electrode and a reference electrode can be measured using a fixed voltage. A current is produced by the oxidation/reduction reaction of an analyte whose concentration is fixed at a predetermined value with respect to the reference electrode. The reduction of DPPH using a carbon electrode is the foundation for the amperometric method of measuring antioxidant activity [52, 54].

12.9.3 Biosensor Technique

Oxidoreductases, being electron-giving enzymes, are widely used biosensors. They are highly stable by nature and do not need any additional chemicals or cofactors to function. The literature has extensively examined and discussed them. More broadly, the polyphenolic components of the extract are responsible for most of its antioxidant qualities. The biosensors based on enzymes can identify these polyphenols. Tyrosinase and lacease are two of these enzymes [52, 55].

12.9.4 Chromatographic Techniques

Chromatographic methods are also utilized to evaluate the extracts' antioxidant potential. Gas chromatography is the method of chromatography that is employed for these objectives. One of the main chromatographic techniques that is frequently used for the analysis and separation of stable but vaporizable compounds is gas chromatography. In this kind of chromatography, the mixture is passed through a liquid stationary phase and a gas mobile phase to complete the process. The stationary phase is a liquid microfilm deposited on a nonreactive solid support, while the mobile phase consists of the inert gases helium and nitrogen. Its value is revealed by the retention time comparison. In gas chromatography, thermal conductivity and flame ionization detectors are typically utilized [52].

12.10 Conclusion

This chapter concluded that lichens are a vital source of antioxidants. Lichens are formed through a unique symbiotic relationship between fungi and algae or cyanobacteria. Moreover, the chapter focused on lichen's constituents with antioxidant qualities such as usnic acid and phenolic compounds, atranorin, lecanoric acid, fumarprotocetraric acid, orcinol, orsellinic acid, and Orsellinates, which have been shown to be strong antioxidant property. These substances aid in shielding our cells from harm brought on by free radicals. However, different techniques are discussed that determine the antioxidant behavior in lichens, which could be used in the near future to treat different complicated diseases.

References

1 Calcott, M.J., Ackerley, D.F., Knight, A. et al. (2018). Secondary metabolism in the lichen symbiosis. *Chemical Society Reviews* 47 (5): 1730–1760. https://doi.org/10.1039/c7cs00431a.

2 Choi, R.-Y., Ham, J.R., Yeo, J. et al. (2017). Anti-obesity property of Lichen *Thamnolia vermicularis* extract in 3T3-L1 cells and diet-induced obese mice. *Preventive Nutrition and Food Science* 22 (4): 285–292. https://doi.org/10.3746/pnf.2017.22.4.285.

3 Kosanic, M., Rankovic, B., Stanojkovic, T. et al. (2014). Biological activities and chemical composition of lichens from Serbia. *Excli Journal* 13: 1226.

4 Molnar, K. and Farkas, E. (2010). Current results on biological activities of lichen secondary metabolites: a review. *Zeitschrift fuer Naturforschung, C: Journal of Biosciences* 65 (3–4): 157–173. https://doi.org/10.1515/znc-2010-3-401.

5 Huang, X., Ma, J., Wei, L. et al. (2018). An antioxidant α-glucan from *Cladina rangiferina* (L.) Nyl. and its protective effect on alveolar epithelial cells from Pb^{2+}-induced oxidative damage. *International Journal of Biological Macromolecules* 112: 101–109. https://doi.org/10.1016/j.ijbiomac.2018.01.154.

6 Zhao, Y., Wang, M., and Xu, B. (2021). A comprehensive review on secondary metabolites and health-promoting effects of edible lichen. *Journal of Functional Foods* 80: 104283.

7 Bayir, Y., Odabasoglu, F., Cakir, A. et al. (2006). The inhibition of gastric mucosal lesion, oxidative stress and neutrophil-infiltration in rats by the lichen constituent diffractaic acid. *Phytomedicine* 13 (8): 584–590. https://doi.org/10.1016/j.phymed.2005.07.002.

8 Grujičić, D., Stošić, I., Kosanić, M. et al. (2014). Evaluation of in vitro antioxidant, antimicrobial, genotoxic and anticancer activities of lichen *Cetraria islandica*. *Cytotechnology* 66: 803–813. https://doi.org/10.1007/s10616-013-9629-4.

9 Gupta, V.K., Verma, S., Gupta, S. et al. (2012). Membrane-damaging potential of natural L-(−)-usnic acid in *Staphylococcus aureus*. *European Journal of Clinical Microbiology & Infectious Diseases* 31: 3375–3383. https://doi.org/10.1007/s10096-012-1706-7.

10 Gulluce, M., Aslan, A., Sokmen, M. et al. (2006). Screening the antioxidant and antimicrobial properties of the lichens *Parmelia saxatilis, Platismatia glauca, Ramalina pollinaria, Ramalina polymorpha* and *Umbilicaria nylanderiana*. *Phytomedicine* 13 (7): 515–521. https://doi.org/10.1016/j.phymed.2005.09.008.

11 Rosentreter, R., Eldridge, D.J., Westberg, M. et al. (2016). Structure, composition, and function of biocrust lichen communities. In: *Biological Soil Crusts: An Organizing Principle in Drylands* (ed. B. Weber, B. Budel, and J. Belnap), 121–138.

12 Shrestha, G. and St. Clair, L.L. (2013). Lichens: a promising source of antibiotic and anticancer drugs. *Phytochemistry Reviews* 12: 229–244.

13 Kumar, G., Murugesan, A.G., and Rajasekara Pandian, M. (2006). Effect of Helicteres isora bark extract on blood glucose and hepatic enzymes in experimental diabetes. *DiePharmazie* 61 (4): 353–355.

14 Karaman, I., Şahin, F., Güllüce, M. et al. (2003). Antimicrobial activity of aqueous and methanol extracts of *Juniperus oxycedrus* L. *Journal of Ethnopharmacology* 85 (2–3): 231–235.

15 Sukalingam, K., Ganesan, K., and Xu, B. (2017). *Trianthema portulacastrum* L. (giant pigweed): phytochemistry and pharmacological properties. *Phytochemistry Reviews* 16: 461–478. https://doi.org/10.1007/s11101-017-9493-5.

16 Lõhmus, A., Motiejūnaitė, J., and Lõhmus, P. (2023). Regionally varying habitat relationships in lichens: the concept and evidence with an emphasis on North-temperate ecosystems. *Journal of Fungi* 9 (3): 341.

17 Boustie, J. and Grube, M. (2005). Lichens—a promising source of bioactive secondary metabolites. *Plant Genetic Resources* 3 (2): 273–287. https://doi.org/10.1079/PGR200572.

18 Kosanić, M., Ranković, B., and Vukojević, J. (2011). Antioxidant properties of some lichen species. *Journal of Food Science and Technology* 48: 584–590.

19 Ranković, B.R., Kosanić, M.M., and Stanojković, T.P. (2011). Antioxidant, antimicrobial and anticancer activity of the lichens *Cladonia furcata*, *Lecanora atra* and *Lecanora muralis*. *BMC Complementary and Alternative Medicine* 11 (1): 1–8.

20 Fabian, A.S., Peter, D.C., Matthew, J.D. et al. (2005). Breeding systems in the lichen-forming fungal genus Cladonia. *Fungal Genetics and Biology* 2005 (42): 554–563. https://doi.org/10.1016/j.fgb.2005.03.006.

21 Kirmizigül, S., Koz, Ö., Anil, H. et al. (2003). Isolation and structure elucidation of novel natural products from Turkish lichens. *Turkish Journal of Chemistry* 27 (4): 493–500.

22 Malhotra, S., Subban, R.A.V.I., and Singh, A. (2008). Lichens-role in traditional medicine and drug discovery. *The Internet Journal of Alternative Medicine* 5 (2): 1–5.

23 Ganesan, K., Sukalingam, K., and Xu, B. (2017). *Solanum trilobatum* L. Ameliorate thioacetamide-induced oxidative stress and hepatic damage in albino rats. *Antioxidants (Basel)* 6 (3): https://doi.org/10.3390/antiox6030068.

24 Kumar, G., Sharmila Banu, G., and Ganesan Murugesan, A. (2008). Effect of Helicteres isora bark extracts on heart antioxidant status and lipid peroxidation in streptozotocin diabetic rats. *Journal of Applied Biomedicine* 6 (2): 89–95. https://doi.org/10.32725/jab.2008.012.

25 Ranković, B., Ranković, D., Kosanić, M., and Marić, D. (2010). Antioxidant and antimicrobial properties of the lichens *Anaptychya ciliaris*, *Nephroma parile*, *Ochrolechia tartarea* and *Parmelia centrifuga*. *Central European Journal of Biology* 5: 649–655.

26 Fernández-Moriano, C., Gómez-Serranillos, M.P., and Crespo, A. (2016a). Antioxidant potential of lichen species and their secondary metabolites. A systematic review. *Pharmaceutical Biology* 54 (1): 1–17. https://doi.org/10.3109/13880209.2014.1003354.

27 Fernández-Moriano, C., González-Burgos, E., Divakar, P.K. et al. (2016b). Evaluation of the antioxidant capacities and cytotoxic effects of ten parmeliaceae lichen species. *Evidence-Based Complementary and Alternative Medicine* 2016: https://doi.org/10.1155/2016/3169751.

28 Odabasoglu, F., Aslan, A., Cakir, A. et al. (2004). Comparison of antioxidant activity and phenolic content of three lichen species. *Phytotherapy Research* 18 (11): 938–941. https://doi.org/10.1002/ptr.1488.

29 Pol, C.S., Savale, S.A., Khare, R. et al. (2017). Antioxidative, cardioprotective, and anticancer potential of two Lichenized Fungi, *Everniastrum cirrhatum* and Parmotrema reticulatum, from Western Ghats of India. *Journal of Herbs, Spices & Medicinal Plants* 23 (2): 142–156. https://doi.org/10.1080/10496475.2017.1280578.

30 Wang, Y., Shu, X., Chen, Y. et al. (2018). Enrichment, purification and in vitro antioxidant activities of polysaccharides from *Umbilicaria esculenta* macrolichen. *Biochemical Engineering Journal* 130: 10–20. https://doi.org/10.1016/j.bej.2017.11.004.

31 Kotan, E., Alpsoy, L., Anar, M. et al. (2011). Protective role of methanol extract of *Cetraria islandica* (L.) against oxidative stress and genotoxic effects of AFB1 in human lymphocytes in vitro. *Toxicology and Industrial Health* 27 (7): 599–605. https://doi.org/10.1177/0748233710394234.

32 Çolak, S., Geyikoglu, F., Türkez, H. et al. (2013). The ameliorative effect of *Cetraria islandica* against diabetes-induced genetic and oxidative damage in human blood. *Pharmaceutical Biology* 51 (12): 1531–1537. https://doi.org/10.3109/13880209.2013.801994.

33 Karakus, B., Odabasoglu, F., Cakir, A. et al. (2009). The effects of methanol extract of *Lobaria pulmonaria*, a lichen species, on indometacin-induced gastric mucosal damage, oxidative stress and neutrophil infiltration. *Phytotherapy Research: An International Journal Devoted to Pharmacological and Toxicological Evaluation of Natural Product Derivatives* 23 (5): 635–639. https://doi.org/10.1002/ptr.2675.

34 Ganesan, K. and Xu, B. (2018b). Anti-obesity effects of medicinal and edible mushrooms. *Molecules* 23 (11): https://doi.org/10.3390/molecules23112880.

35 Watson, R.R. (2014). *Polyphenols in Plants: Isolation, Purification and Extract Preparation*. San Diego: Academic Press.

36 Michalak, A. (2006). Phenolic compounds and their antioxidant activity in plants growing under heavy metal. *Polish Journal of Environmental Studies* 15: 523–530.

37 Bessadottir, M., Egilsson, M., Einarsdottir, E. et al. (2012). Proton-shuttling lichen compound usnic acid affects mitochondrial and lysosomal function in cancer cells. *PLoS One* 7 (12): e51296.

38 Jin, J.Q., Li, C.Q., and He, L.C. (2008). Down-regulatory effect of usnic acid on nuclear factor-κB-dependent tumor necrosis factor-α and inducible nitric oxide synthase expression in lipopolysaccharide-stimulated macrophages RAW 264.7. *Phytotherapy Research: An International Journal Devoted to Pharmacological and Toxicological Evaluation of Natural Product Derivatives* 22 (12): 1605–1609.

39 Melo, M.G.D., dos Santos, J.P.A., Serafini, M.R. et al. (2011). Redox properties and cytoprotective actions of atranorin, a lichen secondary metabolite. *Toxicology In Vitro* 25 (2): 462–468.

40 Marante, F.T., Castellano, A.G., Rosas, F.E. et al. (2003). Identification and quantitation of allelochemicals from the lichen *Lethariella canariensis*: phytotoxicity and antioxidative activity. *Journal of Chemical Ecology* 29: 2049–2071.

41 Dévéhat, F., Tomasi, S., Elix, J.A. et al. (2007). Stictic acid derivatives from the lichen *Usnea articulata* and their antioxidant activities. *Journal of Natural Products* 70: 1218–1220.

42 Thadhani, V.M., Choudhary, M.I., Ali, S. et al. (2011). Antioxidant activity of some lichen metabolites. *Natural Product Research* 25 (19): 1827–1837.

43 Brewer, M.S. (2011). Natural antioxidants: sources, compounds, mechanisms of action, and potential applications. *Comprehensive Reviews in Food Science and Food Safety* 10 (4): 221–247.

44 Lopes, T.I.B., Coelho, R.G., Yoshida, N.C., and Honda, N.K. (2008). Radical-scavenging activity of orsellinates. *Chemical and Pharmaceutical Bulletin* 56 (11): 1551–1554.

45 Jayaprakasha, G.K. and Rao, L.J. (2000). Phenolic constituents from the lichen *Parmotrema stuppeum* (Nyl.) Hale and their antioxidant activity. *Zeitschrift für Naturforschung C* 55 (11–12): 1018–1022.

46 Kosanić, M., Ranković, B., Stanojković, T. et al. (2014). Cladonia lichens and their major metabolites as possible natural antioxidant, antimicrobial and anticancer agents. *LWT-Food Science and Technology* 59 (1): 518–525.

47 Santos, L.C.D., Honda, N.K., Carlos, I.Z., and Vilegas, W. (2004). Intermediate reactive oxygen and nitrogen from macrophages induced by Brazilian lichens. *Fitoterapia* 75 (5): 473–479.

48 Rabelo, T.K., Zeidán-Chuliá, F., Vasques, L.M. et al. (2012). Redox characterization of usnic acid and its cytotoxic effect on human neuron-like cells (SH-SY5Y). *Toxicology In Vitro* 26 (2): 304–314.

49 Gülçin, İ., Oktay, M., Küfrevioğlu, Ö.İ., and Aslan, A. (2002). Determination of antioxidant activity of lichen *Cetraria islandica* (L) Ach. *Journal of Ethnopharmacology* 79 (3): 325–329.

50 Lee, S., Suh, Y.J., Yang, S. et al. (2021). Neuroprotective and anti-Inflammatory effects of evernic acid in an MPTP-induced Parkinson's disease model. *International Journal of Molecular Sciences* 22: 2098. https://doi.org/10.3390/ijms22042098.

51 Torres-Benítez, A., Ortega-Valencia, J.E., Jara-Pinuer, N. et al. (2023). Antioxidant and antidiabetic activity and phytoconstituents of lichen extracts with temperate and polar distribution. *Frontiers in Pharmacology* 14.

52 Ahmad, S., Arshad, M.A., Ijaz, S. et al. (2014). Review on methods used to determine antioxidant activity. *International Journal of Multidisciplinary Research and Development* 1 (1): 35–40.

53 Chevion, S., Roberts, M.A., and Chevion, M. (2000). The use of cyclic voltammetry for the evaluation of antioxidant capacity. *Free Radical Biology and Medicine* 28 (6): 860–870.

54 Milardovic, S., Ivekovic, D., and Grabaric, B.S. (2006). A novel amperometric method for antioxidant activity determination using DPPH free radical. *Bioelectrochemistry* 68: 175–180.

55 Barroso, M.F., De-Los-Santos-Álvarez, N., Delerue-Matos, C., and Oliveira, M.B.P.P. (2011). Towards a reliable technology for antioxidant capacity and oxidative damage evaluation: electrochemical (bio) sensors. *Biosensors and Bioelectronics* 30 (1): 1–12.

13

Antimicrobial Activities of Lichens

Muhammad Zeeshan Ahmed[1], *Tazeen Rao*[1], *Nihad Ashraf Khan*[2], *Muneeba Aslam*[3], *and Yunita Sari Pane*[4]

[1] *Department of Biochemistry, Bahauddin Zakariya University, Multan, Pakistan*
[2] *Department of Biosciences, Faculty of Natural Sciences, Jamia Millia Islamia University, New Delhi, India*
[3] *Department of Biochemistry, Abdul Wali Khan University, Mardan, Pakistan*
[4] *Department of Pharmacology & Therapeutics, Faculty of Medicine, Universitas Sumatera Utara, Medan, Indonesia*

13.1 Introduction

Lichens, despite their inconspicuous stature in the fields of microbiology and pharmacology, represent an intriguing symbiotic amalgamation. Formed through the intricate fusion of a fungus, denoted as the mycobiont, and a photosynthetic partner, usually an alga or cyanobacterium, referred to as the photobiont, lichens inhabit diverse ecosystems globally, harboring a trove of biologically active compounds endowed with potent antimicrobial properties [1–3]. The dynamic interaction between lichens and microorganisms, occurring both in their immediate microenvironment and within the lichen thallus itself, has piqued scientific curiosity for decades. Lichens have evolved an array of secondary metabolites as components of their defense mechanisms, some of which exhibit remarkable antioxidant capabilities. These metabolites, collectively termed lichen substances, exert a plethora of biological effects encompassing analgesic, antibiotic, anti-inflammatory, antimycotic, antiproliferative, antipyretic, antiviral, and cytotoxic actions [4–7].

The ensuing scientific scrutiny has been intense, driven by the vast potential of lichen substances in diverse domains, including medicine, agriculture, and industry. Within the context of this investigation into the antimicrobial attributes of lichens, we embark upon an exploration of these composite organisms' ability to synthesize a broad spectrum of bioactive molecules [8]. From the historical applications in traditional remedies to pioneering research within the pharmaceutical realm, the antimicrobial promise held by lichens emerges as a beacon of hope for combating infectious diseases, preserving perishables, and confronting the modern conundrum of antibiotic resistance [9, 10]. Experimental findings reveal the substantial antimicrobial activity of acetone, ethanol, and methanol extracts derived from various lichen specimens, effectively targeting both

Chemistry, Biology and Pharmacology of Lichen, First Edition. Edited by Ashoke Kumar Das, Ajay Sharma, Deepika Kathuria, Mohammad Javed Ansari, and Garima Bhardwaj.

gram-positive (G+) and select gram-negative (G−) bacteria. This antimicrobial efficacy extends to diverse microorganisms, further affirming the potency of lichen compounds in this regard, particularly against G+ bacterial growth inhibition [11–13].

The multifaceted potential of lichens as natural antimicrobial agents finds resonance in biopharmaceutical applications, as lichen-derived bioactive compounds exhibit considerable promise. Notably, lichens inherently produce a spectrum of distinctive defense chemicals, harking back to their historical medicinal use. These delve into the extraction and purification of eight lichen substances of diverse chemical origins, providing a comparative analysis of their antimicrobial properties. Furthermore, these unveil the intricate mechanisms underpinning lichen substances' antimicrobial activity [4, 8, 13–15]. Notably, the antibacterial efficacy of usnic acid, a prominent lichen secondary metabolite, primarily hinges upon its role in inhibiting RNA and DNA synthesis in targeted microorganisms. These findings underscore the multifaceted nature of lichens and their metabolites, showcasing their diverse biological activities encompassing antibiotic, antitumor, and antiviral properties, allergenic attributes, enzyme inhibition, plant growth regulation, and ecological roles [16, 17].

As this chapter unfolds, it guides you through the rich tapestry of lichen utilization in historical medicine and agriculture. It sheds light on the captivating chemistry governing their antimicrobial properties while spotlighting contemporary research endeavors aimed at harnessing the latent potential of these often-overlooked organisms. Our journey through the world of lichens and their antimicrobial prowess reveals a compelling narrative wherein ancient wisdom converges with modern scientific pursuits in the battle against microbial adversaries.

13.2 Antimicrobial Compounds of Lichen

Lichen, a type of plant, serves several important roles. Its prevalence in an area can be a reliable indicator of relatively low air pollution levels. Additionally, lichen possesses valuable properties for traditional medicine [18]. Within the healthcare sector, lichen finds extensive use, primarily owing to its chemical constituents. These compounds have applications as antibiotics, antifungals, antivirals, anti-inflammatories, pain relievers, fever reducers, inhibitors of abnormal cell growth, and substances toxic to certain cells [19]. Lichens produce various secondary metabolites (Table 13.1), encompassing aliphatic acids, amino acid derivatives, anthraquinones, carotenoids, depsides, dibenzofurans, monocyclic phenols, peptides, pulvinic acid, steroids, sugar alcohols, terpenoids, usnic acid, xanthones, and other substances [6].

13.2.1 Dibenzofuran and Its Derivatives

Dibenzofuran, a secondary metabolite found in lichen [20], is among a group of approximately 32 dibenzofurans that have been identified in lichens [39]. Usnic acid, which is a type of dibenzofuran, was initially isolated from lichen-producing secondary metabolites. It is renowned for its antibiotic properties and possesses several other intriguing characteristics. Newly synthesized compounds derived from usnic acid exhibit strong antibacterial activity, with Minimum Inhibitory Concentrations (MIC) ranging from 1.02 to 50.93 × 10–2 mmol/mL and Minimum Bactericidal Concentrations (MBC) from 2.05 to 70.57 × 10–2 mmol/mL [21].

Table 13.1 Summarizes the diverse secondary metabolites found in lichens, their descriptions, and their respective antimicrobial activities.

Secondary metabolite	Description	Antimicrobial activity	References
Dibenzofuran and its derivatives	Dibenzofuran is found in lichen and has various derivatives. Usnic acid, a type of dibenzofuran, is renowned for its antibiotic properties.	Strong antibacterial activity, especially against *Staphylococcus aureus*, *Enterococcus* spp., and *Mycobacteria* spp.	[20, 21]
Depsidones	Lichen depsidones are effective against a wide spectrum of microorganisms, including fungi and bacteria.	Strong antifungal activity; weaker antimicrobial efficacy compared to usnic acid	[22, 23]
Depsides	Depsides are compounds formed by the linkage of two phenyl carboxylic acids through an ester bond.	Significant antimicrobial activity, especially against *Staphylococcus* species (*Stereocaulon halei* Lamb)	[23]
Quinones and its derivatives	Naphthoquinones are prevalent quinones in nature and are being explored for antimicrobial properties	Increased interest in exploring quinones for potential antimicrobial properties	[24, 25]
Xanthones	Xanthones are secondary metabolites found in select plants, fungi, and lichens. They exhibit diverse bioactivities.	More than 250 xanthones have demonstrated significant bioactivities, including antimicrobial properties	[26]
Chromones and pyrones	Chromones and 2-pyrones are found in lichens and exhibit biological effects.	Certain chromones have potential as anticancer and antibacterial agents. 2-pyrones display antimicrobial activity	[4, 27, 28]
Nitrogenous and sulfur-containing compounds	Bacteria produce VOCs, including sulfur- and nitrogen-containing molecules, with diverse roles	VOCs, including sulfur- and nitrogen-containing compounds, have been discovered in bacterial VOC production	[29, 30]
Steroids and terpenoids	Terpenoids and steroids in lichens act as antimicrobial agents, disrupting bacterial cell walls	Terpenoids act by interacting with bacterial porins, reducing permeability and inhibiting growth	[31, 32]
Ambuic acid and its derivatives	Ambuic acid and derivatives serve as antibiotics and antifungal agents, inhibiting quorum sensing in G+ bacteria	Antibacterial effects *in vivo*, including inhibition of quorum sensing and gelatinase production	[33]
Phenolic compounds and their derivatives	Lichen extracts contain phenolic compounds with antimicrobial activity	Phenolic compounds exhibit antimicrobial activity against multidrug-resistant bacteria and MRSA	[34]

(Continued)

Table 13.1 (Continued)

Secondary metabolite	Description	Antimicrobial activity	References
Miscellaneous compounds	Lichens have diverse uses in various cultures, including medicine, food, perfumes, and more. Some lichen species have compounds with biological activities	Lichens are used for various purposes, including traditional medicine, spices, and perfumes	[35–38]

It was noted that both enantiomers of usnic acid inhibit the growth of multi-resistant strains of bacteria such as *Staphylococcus aureus*, *Enterococcus* spp., and *Mycobacteria* spp. [40]. Furthermore, it has been documented that dibenzofuran derivatives were initially isolated from a complete lichen organism. Several of these isolated metabolites have demonstrated notable antibacterial activity against oral pathogens such as *Porphyromonas gingivalis* and *Streptococcus mutans* [41]. *Usnea diffracta*, along with its primary constituent usnic acid, displays potent antimicrobial properties, effectively inhibiting the growth of various planktonic bacterial and fungal strains [42].

13.2.2 Depsidones

Lichens produce secondary metabolites with antibiotic properties as a natural defense mechanism against pathogens in their environment. These compounds, known as lichen depsidones, have demonstrated significant effectiveness against a wide spectrum of microorganisms, including fungi, algae, yeast, and both G+ and G− bacteria. Research suggested that lichen is particularly potent as an antifungal agent, while its antimicrobial efficacy is comparatively weaker. Among these compounds, physodic acid exhibits the lowest antimicrobial activity when compared to the activities of usnic acid (a dibenzofuran) and the depsides antranorin and gyrophoric acid [22, 23].

13.2.3 Depsides

A depside represents a chemical compound resulting from the connection of two phenylcarboxylic acids via an ester bond. In this study, we conducted antimicrobial assays on various fractions of *Stereocaulon halei* Lamb, including *n*-hexane, ethyl acetate, acetone, and methanol fractions. The findings revealed noteworthy antimicrobial efficacy against a spectrum of bacteria, with a particular emphasis on *Staphylococcus* strains. Notably, the ethyl acetate fraction of *S. halei* Lamb exhibited robust antimicrobial properties, successfully inhibiting the proliferation of all examined bacteria, with the exception of *Enterococcus faecalis*. Specifically, this fraction displayed moderate activity against bacteria such as *Escherichia coli*, *Pseudomonas aeruginosa*, *Salmonella typhosa*, and *Salmonella typhimurium*, while demonstrating potent activity against *S. aureus*. Among the tested fractions, the ethyl acetate fraction of *S. halei* Lamb showcased the most extensive inhibitory zone against *S. aureus* [23].

13.2.4 Quinones and Its Derivatives

Naphthoquinones are the predominant class of quinones identified in the natural environment. These substances constitute a diverse group of secondary metabolites present in plants, lichens, and diverse microorganisms. The naphthoquinone family continues to expand as novel natural products are uncovered, and innovative methods are employed to synthesize new compounds. Recently, there has been a heightened fascination with quinones, particularly centered around investigating their potential as antimicrobial agents [24, 25].

13.2.5 Xanthones

Xanthones are a recognized group of secondary metabolites discovered in a select group of higher plants, fungi, and lichens. The biological activities demonstrated by these compounds depend on both their tricyclic core structure and the type and location of their remarkably diverse substituents, making them a "privileged structure" capable of engaging with a wide array of molecular targets. Remarkably, more than 250 xanthones have demonstrated significant bioactivities, encompassing antimicrobial, antioxidant, and cytotoxic properties. Although xanthones may be considered less specific in lichens compared to other structural classes of compounds, the distinct biosynthetic pathway leading to fungal xanthones results in novel substitution patterns. This underscores the scientific interest in exploring xanthones derived from lichens [26].

13.2.6 Chromones and Pyrones

Chromones are one of the various secondary metabolites found in lichens. Some complexes of alkynyl chromones have been investigated for their potential as antibacterial and anticancer agents in human cancer cell lines and against pathogenic bacterial strains. Additionally, certain simple compounds known as 2-pyrones exhibit notable biological effects. For instance, 6-pentyl-2-pyrone, which has been isolated as a fungal product from various natural sources, including *Trichoderma viride*, displays antimicrobial activity against *Botrytis cinerea*, *Gaeumannomyces graminis*, and *Rhizoctonia cerealis* [4, 27, 28].

13.2.7 Nitrogenous and Sulfur-Containing Compounds

Microorganisms are prolific producers of diverse secondary metabolites, encompassing antibiotics, toxins, pigments, and various other compounds. Interestingly, small-molecular-mass metabolites like volatile organic compounds (VOCs) were historically neglected for a significant period. However, research conducted in recent decades has unveiled that bacteria, in particular, generate a wide array of VOCs. These compounds include hydrocarbons, ketones, alcohols, sulfur- and nitrogen-containing molecules, terpenes, and others. Despite this newfound awareness of VOC production, our understanding of their precise biological and ecological roles remains limited [29, 30].

13.2.8 Steroids and Terpenoids

The identification of triterpenoid/steroid compounds in lichen extracts is confirmed through the development of a distinctive purple-red color. Saponins are compounds with glycosyl groups serving as polar elements and steroid/triterpenoid groups acting as nonpolar components. Compounds possessing both polar and nonpolar characteristics are surface-active, allowing saponins to form micelles when agitated with water. Steroids, triterpenoids, saponins, and tannins within lichens function as antimicrobial agents. Lichens within the same family likely share similarities in their compound composition or contain other characteristic constituents that are structurally related. The mechanism by which terpenoids act as antibacterials involves their interaction with porins, which are transmembrane proteins located on the outer membrane of bacterial cells. This interaction results in the formation of robust polymer bonds, causing damage to the porins. Impaired porins, responsible for the entry and exit of compounds, reduce the permeability of the bacterial cell wall. Consequently, the bacterial cells face nutrient scarcity, leading to inhibited growth or cell death [31, 32].

13.2.9 Ambuic Acid and Its Derivatives

Ambuic acid and its derivatives serve as antibiotics and antifungal agents. They function by inhibiting the biosynthesis of cyclic peptide quorum-sensing molecules in G+ bacteria. Additionally, they can hinder the production of gelatinase mediated by quorum sensing. These compounds have demonstrated antibacterial effects *in vivo* [33].

13.2.10 Phenolic Compounds and Its Derivatives

Phenolic profiles were determined using HPLC-DAD-ESI/MS, and the antimicrobial activity was assessed against clinical isolates of multidrug-resistant bacteria, including *E. coli*, extended-spectrum β-lactamases (ESBL)-producing *E. coli*, *Proteus mirabilis*, methicillin-resistant *S. aureus* (MRSA), and *P. aeruginosa*. Artichoke syrup was found to contain vanillic acid and luteolin-7-O-glucoside, while the pills had higher concentrations of 4-O-caffeoylquinic, 5-O-caffeoylquinic, and 1,3-O-dicaffeoylquinic acids, with the latter inhibiting MRSA growth. Overall, all the formulations studied were rich in phenolic compounds. Milk thistle syrup, in particular, had the highest variety and concentration of flavonoids, likely contributing to its potent antimicrobial activity [34].

The methanol extract obtained from *C. pinastri* exhibited a remarkable level of antioxidant activity, surpassing other lichen extracts. Meanwhile, extracts from *Fulgens fulgens*, *Cladonia fimbriata*, and *Parmelia crinita* demonstrated moderate antioxidant activity, whereas those from *Ochrolechia parella* and *C. digitata* exhibited relatively weaker activity. It is worth noting that *C. pinastri's* methanol extract contained the highest total phenol content, measured at 32.9 mg/g of dry extract. Notably, a positive correlation was observed between the antioxidant activity and the total phenol content in all the tested lichen extracts, a phenomenon documented in previous studies [43, 44]. Furthermore, this study delved into the antimicrobial properties of methanol extracts derived from various lichen species, assessing their effectiveness against six bacterial and eleven fungal species using both disc diffusion and MIC methods. Among these extracts, those from *C. pinastri* and *P. crinita* displayed the most potent antibacterial and antifungal activity against a majority of the tested microorganisms. These findings underscore the potential of *C. pinastri* lichens as promising sources of natural antioxidants and antimicrobial agents, holding significance in various applications [45].

13.2.11 Ether Derivatives

The ability of various extracts (acetone, chloroform, diethyl ether, ethanol, and petroleum ether) obtained from the lichen *Xanthoparmelia pokornyi*, as well as its constituents, stenosporic acid, and gyrophoric acid, was evaluated to combat several foodborne bacteria and fungi. The findings revealed that both the extracts and the individual acids displayed antimicrobial properties against a range of microorganisms, including *Yersinia enterocolitica*, *Streptococcus faecalis*, *S. aureus*, *Proteus vulgaris*, *Listeria monocytogenes*, *Candida glabrata*, *Candida albicans*, *Bacillus subtilis*, *Bacillus cereus*, and *Aeromonas hydrophila* [35, 36].

13.2.12 Miscellaneous Compounds

Lichens are valuable resources in various cultures worldwide, serving roles in medicine, food, animal feed, perfumes, spices, dyes, and other applications. Lichens are renowned for the rich diversity of secondary metabolites and chemical compounds they produce. Many different lichen species have yielded compounds with diverse biological activities. In this chapter, we delve into the medicinal potential of these lichen substances, aiming to uncover their pharmaceutical uses. Throughout history, lichens have been found to be used in multiple domains, including as dyes, perfumes, and remedies in traditional folk medicines. In the Indian spice market, lichens are traded under the name "Chharila," typically comprising a mixture of several species like *Parmelia*, *Usnea longissima*, *Ramalina subcomplanata*, and *Heterodermia tremulans*. Chharila is known for its astringent, resolvent, laxative, and carminative properties and is even believed to possess aphrodisiac qualities [37, 38].

13.3 Lichen Species Having Antimicrobial Properties

Globally, lichens comprise an extensive group of approximately 20,000 species, exhibiting adaptability to a wide range of substrates, including rocks, walls, exposed soil surfaces, and epiphytes residing on the bark of trees and leaves [46]. These resilient organisms have acclimated to thrive across diverse geographical zones, spanning from sea level to high-altitude regions and from equatorial to polar latitudes. Lichens represent intricate biological entities formed through a mutualistic symbiosis between fungi (the mycobiont) and photosynthetic green algae and/or cyanobacteria (the photobiont) [47]. Contemporary investigations have unveiled the presence of additional bacterial partners residing within lichen thalli, suggesting their active participation in mutualistic relationships. These multifaceted mutualisms empower lichens to flourish in environments characterized by extreme conditions, including severe temperature fluctuations, intermittent desiccation, intense UV radiation, and salinity [5]. To endure such challenging circumstances, lichens have evolved mechanisms for the production of various metabolites, including UV protectants, cryoprotectants, and osmolytes, thereby positioning them as valuable sources for pioneering biotechnological applications. Over the course of history, lichens have served as reservoirs for pigments, cosmetics, and therapeutic remedies, and their capacity to synthesize a diverse array of unique chemical compounds underscores their potential for a broad spectrum of applications [48].

Recent discoveries indicate that lichens, being composite organisms comprised of a fungal component and photosynthetic organisms, possess a multitude of biological properties, including analgesic, anti-inflammatory, antimicrobial, antiproliferative, antipyretic, antiprotozoal, antitumor, and antiviral. Given the increasing resistance of microorganisms to conventional antibiotics, the antimicrobial potential of lichens has been extensively explored [15]. There are several species of lichens that have antimicrobial properties like *Alectoria sarmentosa*, *Bryoria fuscescens*, *C. pinastri (Scop.) Gray*, *Cetraria aculeata*, *C. digitata (L.) Hoffm.*, *C. fimbriata (L.) Fr.*, *Evernia divaricate*, *Lecanora frustulosa*, *O. parella (L.) A. Massl.*, *Parmelia crinite Ach*, *Parmeliopsis hyperopta*, *Platismatia glauca*, *Ramalina farinacea*, *Usnea*, and others, so a few of them are explained below. Table 13.2 also depicts examples of lichen species with potential antimicrobial activity.

13.3.1 *Ramalina farinacea*

R. farinacea is a type of lichen that has been traditionally used for medicinal purposes in different parts of the world. Some studies have shown that *R. farinacea* has the ability to inhibit the growth of or kill microorganisms such as bacteria, fungi, and viruses and also has low cytotoxicity toward human cells, indicating their potential use as an antimicrobial agent. The ethanolic extracts had significant inhibitory effects on the growth of different bacterial strains, including *E. coli*, *S. aureus*, and *P. aeruginosa*. The extracts also showed antifungal activity against *C. albicans* and *Aspergillus niger* [49].

13.3.2 *Platismatia glauca*

P. glauca, a commonly found lichen, has been traditionally utilized as a spice and exhibits noteworthy anti-cancer properties [48]. The methanol extract obtained from *P. glauca* exhibited remarkable antibacterial activity against both G+ and G− bacteria. In particular, the extract demonstrated potent antibacterial effects against *Staphylococcus saprophyticus*, which is commonly associated with urinary tract infections in women. Given that both *C. albicans and S. saprophyticus* are causative agents of UTIs, *P. glauca* lichen thalli may hold potential as a source for developing drugs to treat UTIs [50].

13.3.3 *Evernia divaricata*

Notable antimicrobial properties are exhibited by *Evernia divaricata*. Extracts derived from this lichen have been demonstrated to display antibacterial activity against a diverse range of G+ and G− bacteria, including those that are resistant to conventional antibiotics. Besides,

Table 13.2 Antimicrobial activity of lichen species.

Lichen species	Description	Antimicrobial activity	References
Ramalina farinacea	Traditionally used for medicinal purposes; exhibits antimicrobial potential.	Inhibits the growth of bacteria, fungi, and viruses. Low cytotoxicity to human cells. Significant inhibitory effects on bacterial strains including *Escherichia coli*, *S. aureus*, and *Pseudomonas aeruginosa*. Also, antifungal activity against *Candida albicans* and *Aspergillus niger*.	[49]
Platismatia glauca	Commonly used as a spice; possesses anticancer properties.	Remarkable antibacterial activity against G+ and G− bacteria, particularly effective against *Staphylococcus saprophyticus*. Potential for UTI treatment.	[48, 50]
Evernia divaricata	Demonstrates notable antimicrobial properties.	Exhibits antibacterial activity against diverse G+ and G− bacteria, including antibiotic-resistant strains. Possesses antifungal properties. Potential for novel antimicrobial agents.	[50]
Bryoria fuscescens	Recognized for medicinal use; utilized as an antimicrobial agent.	Ethyl acetate extract shows significant antibacterial activity against *E. coli* and *S. aureus*. Antifungal activity against *C. albicans* and *A. niger*.	[51]
Alectoria sarmentosa	Utilized traditionally for medicinal purposes.	Displays moderate to strong antimicrobial activity against bacteria like *S. aureus*, *E. coli*, and fungi including *C. albicans* and *C. neoformans*.	[52, 53]
Cetraria pinastri (Scop.) Gray	Thrives in cool and damp environments, used in traditional medicine.	Effective against various bacteria and fungi, including *S. aureus*, *E. coli*, *C. albicans*, and *A. niger*. A high concentration of bioactive compounds contributes to antimicrobial properties.	[15, 45, 54]
Cladonia digitata (L.) Hoffm	Contains bioactive compounds with antimicrobial properties.	Inhibits the growth of pathogenic bacteria such as *S. aureus*, *P. aeruginosa*, and *E. coli*. Antifungal activity against *C. albicans* and *Aspergillus fumigatus*.	[45, 55]
Cladonia fimbriate (L.) Fr.	Found in various regions globally; possesses antimicrobial potential.	Methanol extract exhibits potent antimicrobial activity. Moderate activity against *E. coli* and *S. aureus*. Moderate activity against bacteria including *P. aeruginosa* and *S. aureus*.	[14, 56]
Ochrolechia parella (L.) A. Massal.	Common in temperate regions; antimicrobial properties explored.	Moderate to strong antibacterial effects against G+ bacteria such as *S. aureus* and *Bacillus subtilis*. Moderate antifungal activity against specific fungal strains including *C. albicans* and *A. fumigatus*.	[5, 57, 58]

(Continued)

Table 13.2 (Continued)

Lichen species	Description	Antimicrobial activity	References
Parmeliopsis hyperopta	Investigated for its antimicrobial properties.	Significant antibacterial activity against G+ bacteria including *S. aureus* and *Streptococcus pyogenes*. Moderate antifungal activity against specific fungi strains, including *C. albicans* and *A. fumigatus*.	[15, 59, 60]
Lecanora frustulosa	Utilized for medicinal purposes; possesses antimicrobial potential.	Exhibits antibacterial and antifungal activity against various microorganisms, including *Bacillus* spp., *S. aureus*, *E. aerogenes*, *E. coli*, *C. albicans*, and *A. niger*. Contains divaricatic acid and zeorin with antimicrobial properties.	[6, 60–62]

E. divaricata extracts possess antifungal properties, making them effective against various fungal pathogens. Consequently, the antimicrobial properties exhibited by *E. divaricata* render it a promising natural source for the development of novel antimicrobial agents [50].

13.3.4 *Bryoria fuscescens*

B. fuscescens, a type of lichen, is distributed across various regions globally, including North America, Europe, and Asia. It is renowned in certain cultures for its medicinal attributes and is employed as an antimicrobial agent. Several investigations have examined various extracts from this lichen for their antibacterial potential against different bacterial strains [63]. The results indicated that the ethyl acetate extract exhibited substantial antibacterial activity against both *E. coli* and *S. aureus*. Furthermore, research into *B. fuscescens* extracts highlighted their antifungal properties, with the ethyl acetate extract displaying notable effectiveness against *C. albicans* and *A. niger* [51].

13.3.5 *Alectoria sarmentosa*

A. sarmentosa, a type of lichen, has a history of traditional use for its medicinal qualities in indigenous communities. Some studies suggest its potential antimicrobial properties, demonstrating moderate to strong activity against bacteria such as *S. aureus* and *E. coli*, as well as fungi like *C. albicans* and *Cryptococcus neoformans* [52]. In another investigation, the chemical composition and antimicrobial activity of *A. sarmentosa* extracts were explored. Several compounds, including usnic acid commonly found in lichens, exhibited antimicrobial effects. The extract proved effective against various bacterial strains, including *S. aureus* and *E. coli*, and fungi such as *A. niger* and *C. albicans* [53].

13.3.6 *Cetraria pinastri* (Scop.) Gray

The gray reindeer lichen, also known as *C. pinastri* (Scop.) Gray, is a type of lichen thriving in cool, damp environments like boreal forests. It has a longstanding history in traditional medicine due to its anti-inflammatory and antimicrobial properties [54]. Extracts from *C. pinastri* have demonstrated effectiveness against a variety of bacteria and fungi, including *S. aureus*, *E. coli*, *C. albicans*, and *A. niger*. This lichen is particularly potent against pathogenic bacterial strains like MRSA and vancomycin-resistant Enterococcus (VRE) due to its high concentration of bioactive compounds such as usnic acid, atranorin, and protocetraric acid, all of which possess potent antimicrobial, antioxidant, and anti-inflammatory properties [45].

13.3.7 *Cladonia digitata* (L.) Hoffm

C. digitata (L.) Hoffm is a lichen containing various bioactive compounds, including usnic acid, fumarprotocetraric acid, and diffractaic acid, all with antimicrobial properties [45]. Studies have confirmed that extracts of *C. digitata* can inhibit the growth of pathogenic bacteria such as *S. aureus*, *P. aeruginosa*, and *E. coli*, as well as demonstrate antifungal activity against *C. albicans* and *Aspergillus fumigatus*. These bioactive compounds disrupt the cell membranes and metabolic processes of microorganisms, contributing to *C. digitata*'s antimicrobial activity [55]. This lichen holds promise as a natural source of valuable antimicrobial agents for medical and other applications.

13.3.8 *Cladonia fimbriate* (L.) Fr

C. fimbriata (L.) Fr. is a lichen species found in various regions of the world, including Asia, Europe, North America, and South America. This lichen has demonstrated the ability to inhibit the growth of microorganisms, including viruses, bacteria, and fungi. A study published in the Microbiology journal reported potent antimicrobial activity in the methanol extract of *C. fimbriata* [14]. Another study investigating the chemical composition and biological activities of *C. fimbriata* collected from the Brazilian Amazon rainforest, revealed moderate activity against *E. coli* and *S. aureus* bacteria. Furthermore, *C. fimbriata* collected from the Brazilian Caatinga biome exhibited moderate activity against several bacteria, including *P. aeruginosa* and *S. aureus* [56].

13.3.9 *Ochrolechia parella* (L.) A. Massal

O. parella is a lichen-forming fungus commonly found on trees and rocks in temperate regions. In 2018, a research study explored the antimicrobial properties of various lichen species, including *O. parella* [5]. The study revealed that extracts from *O. parella* exhibited moderate to strong activity against G+ bacteria, such as *S. aureus* and *B. subtilis*. Additionally, the extracts displayed moderate antifungal activity against certain fungal strains, including *C. albicans* and *A. fumigatus* [57]. Another investigation demonstrated that an ethyl acetate extract from the lichen had significant antibacterial effects against different G+ bacteria, such as *S. aureus* and *Streptococcus pyogenes*, as well as moderate antifungal effects against specific fungal strains, including *C. albicans* and *C. neoformans* [58].

13.3.10 *Parmeliopsis hyperopta*

P. hyperopta, a type of lichen, has been researched for its antimicrobial properties. Studies have demonstrated that extracts from *P. hyperopta* can impede the growth of certain bacteria and fungi [59]. The antibacterial and antifungal properties of *P. hyperopta* were evaluated by examining the effects of its acetone, methanol, and aqueous extracts. Notably, the lichen displays significant antibacterial activity against various strains of G+ bacteria, including *S. aureus* and *S. pyogenes*, as well as some G− bacteria [15]. Moreover, *P. hyperopta* has demonstrated moderate antifungal activity against specific fungi strains, including *C. albicans* and *A. fumigatus*. The precise mechanisms through which *P. hyperopta* exerts its antimicrobial effects remain unclear and require further investigation [60]. Nevertheless, these results suggest that *P. hyperopta* may hold potential as a source of novel antimicrobial agents for treating infectious diseases.

13.3.11 *Lecanora frustulosa*

L. frustulosa, a type of lichen, has been used for its medicinal properties by various cultures [60]. According to a study, the acetone, methanol, and aqueous extracts of *L. frustulosa* exhibit antibacterial and antifungal activity against a variety of microorganisms, including *Bacillus mycoides*, *B. subtilis*, *S. aureus*, *Enterobacter aerogenes*, *E. coli*, *P. aeruginosa*, *C. albicans*, and *A. niger* [61]. The antimicrobial properties of divaricatic acid and zeorin, which are constituents of *L. frustulosa*, have also been identified. Extracts obtained from this lichen have been shown to hinder the growth of certain bacteria and fungi [6]. Specifically, *L. frustulosa* has considerable antibacterial effects against various G+ bacteria and some G− bacteria, as well as moderate antifungal activity against specific fungi strains. Nevertheless, further research is necessary to fully comprehend how *L. frustulosa* exerts its antimicrobial effects. These findings suggest that *L. frustulosa* may have potential as a natural source of antimicrobial agents to combat infectious diseases [62].

13.4 Antibacterial Properties

The metabolites extracted from lichens exhibit a wide variety of biological activities such as antibacterial, antiviral, anti-inflammatory, antifungal, antipyretic, antiproliferative, and cytotoxic effects. The increasing resistance to bacterial infections has posed a major public health problem throughout the world, which is accelerating at an alarming rate [64]. Therefore, an immediate requirement for new therapeutic agents derived from plants or other sources is of great importance. Lichens, a complex life form having a symbiotic partnership with a fungus and an alga synthesize numerous metabolites comprising aliphatic, aromatic, cycloaliphatic, and terpenic compounds. Thus, metabolite extracts from different lichen species stand out as new therapeutic agents thereby, providing a promising scope as antibacterial agents. These agents possess antibacterial properties and efficiently destroy bacterial strains of various species [65].

13.4.1 Properties

Agents or a group of compounds able to fight against pathogenic bacteria are known as antibacterial agents. They kill or reduce the metabolic activities of bacteria thereby, minimizing their pathogenic effect on the environment. Thus, manufacturing products containing antibacterial agents, or coating materials with antibacterial compounds has become an interesting topic of research. Lichen is a symbiotic organism comprising of fungus that forms a thallus or lichenized stroma containing secondary metabolites and alga, which contain photosynthetic pigments [13]. Being a valuable plant resource, it is used as medicine, food, fodder, dyes, perfume, spice, and other miscellaneous purposes. Several lichen species used for medicinal purposes throughout the ages *Cetraria islandica* and *Lobaria pulmonaria* show significant and effective treatment against *Mycobacterium tuberculosis* [13]. Lichens harbor a diverse array of biologically active compounds that manifest antimicrobial properties, encompassing antibacterial, antifungal, and antiviral activities. The protracted and extensive utilization of antibiotics has culminated in the emergence of antibiotic resistance and the proliferation of multi-resistant microorganisms. The ubiquity and prevalence of lichens, represented by the distribution of approximately 20,000 lichen species worldwide, align with their abundant presence in terrestrial ecosystems. Nevertheless, the full extent of their biological activities and the biologically active compounds synthesized by lichens, particularly in the context of their antimicrobial potential, remains largely uncharted [66].

Usnea ghattensis belongs to the group of fruticose lichen mostly grown on different trees and shrubs in the Northern Western Ghats of India contains usnic acid, a major chemical constituent that is traditionally used to cure upper respiratory infections, and skin infections has been extensively studied affecting *S. aureus*, *S. faecalis*, *B. cereus*, *E. coli*, *P. aeruginosa*, and *S. typhimurium* [38]. Additionally, several other crude extracts from different groups of lichens have been extracted to study the antibacterial effect.

13.4.2 Lichens Crude Extracts

Lichens, being an important pharmacological agent, are not extensively explored due to their slow growth rate of 1 cm per year and difficulties in culturing, which is regarded as a major problem for the production of metabolites. However, culturing of lichen mycobionts can be optimized to support the synthesis of secondary compounds. The crude extracts of lichens are usually taken in the presence of acetone, ethanol, and methanol. An *in vitro* antibacterial test is carried out using the *Kirby-Bauer* technique of disc diffusion method, and MIC is assigned using *Broth microdilution*. The cosmopolitan nature of lichens shows a distribution from arctic to tropical regions as well as from plains to the highest mountains. These lichens synthesize various classes of metabolites such as xanthones, pulvinic acid derivatives, naphthoquinones, depsidones, depsides, diterpenes, anthraquinones, aliphatic acids, etc. Thus, assisting to overcome the multiple drug-resistant bacterial strains by imparting antibacterial properties to inhibit the pathogenic attributes of various bacterial species [67].

13.4.3 Lichenic Acids

Generally, lichen is divided into three groups, i.e. crustose, foliose, and fruticose. The symbiotic association of lichen has the potential to produce secondary metabolites for therapeutic effects against bacterial, viral, and fungal infections. Usnea is a popular genera of lichens, among which several species such as *Usnea articulate*, *Usnea dasypoga*, *Usnea longissimi*, *Usnea sternaii*, *Usnea vireseana*, *Usnea comosa*, *Usnea blepharea*, *Usnea baileyi*,

Usnea javanica, and *Usnea flexuosa*. A study conducted by Kosanic et al. demonstrated that lichens can potentially treat disease caused by the bacteria, and its pathogenic attributes can be inhibited by metabolites such as lauric acid derived from (*Usnea barbata*), usnic acid (*U. longissima*), protolichesterinic acid (*Usnea albopunctata*) and atranorin (*Usnea rubrotincta*) [68]. A secondary metabolite extracted from *U. blepharea* is a diffractaic acid that actively inhibits the growth of *S. aureus* and *E. coli* at concentrations 750 and 1000 ppm [69]. Usnic acid, a lichenic acid, is a potent compound that causes rapid and strong inhibition of RNA and DNA synthesis in G+ bacteria, whereas it does not cause inhibition in *E. coli*, thereby demonstrating resistant traits of *E. coli* to even high doses of usnic acid. However, inhibition of RNA synthesis was seen in the G− bacterium, *Vibrio harveyi*. Confirming two main mechanisms of usnic acid: (i) inhibition of RNA synthesis and (ii) causing impairment of DNA replication [17]. Additionally, protolichesterinic acid has significant broad-spectrum antimicrobial properties against *Helicobacter pylori*, whereas vulpinic acid showed some activity against aerobic and anaerobic microorganisms. Hence, lichenic acids are being predominantly explored for their antibacterial activities against various bacterial strains as potential therapeutic drugs in the coming years. The promising scope of lichenic acids for a variety of antimicrobial activities is a boon for the public health sector due to increasing resistance against pharmaceutical drugs.

13.4.4 Bioactive Compounds

Since ancient times, lichens have been used as a source of natural drugs as well as an important source of biologically active compounds. For centuries, humans have used lichens as a source of food, dye, and therapeutic agents in traditional medicines. The efficacy and effectiveness of lichens as a therapeutic drug are mainly due to the synthesis of secondary metabolites [70]. Usnic acid, an active metabolite of lichen, has several applications in pharmaceutical preparations. Additionally, atranorin, evernic, and physodes are powerful inhibitors of some metabolic enzymes. Several lichens, such as *Cladonia foliacea*, *Dermatocarpon miniatum*, *E. divaricate*, *Evernia prunastri*, and *Neofuscella pulla*, revealed antioxidant and antimicrobial properties [38].

13.4.5 Secondary Metabolites of Lichens and Standard Antibiotics

Lichens are recognized as prolific producers of an extensive array of secondary metabolites, with more than 1000 distinct compounds identified thus far. Among these compounds, depsidones, depsides, dibenzofurans, and xanthones predominate and are primarily synthesized through the acetate-malonate pathway. Their accumulation within the thallus exhibits notable variability, spanning from 0.1% to 30% of the thallus' dry weight, with a predominant deposition in the cortex and medullary layers. Of particular significance are depsidones, which have garnered substantial attention in pharmacological research due to their pronounced antioxidative, antimicrobial, and cytotoxic properties [22]. Prominent among the studied depsidones are fumarprotocetraric acid, lobaric acid, norstictic acid, physodic acid, salazinic acid, and stictic acid. These compounds have demonstrated remarkable antimicrobial efficacy against a broad spectrum of microorganisms, encompassing both G+ and G− bacteria, as well as fungi, with a notable impact on

Table 13.3 Antibacterial activities of various despidones produced by lichens.

Depsidones	Lichen spp.	Experimental model	Activity
Fumarprotocetraric acid	*Cladonia furcate*	G+ and G− bacteria	Antibacterial activity
Pannarin	*Psoroma* spp.	Methicillin-resistant *S. aureus*	Antibacterial activity
Physodic acid	*Hypogymnia physodes*	G+ (*Bacillus mycoides, B. subtilis, S. aureus*) and G− bacteria (*E. coli, K. pneumoniae*)	Antibacterial activity
Protocetraric acid	*Flavoparmelia caperata* *Parmelia caperata*	*S. aureus*	Antibacterial activity
Psoromic acid	*Squamarina cartilaginea*	*S. gordonii, P. gingivalis*	Antibacterial activity
Salazinic acid	*Parmelia sulcata*	*B. subtilis, L. monocytogenes, S. aureus, Streptococcus faecalis, P. vulgaris, Y. enterocolitica*	Antibacterial activity

Candida species. Their mechanisms of action include the potentiation of bactericidal activity and the reduction of antibiotic resistance, attributed in part to their RecA inhibitory properties. Collatolic acid, in particular, has exhibited potent antimicrobial activity against methicillin-resistant strains of *S. aureus*. Furthermore, fumarprotocetraric acid has displayed antimicrobial effectiveness against G+ bacteria such as *B. cereus* and *B. subtilis*, as well as G− bacteria like Listeria monocytogenes and *C. albicans*. Additional antibacterial activities of several other depsidones are detailed in Table 13.3.

13.5 Antifungal Properties

Lichens are used worldwide as a plant resource for producing a vast range of bioactive compounds. Lichens are well known for their use in folk medicine as antibiotics [71]. A wide range of lichen isolates have been used pharmaceutically for humans [72]. Lichens are also known for their medicinal properties, including their antifungal properties. In recent years, scientific studies have confirmed the antifungal activity of lichens and have identified several compounds responsible for this activity. Lichen crude extracts have been found to have antimicrobial activity against various microorganisms, including bacteria, fungi, and viruses [64]. The production of a diverse array of secondary metabolites has shown antifungal properties. Lichens have not been used widely for the treatment of fungal infections, due to the difficulty in obtaining adequate amounts of secondary metabolites for drug development. However, recent advances in biotechnology and synthetic chemistry may provide new opportunities for the development of lichen-based antifungal drugs. One of the most well-known lichens with antifungal properties is *Usnea* spp. Usnea is a genus of lichens commonly found in temperate and arctic regions. It has been used in traditional medicine to treat fungal infections, wounds, and respiratory diseases [73]. Usnea contains usnic acid, a potent antifungal compound that has been shown to inhibit the growth of a wide range of fungal species, including *C. albicans*, *A. fumigatus*, and *C. neoformans*. Another lichen with strong antifungal properties is Cladonia rangiferina. This lichen is commonly found in arctic and subarctic regions and is used in traditional medicine to treat skin infections, such as ringworm. *Cladonia rangiferina* contains several bioactive compounds, including atranorin, chloroatranorin, and usnic acid, which have been shown to have antifungal activity against several fungal species [74].

Parmotrema tinctorum is another lichen with potent antifungal properties. It is commonly found in tropical and subtropical regions and is used in traditional medicine to treat fungal infections of the skin and nails. Parmotrema tinctorum contains several compounds, including parmotrein, which has been shown to have strong antifungal activity against several fungal species, including *C. albicans* and Trichophyton rubrum. In addition to these lichens, many other species have been found to have antifungal properties, including *Cladonia convoluta*, *L. pulmonaria*, and *Peltigera canina*. These lichens contain a variety of bioactive compounds, including lichen acids, depsides, and depsidones, which have been shown to have antifungal activity against various fungal species. The antifungal properties of lichens have several potential applications. They could be used as natural sources of antifungal agents for the development of new drugs. They could also be used in the development of antifungal coatings for surfaces, such as medical devices, to prevent fungal infections. Finally, lichens could be used in the development of new topical antifungal agents for the treatment of skin and nail infections but the literature available is still insufficient [75].

Fungal infections represent a substantial global public health challenge, characterized by a growing incidence attributed to multiple factors, including the expanding population of immunocompromised individuals, the extensive utilization of antibiotics, and the emergence of drug-resistant fungal strains. Therefore, the search for new antifungal agents has become a major focus of research in the pharmaceutical industry. The increasing incidence of fungal infections could be prevented with the help of antifungal drug development.

13.5.1 Lichens Crude Extracts

Crude extracts of lichens have been historically used in traditional medicine for the treatment of several diseases, for example, skin diseases, digestive disorders, respiratory system infections, jaundice, and various ailments [28]. Lichens crude extracts have also been used as antiviral and anticarcinogenic [4].

Antioxidant activity is also one of the crucial biological activities of lichen extracts. Phenolic compounds like dibenzofurans, depsidones, and depsides are examples of potential antioxidants. These compounds can act as scavengers of free radicals and prevent cells and tissues from any oxidative damage [5, 11]. *Cladonia rangiferina* and *Parmotrema tinctorum* are the crude extracts of lichens, having the highest antioxidant activity and are used as natural antioxidant sources in nutraceuticals and functional foods.

Lichen extracts are well known for possessing antifungal and antibacterial properties [4]. A well-known lichen with potent antibacterial and antifungal properties is *Usnea* spp. [76]. Many bacterial and fungal species, including *S. aureus*, *E. coli*, *C. albicans*, and *A. fumigatus*, have been demonstrated to be inhibited in growth by its crude extract [8].

Several lichens with antibacterial and antifungal characteristics have also been discovered, including *C. convoluta* and *L. pulmonaria*.

Lichen crude extracts have anti-inflammatory and immunomodulatory potential. Polysaccharides like lichenan and isolichenan, which have been demonstrated to stimulate the immune system and alter the generation of cytokines, are primarily responsible for these actions. *In vitro* and *in vivo* studies have revealed that crude extracts of lichens like *Parmelia sulcata* and *Ramalina fraxinea* have anti-inflammatory and immunomodulatory properties [77].

Lichen extracts have potential use in the industrial and agricultural sectors. It is well known that lichens produce several distinctive substances with antibacterial and antifouling effects, including depsidones and usnic acid. New insecticides and antifouling agents might be created using these substances. Lichen extracts have also been demonstrated to have dyeing abilities, and some lichens, including *Xanthoria parietina* and *Parmelia saxatilis*, have been utilized as natural dyes for leather and textiles [78].

Lichen crude extracts have shown antifungal activity towards various pathogenic fungi, including *Trichophyton mentagrophytes*, *A. fumigatus*, and *C. albicans*. It has been discovered that the methanolic extract of *Parmotrema praesorediosum* has strong antifungal activity against *A. fumigatus* and *C. albicans* [79]. Antifungal properties of crude extracts of lichens have been recognized in various secondary metabolites, including salazinic acid, atranorin, and usnic acid.

Table 13.4 provides a comprehensive overview of the MIC of lichen extracts against various fungal strains. Among the tested lichen species, *Cladonia macilenta*, extracted using acetone, exhibited a MIC of 50 μg/mL against *T. mentagrophytes*. Additionally, *P. praesorediosum*, extracted with methanolic solvent, demonstrated notable inhibitory activity against *C. albicans* and *A. fumigatus*, with MIC values ranging from 12.5 to 50 μg/mL. These findings highlight the potential of lichen-derived compounds as effective antifungal agents, suggesting their utility in the development of treatments for fungal infections caused by a diverse range of pathogens. Further research into the mechanisms underlying these inhibitory effects and the isolation of active compounds from lichen extracts may pave the way for innovative therapeutic strategies in the field of mycology.

Lichen crude extracts are well known for their biological activities and potential uses in agriculture, medicine, and industry. To understand and explore the therapeutic effects of lichen crude extracts, further research is needed. Lichen extracts may offer a viable and eco-friendly substitute for synthetic chemicals in a variety of applications.

Table 13.4 Minimum inhibitory concentration (MIC) of lichen extracts against fungal strains.

Lichen species	Extract type	Fungal strain	Minimum inhibitory concentration (MIC) (μg/mL)
Cladonia macilenta	Acetone	*Trichophyton mentagrophytes*	50
Parmotrema praesorediosum	Methanolic	*C. albicans, A. fumigatus*	12.5–50

13.5.2 Secondary Metabolites of Lichens and Standard Antibiotics

The chemical substances produced by lichens are known as lichen acids or secondary metabolites. Lichens are known to produce a variety of secondary metabolites that have shown potential antimicrobial activity, including against bacterial strains. Some of these secondary metabolites have been identified and characterized as standard antibiotics. Here, we discuss a few examples of standard antibiotics produced by lichens (Table 13.5).

13.5.2.1 Usnic Acid

The phenolic compound Usnic acid is uniquely found in many genera of lichens, such as *Usnea*, *Lecanora*, *Alectoria*, *Evernia*, *Cladonia*, and *Ramalina*. Usnic acid or dibenzofuran is a crucial metabolite produced by the lichens. The chemical formula of Usnic acid is $C_{18}H_{16}O_7$, [UA = 2,6-diacetyl-7,9-dihydroxy-8-9b-dimethyl-1,3(2H, 9b/aH)-dibenzofurandione] [80]. It has been found to be highly effective in wound healing due to its anti-inflammatory potential. Usnic acid has been shown to have potent antifungal activity against a wide range of fungal species, including *C. albicans*, *A. fumigatus*, and *C. neoformans*. In higher plants, usnic has shown growth regulator activity [72]. Usnic acid isolated from *Ramalina celastri* shows antiviral activity toward the Junin virus [81]. Despite the pharmacological potential of usnic acid,

Table 13.5 Lichen secondary metabolites and their biological activities.

Lichen metabolite	Source lichen genera	Biological activities	References
Usnic acid	*Usnea, Lecanora, Alectoria, Evernia, Cladonia, Ramalina*	Wound healing, antifungal (*C. albicans, A. fumigatus, C. neoformans*), growth regulator, antiviral (Junin virus), antimicrobial (including MRSA), toxic effects	[64, 72, 80–82]
Atranorin	*P. sulcata, Stereocaulon caespitosum, Evernia prunastri, Cladina kalbii, Lethariella canariensis, Stereocaulon alpinum*	Antimicrobial, antitumor, cytotoxic, immunomodulatory, antiviral, anti-inflammatory, antioxidant, antinociceptive, antifungal, antiprotozoal	[83–85]
Evernic acid	*Usnea, Evernia, Ramalina, Hypogymnia*	Antibacterial (*B. mycoides, B. subtilis, K. pneumoniae*), antifungal (*C. albicans, A. flavus, A. fumigatus, P. purpurescens, P. verrucosum*), anti-quorum sensing (*P. aeruginosa*), antioxidant	[86, 87]
Lobaric acid	Not specified	Antifungal, antibacterial (*B. subtilis, S. aureus*), anticancer	[88]
Vulpinic acid	*Umbilicaria sp., Parmelia saxatilis*	Antibacterial (*Streptococcus mutans, S. aureus*), disrupts the cell membrane, induces oxidative stress	[89]
Physodic acid		Cytotoxic effects in cancer cells	[90]

it has also been reported as a toxic compound. Usnic acid has the capacity to detach the electron transport chain taking place in the mitochondria, which eventually causes stress in the cell [82]. Usnic acid has shown antimicrobial activity against MRSA in patients with cystic fibrosis. The activity of usnic acid in tumor inhibition was first identified by Kupchan and Kopperman [64].

13.5.2.2 Atranorin

After Usnic acid, ATR is a corticol secondary metabolite of lichen, specifically *P. sulcata*, *Stereocaulon caespitosum*, *Evernia prunastri*, *Cladina kalbii*, *Lethariella canariensis*, and *Stereocaulon alpinum* [83]. In 1898, Hesse first isolated ATR, which is the second most studied metabolite. Different effects of ATR have been studied only *in vivo*, and the molecular studies are poorly understood to date. ATR has antimicrobial, antitumor, cytotoxic, immunomodulatory, antiviral, anti-inflammatory, antioxidant, antinociceptive, and antifungal activities. The molecular formula of ATR is $C_{19}H_{18}O_8$. *In silico* studies have been carried out to understand the interaction of ATR with breast cancer targets [84]. The color of ATR ranges from white to pale yellow and is soluble in various organic solvents. It is also used as a dye, in traditional medicine and as an air pollution marker. ATR also has antiprotozoal activity. Unlike other lichen metabolites, ATR is active against G− bacteria [85]. The molecular interaction studies of ATR are yet not fully understood. However, the cytotoxic effect of ATR was studied in breast cancer cells that showed inhibition of the target cells. Further research is needed to fully understand the molecular mechanism of action of ATR.

13.5.2.3 Evernic Acid

Evernic acid is specifically isolated for the first time from *U. longissima* and is found in *Evernia*, *Ramalina*, and *Hypogymnia*. The molecular formula of evernic acid is $C_{17}H_{16}O_7$. It is the most studied compound [86]. *B. mycoides*, *B. subtilis*, *Klebsiella pneumoniae*, and *C. albicans* at MIC 0.25 mg/mL, against *E. coli* at MIC 0.5 mg/mL, and against *Aspergillus flavus*, *A. fumigatus*, *Penicillium purpurescens* and *Philodendron verrucosum* at MIC 1 mg/mL as well as against *P. aeruginosa* quorum-sensing systems. Like other lichenic compounds, Evernic acid also has antioxidant properties. The structural similarity of Evernic acid was studied to understand its mechanism of action [87].

13.5.2.4 Lobaric Acid

Lobaric acid is a well-known antifungal lichenic secondary metabolite. It also has antibacterial activity against *B. subtilis* and *S. aureus*, which are G+ bacteria. Lobaric acid inhibits the growth of bacteria by erupting the cell membrane and induces oxidative stress on the cell membrane. Lobaric acid also showed anticancer activity [88].

13.5.2.5 Vulpinic Acid

The depsidone compound Vulpinic acid has been specifically isolated from some of the lichen species, including *Umbilicaria sp* and *P. saxatilis*. Vulpinic acid has antibacterial properties against G+ bacteria like *S. mutans* and *S. aureus*. Like lobaric acid, vulpinic acid disrupts the cell membrane by inducing oxidative stress [89].

13.5.2.6 Physodic Acid

Physodic acid is a secondary metabolite produced by the lichens as a defense mechanism against herbivores. It belongs to the class depsidones, which are distinguished by having molecules with two depsides connected by an ester link. The chemical formula of physodic acid is $C_{20}H_{22}O_{10}$, and it has a molecular mass of 422.38 g/mol. Physodic acid has been found to have cytotoxic effects in cancer cells [90].

13.6 Conclusion

In conclusion, lichens represent a valuable source of antimicrobial compounds derived from various species. These compounds exhibit a broad spectrum of activity against bacteria and fungi. For instance, *R. farinacea* demonstrates potent antibacterial and antifungal effects with low cytotoxicity to human cells. *P. glauca* not only possesses remarkable antibacterial properties but also shows promise for treating urinary tract infections. Additionally, *E. divaricata* displays notable antimicrobial properties, particularly against antibiotic-resistant strains. *B. fuscescens* serves as an effective antimicrobial agent, especially against *E. coli* and *S. aureus*. Furthermore, lichen species like *A. sarmentosa*, *C. pinastri*, *C. digitata*, *C. fimbriate*, *O. parella*, *P. hyperopta*, and *L. frustulosa* exhibit significant antibacterial and antifungal activity, further highlighting the potential of lichens as sources of novel antimicrobial agents. Their diverse bioactive compounds offer promising avenues for future research and development in the field of antimicrobial therapeutics.

References

1 Morillas, L., Roales, J., Cruz, C., and Munzi, S. (2022). *Encyclopedia* 2: 1421–1431.

2 Armaleo, D., Müller, O., Lutzoni, F. et al. (2019). *BMC Genomics* 20: 1–33.

3 Grimm, M., Grube, M., Schiefelbein, U. et al. (2021). *Frontiers in Microbiology* 12: 623839.

4 Ranković, B. and Kosanić, M. (2015). *Lichen Secondary Metabolites Bioactive Properties and Pharmaceutical Potential*, 1–26. Springer International Publishing.

5 Calcott, M.J., Ackerley, D.F., Knight, A. et al. (2018). *Chemical Society Reviews* 47: 1730–1760.

6 Goga, M., Elečko, J., Marcinčinová, M. et al. (2020). *Reference Series in Phytochemistry*, 175–209. Springer Science And Business Media B.V.

7 Karagöz, Y. and Karagöz, B.Ö. (2022). *Eurasian Journal of Medicine* 54: S195–S208.

8 Paguirigan, J.A., Liu, R., Im, S.M. et al. (2022). *The Plant Pathology Journal* 38. https://doi.org/10.5423/PPJ.OA.12.2021.0176.

9 Ahmed, S., Ahmed, M.Z., Rafique, S. et al. (2023). *BioMed Research International* 2023: 1–27.

10 Ahmed, M.Z., Rao, T., Ahmad, W. et al. (2021). *Biochemistry of Drug Resistance*, 33–56. Springer International Publishing, Cham.

11 Sargsyan, R., Gasparyan, A., Tadevosyan, G., and Panosyan, H. (2021). *AMB Express* 11: 110.

12 Felczykowska, A., Pastuszak-Skrzypczak, A., Pawlik, A. et al. (2017). *BMC Complementary and Alternative Medicine* 17: 1–12.

13 Srivastava, P., Upreti, D.K., Dhole, T.N. et al. (2013). *Interdisciplinary Perspectives on Infectious Diseases* 2013: https://doi.org/10.1155/2013/709348.

14 Kosanić, M., Ristić, S., Stanojković, T. et al. (2018). *Farmácia* 66: 644–651.

15 Kosanić, M. and Ranković, B. (2011). *Journal of Medicinal Food* 14: 1624–1630.

16 Noël, A., Garnier, A., Clément, M. et al. (2021). *Phytochemistry* 181: 112535.

17 Maciag-Dorszyńska, M., Wegrzyn, G., and Guzow-Krzemińska, B. (2014). *FEMS Microbiology Letters* 353: 57–62.

18 Septiana, E. (2011). *Journal of Biology Udayana* 15: 2.

19 Manojlović, N.T., Vasiljević, P., Jusković, M. et al. (2010). *Journal of Medicinal Plants Research* 4: 817–823.

20 Balaji, G. and Hariharan, P. (2007). *Research Journal of Botany* 2: 54–59.

21 Victor, K., Boris, L., Athina, G. et al. (2018). *MedChemComm* 9: 870–882.

22 Ureña-Vacas, I., González-Burgos, E., Divakar, P.K., and Gómez-Serranillos, M.P. (2022). *Planta Medica* 88: 855–880.

23 Ureña-Vacas, I., González-Burgos, E., Divakar, P.K., and Gómez-Serranillos, M.P. (2023). *Journal of Fungi* 9: 116.

24 Chan-Zapata, I., Borges-Argáez, R., and Ayora-Talavera, G. (1981). *Molecules* 2023: 28.

25 Futuro, D.O., Ferreira, P.G., Nicoletti, C.D. et al. (2018). *Anais da Academia Brasileira de Ciências* 90: 1187–1214.

26 Le Pogam, P. and Boustie, J. (2016). *Molecules* 21: 294.

27 Shahid, M., Rasool, A., Anjum, F., and Rehman, M.T. (2020). *Lichen-Derived Products*, 263–276. Wiley.

28 Solárová, Z., Liskova, A., Samec, M. et al. (2020). *Biomolecules* 10: 87.

29 Kai, M. (2020). *Frontiers in Microbiology* 11: 559.

30 Fouillaud, M. and Dufossé, L. (2022). *Microorganisms* 10: 123.

31 Gadouche, L., Alsoufi, A.S.M., Pacholska, D. et al. (2023). *Molecules* 28: 697.

32 Kiranmayee, P., Anitha, K., and Usha, R. (2016). *International Journal of Pharmacognosy and Phytochemical Research* 8: 912–929.

33 Desouky, S.E., Abu-Elghait, M., Fayed, E.A. et al. (2022). *Metabolites* 12: 246.

34 Pandit, R., Awal, B., Shrestha, S.S. et al. (2020). *Interdisciplinary Perspectives on Infectious Diseases* 2020: 6525826.

35 Ahmed, M., Marrez, D.A., Mohamed Abdelmoeen, N. et al. (2023). *Processes* 11: 1644.

36 Karaahmet, Z., Kinalioğlu, K., and Aydin, S. (2019). *Gümüşhane Üniversitesi Fen Bilimleri Enstitüsü Dergisi* 9: 415–424.

37 Yang, M.X., Devkota, S., Wang, L.S., and Scheidegger, C. (2021). *Diversity* 13: 330.

38 Shukla, V., Joshi, G.P., and Rawat, M.S.M. (2010). *Phytochemistry review* 9: 303–314.

39 Ureña-Vacas, I., Burgos, E.G., Divakar, P.K., and Gómez-Serranillos, M.P. (2021). *Current Topics in Medicinal Chemistry* 21: 2397–2408.

40 Wang, H., Xuan, M., Huang, C., and Wang, C. (2022). *Molecules* 27: 7469.

41 Carpentier, C., Queiroz, E.F., Marcourt, L. et al. (2017). *Journal of Natural Products* 80: 210–214.

42 Hao, Y.M., Yan, Y.C., Zhang, Q. et al. (2023). *Frontiers in Chemistry* 10: 1063645.

43 Ali, H.M., Salem, M.Z.M., and Al Sahli, A.A. (2013). *Journal of Pure Applied Microbiology* 7: 131–137.
44 Özen, T. and Kinalioğlu, K. (2008). *Biologia (Bratisl)* 63: 211–216.
45 Ranković, B., Ranković, D., and Marić, D. (2010). *Mikrobiologiia* 79: 812–818.
46 Ellis, C.J. (2012). *Perspectives in Plant Ecology, Evolution and Systematics* 14: 131–152.
47 Pankratov, T.A., Kachalkin, A.V., Korchikov, E.S., and Dobrovol'skaya, T.G. (2017). *Microbiology* 86: 293–309.
48 Šeklić, D.S. and Jovanović, M.M. (2022). *Journal of Food Biochemistry* 46: e14096.
49 Esimone, C.O. and Adikwu, M.U. (1999). *Fitoterapia* 70: 428–431.
50 Abdallah, E.M. (2019). *Advanced Life Science* 6: 110–115.
51 Milovanovic, N., Kostic, D., Milovanovic, V. et al. (2018). *Facta Universitatis, Series Physics, Chemical Technology* 16: 183–189.
52 Çobanoğlu, G., Sesal, C., Gökmen, B., and Çakar, S. (2010). *South Western Journal of Horticulture Biology Environment* 1: 153–158.
53 Ranković, B. and Mišić, M. (2007). *Mikologiya I Fitopatologiya* 41: 276–281.
54 Mitrović, T., Stamenković, S., Cvetković, V. et al. (2011). *International Journal of Molecular Sciences* 12: 5428–5448.
55 Dzomba, P. (2012). *AFRICAN Journal of Biotechnology* 11: 7995–7999. https://doi.org/10.5897/ajb12.4130.
56 Yoshikawa, K., Kokudo, N., Tanaka, M. et al. (2008). *Chemical & Pharmaceutical Bulletin* 56: 89–92.
57 Cintas, P., Calcio-Gaudino, E., and Cravotto, G. (2013). *Food Engineering Series*, 181–206. Springer.
58 Ranković, B., Ranković, D., Kosanić, M., and Marić, D. (2010). *Central European Journal of Biology* 5: 649–655.
59 Marijana, K., Branislav, R., and Slobodan, S. (2010). *Journal of Microbiology* 4: 885–890.
60 Kosanić, M. and Ranković, B. (2019). *Lichen Secondary Metabolites*, 99–127. Springer, Cham.
61 Oh, J.M., Kim, Y.J., Gang, H.S. et al. (2018). *Molecules* 23: 3068.
62 Nasisri Semnani, S., Rahnema, M., Ghasempou, H. et al. (2014). *Journal of Fasa University of Medical Science* 4: 318–326.
63 Anar, M., Aslan, A., Agar, G., and Ozgencli, I. (2016). *Journal of Medicinal and Aromatic Plants* 5: 2167–0412. https://doi.org/10.4172/2167-0412.1000233.
64 Basile, A., Rigano, D., Loppi, S. et al. (2015). *International Journal of Molecular Sciences* 16: 7861–7875.
65 Micheletti, A.C., Honda, N.K., Ravaglia, L.M. et al. (2021). *Anais da Academia Brasileira de Ciências* 93. https://doi.org/10.1590/0001-3765202120191194.
66 Molnár, K. and Farkas, E. (2010). *Zeitschrift fur Naturforsch. – Section C Journal of Biosciences* 65: 157–173.
67 Jha, B.N., Shrestha, M., Pandey, D.P. et al. (2017). *BMC Complementary and Alternative Medicine* 17: 1–8.
68 Kosanić, M., Ranković, B., Stanojković, T. et al. (2014). *EXCLI Journal* 13: 1226–1238.
69 Zambare, V.P. and Christopher, L.P. (2012). *Pharmaceutical Biology* 50: 778–798.
70 Dar, T.U.H., Dar, S.A., Islam, S.U. et al. (2022). *Seminars in Cancer Biology* 86: 1120–1137.
71 Thakur, M. and Chander, H. (2021). *Biological Forum – An International Journal* 13: 39.

72 Nunes, P.S., Albuquerque-Jnior, R.L.C., Cavalcante, D.R.R. et al. (2011). *BioMed Research International* 2011: https://doi.org/10.1155/2011/761593.

73 Sepahvand, A., Studzińska-Sroka, E., Ramak, P., and Karimian, V. (2021). *Journal of Ethnopharmacology* 268: 113656.

74 Egbert, S., Hoffman, J.R., McMullin, R.T. et al. (2022). *Fungal Biology* 126: 697–706.

75 Iorizzo, M., Tosti, A., Starace, M. et al. (2020). *Journal of the American Academy of Dermatology* 83: 1717–1723.

76 Yadav, H., Nayaka, S., and Dwivedi, M. (2021). *Journal of Pure Applied Microbiology* 15: 701–708.

77 Hamida, R.S., Ali, M.A., Abdelmeguid, N.E. et al. (2021). *Journal of Fungi* 7: 291.

78 Karadag, R. (2023). *Journal of Natural Fibers* 20: 2162187.

79 Okla, M.K., Alatar, A.A., Al-Amri, S.S. et al. (2021). *Plants* 10: 1–13.

80 Macedo, D.C.S., Almeida, F.J.F., Wanderley, M.S.O. et al. (2021). *Phytochemistry Reviews* 20: 609–630.

81 Fazio, A.T., Adler, M.T., Bertoni, M.D. et al. (2007). *Zeitschrift fur Naturforsch. – Section C Journal of Biosciences* 62: 543–549.

82 Han, D., Matsumaru, K., Rettori, D., and Kaplowitz, N. (2004). *Biochemical Pharmacology* 67: 439–451.

83 Urbanska, N., Simko, P., Leskanicova, A. et al. (1850). *Life* 2022: 12.

84 Harikrishnan, A., Veena, V., Lakshmi, B. et al. (2021). *Journal of Biomolecular Structure & Dynamics* 39: 1248–1258.

85 Studzinska-Sroka, E., Galanty, A., and Bylka, W. (2017). *Mini-Reviews Medicinal Chemistry* 17: 1633–1645.

86 Staples, R., LaDuca, R.L., Roze, L.V. et al. (2020). *Chemistry & Biodiversity* 17: e1900465.

87 Shcherbakova, A., Strömstedt, A.A., Göransson, U. et al. (2021). *World Journal of Microbiology and Biotechnology* 37: 129.

88 Hong, J.M., Suh, S.S., Kim, T.K. et al. (2018). *Molecules* 23: 658.

89 Sahin, E., Dabagoglu Psav, S., Avan, I. et al. (2019). *Human & Experimental Toxicology* 38: 675–684.

90 Studzińska-Sroka, E., Piotrowska, H., Kucińska, M. et al. (2016). *Pharmaceutical Biology* 54: 2480–2485.

14

Lichens: A Source of Anticancer Drugs

Ari Satia Nugraha[1,2], Thi Hai Yen Lam[3], Hendris Wongso[4,5], Lilla Nur Firli[1], and Paul A. Keller[2]

[1] *Drug Utilisation and Discovery Research Group, Faculty of Pharmacy, Universitas Jember, Jember, Indonesia*
[2] *School of Chemistry and Molecular Biosciences, Molecular Horizons, University of Wollongong, Wollongong, Australia*
[3] *Department of Bioorganic Chemistry, Leibniz Institute of Plant Biochemistry, Halle (Saale), Germany*
[4] *Research Center for Radioisotope, Radiopharmaceutical, and Biodosimetry Technology, Research Organization for Nuclear Energy, National Research and Innovation Agency, Banten, Indonesia*
[5] *Research Collaboration Center for Theranostic Radiopharmaceuticals, National Research and Innovation Agency, Sumedang, Indonesia*

14.1 Introduction

Lichens are composed organisms, a symbiotic relationship between algae or cyanobacteria (photobiont) and fungi (mycobiont) in which their taxonomical classification is based on the fungal species [1]. This cosmopolitan organism is found in an array of environments from polar to tropical, and from coastline to mountain range regions, mostly as they are able to adapt to harsh conditions with inadequate nutrition sources such as stone and tree bark (Figure 14.1) [2]. Despite limited biomass, several lichens are well known for their traditional uses, for example, *Usnea misaminensis* (Vain.) Not., *Parmelia perlata* (Huds.) Ach., *Lobaria pulmonaria* (L.) Hoffm, *Usnea barbata* (L.) Weber ex F. H. Wigg, and *Heterodermia galactophylla* (Tuck.) Culb. Specifically, *U. misaminensis* (Vain.) Not., which is locally named as Kayu Angin (wind wood) by the Indigenous people of Java Island—Indonesia, is commonly prepared as a traditional medicament to treat cold, dysentery, blister and is used as a diuretic agent (Figure 14.1) [3], whereas *P. perlata* (Huds.) Ach. was prepared in India as an oral supplement to treat wounds, infections, and inflammation [4]. Likewise, the thallus of *L. pulmonaria* (L.) Hoffm has been prepared to treat cuts in southern Italy [4], while *U. barbata* (L.) Weber ex F. H. Wigg has been used in South Africa in mammary infection therapy for cattle fresh wounds, cuts, and moles [5]. Similarly, *H. galactophylla* (Tuck.) Culb. has been used in the Amazon to treat epilepsy and digestive disorders such as diarrhea and nausea [6]. In addition, there are a few lichen tinctures available on the market, such as Lichen Islandica (*Cetraria islandica* (L.) Acharius s.l.) tincture, which was described in European Union herbal monographs to treat oral or pharyngeal irritation [7].

Chemistry, Biology and Pharmacology of Lichen, First Edition. Edited by Ashoke Kumar Das, Ajay Sharma, Deepika Kathuria, Mohammad Javed Ansari, and Garima Bhardwaj.

Figure 14.1 (a)–(d) *Usnea misaminensis* (Vain) and *Parmelia dilatata* Vain. grown on *Swietenia mahagoni* bark at Bondowoso District, *Parmelia aurulenta* (Tuck.) grown on a palm tree (*Adonidia merrillii*) at Jember District Indonesia, *Phyloporon aciculare* (Ach.) Nyl grown on stone at Magetan District-Indonesia.

Cancer is one of the most serious health problems worldwide, with 19 million people affected in 185 countries and causing almost 10 million deaths in 2020 (one in six deaths) [8]. Despite the progress in cancer therapy, the need for effective and safe therapeutic agents remains a challenging task. Around 60% of anticancer drugs were discovered as natural-based chemicals such as docetaxel, vincristine, vinblastine, epirubicin hydrochloride, cytarabine ocfosfate, etoposide, dromostanolone, estramustine, and fluoxymesterone [9]. Empirical knowledge about the uses of lichen species as medicament has allocated this organism as one of the interesting sources for biodiscoveries, including searching for anticancer agents.

14.2 Lichen Extracts with Anticancer Activities

Drug discovery from natural sources commonly starts with the *in vitro* or *in vivo* bioactivity screening of mixtures of crude extracts or fractions, with the former becoming more widely used in large-scale bioprospecting. In searching for new anticancer agents, arrays of lichen species have been tested against several cancer cell lines as well as normal cells for cytotoxicity determination (Table 14.1). Several secondary metabolites-rich lichen genera include *Usnea*, *Parmelia*, *Cladonia*, *Evernia*, and *Ramalina* (Figure 14.1). The most studied genus, *Usnea*, belongs to Parmeliaceae, the largest family of lichenized fungi, which is composed of 2500 species spread over 80 genera [26]. This fruticose genus is well-known as beard lichen, and its high available biomass has made it an accessible source for traditional medicine treatments through the generations. It is used to treat both infectious and non-infectious ailments including cough and epilepsy as well as being used as a tonic and diuretic agent. In the wildlife, the dried lichens are eaten by animals or collected as fodder. Due to the high accessibility, the *Usnea* lichen has been one of the most bio-prospected lichens with the crude extracts and fractions of many *Usnea* species being reported as having low to high activity against cancer cell lines (Table 14.1). Significant anticancer activity was indicated for the less-polar fraction of *Usnea rubicunda* against human prostate cancer DU145 (IC_{50} value of 8.1 μg/mL).

Another important genus is *Parmelia*, which comprises 41 different species. The *Parmelia* lichens were considered to have large biomass in the form of large foliose thallus [27]. These lichens are considered powerful tools for biomonitoring environmental quality and are pharmacologically important compound producers. Investigations on *P. perlata* of British origins revealed the *n*-hexane crude extract to possess significant toxicity against the L1210 (mouse lymphocytic leukemia) cell line (IC_{50} value of 4.9 μg/mL) [12]. Lichen fruticose, such as the *Ramalina* genus, presents high biomass compared to the other genera. This genus consists of 200 reported species worldwide [28]. Studies on *Ramalina cuspidate* from Great Britain showed the *n*-hexane crude extract to possess significant cytotoxicity against the 3LL (lung) carcinoma cell line [12].

14.3 Lichen Compounds with Anticancer Activities

14.3.1 Depside

Depsides are polyphenolic benzoic acid derivatives commonly present in lichens of which the building block unit, *o*-orsellinic acid (2,4-dihydroxy-6-methylbenzoic acid) is formed through a polyketide pathway starting from acetyl-CoA reaction with malonyl-CoA forming the intermediate (methyl tetra-β-ketide) which undergoes aldol cyclization at C2-7 forming the 2,4-dihydroxy-6-methylbenzoic acid (**1**) depside monomer (Figure 14.2) [29].

Apart from this simple pathway, arrays of intermediates with various starter loading stages followed by cyclization and esterification are able to produce a number of depside-related derivatives [30]. In addition, the size of alkyl side chains attached to C-6 depends on the carbon skeleton brought by fatty acid synthetase, which was then carried through the acid biosynthesis [31] as illustrated in the biosynthesis of olivetoric acid (**2**) (Figure 14.3) [31].

Table 14.1 *In vitro* anticancer activity of crude extract or fractions of lichens.

Serial number	Species	Location/date collected	Solvent	Anticancer activity	References
1	*Bulbothrix setschwanensis* (Zahlbr.) Hale	Uttarakhand, Uttarkashi district, India September 2014	Acetone, chloroform, methanol	The extracts exhibited no significant cytotoxic activity against human cervical cancer HeLa and human normal embryonic kidney HEK-293cells.	[10]
2	*B. setschwanensis* (Zahlbr.) Hale	Uttarakhand, Uttarkashi district, India November 2012	Methanol	The extract showed low cytotoxicity against breast MCF-7 and liver HepG2 cell lines with LD_{50} values of 91 and 92 μg/mL, respectively.	[11]
3	*Cladonia convoluta*	Quercy (near Cressensac), France	*n*-hexane, diethyl ether and methanol	Most of the extracts possessed significant cytotoxicity (IC_{50} <20 μg/mL) against arrays of cell lines, such as L1210 (leukemia), 3LL (lung), DU145 (prostate), MCF-7 (breast), K562 (leukemia), and U251 (glioblastoma) with the exception of diethyl ether extract against K562 and methanol extracts for both MCF-7 and K562 (IC_{50} >90 μg/mL). The diethyl ether fractions showed promising cytotoxic activity against various cell lines, especially Lewis lung carcinoma.	[12]
4	*Cladonia furcata* (Hudson) Schrad.	Kopaonik, Serbia May 2013	Acetone	The extract displayed good cytotoxic activity against Hela cell lines with an IC_{50} value of 11.69 ± 0.45 μg/mL.	[13]
5	*C. furcata* (Hudson) Schrad.	Kopaonik, Serbia September of 2010	Acetone	The extract manifested a weak cytotoxic activity against FemX and LS174 (colon) cells with IC_{50} values of 23.52 and 40.22 μg/mL, respectively.	[14]
6	*Cladonia rangiformis*	Brittany	*n*-hexane, diethyl ether and methanol	The diethyl ether fractions showed promising cytotoxic activity against various cell lines, especially Lewis lung carcinoma.	[12]

7	*Evernia prunastri*	Brittany	*n*-hexane, diethyl ether, and methanol	The majority of the extracts showed low to moderate cytotoxicity for L1210 (leukemia), 3LL (lung), DU145 (prostate), MCF-7 (breast), K562 (leukemia), and U251 (glioblastoma) with IC_{50} values in the range of 18.5–100 μg/mL.	[12]
8	*E. prunastri*	Tuscany, central Italy	Water, methanol, ethanol	The extracts displayed no substantial cytotoxicity against murine myeloma cells (P3X63-Ag8.653).	[15]
9	*Flavoparmelia caperata* (L.) Hale	Gran Canaria, Canary Islands, Spain June 2009	Methanol	The extract showed low cytotoxicity against breast MCF-7 and liver HepG2 cell lines with LD_{50} values of >100 μg/mL.	[11]
10	*Flavoparmelia euplecta* (Stirt.) Hale	New South Wales, Australia February 2004	Methanol	The extract displayed relatively strong cytotoxicity against breast MCF-7 and liver HepG2 cell lines with LD_{50} values of 67 μg/mL, for both cell lines.	[11]
11	*Flavoparmelia haysomii* (CW Dodge) Hale	Canberra, Australia September 1999	Methanol	The extract showed low cytotoxicity against breast MCF-7 and liver HepG2 cell lines with LD_{50} values of >100 μg/mL.	[11]
12	*Hypogymnia physodes*	Uludag Mountain, Bursa, Turkey May 2010	Methanol	The extract possessed an antiproliferation against breast cancer MCF-7 and MDA-MB-231 cells with IC_{50} values of 50 and 44 μg/mL, respectively. Their genotoxic activity occurred at higher concentrations.	[16]
13	*Hypotrachyna cirrhata* (Fr.) Hale	Uttarakhand, Uttarkashi district, Kedarkantha, India November 2012	Methanol	The extract showed no activity against MCF-7, but exhibited potent cytotoxicity against liver HepG2 cell lines with LD_{50} values of 41 μg/mL.	[11]
14	*Lecanora atra* (Hudson) Ach.	Kopaonik, Serbia September of 2010	Acetone	The extract exhibited potent cytotoxicity against FemX and LS174 with IC_{50} values of 8.51 and 10.29 μg/mL, respectively.	[14]
15	*Lecanora muralis* (Schreber) Rabenh.	Kopaonik, Serbia September of 2010	Acetone	The extract are significantly cytotoxic against FemX and LS174 cell lines, with IC_{50} values of 9.58 μg/mL and 12.23 μg/mL, respectively.	[14]

(*Continued*)

Table 14.1 (Continued)

Serial number	Species	Location/date collected	Solvent	Anticancer activity	References
16	*Lethariella zahlbruckneri*		Acetone, methanol	Both the acetone extract (AEL) and methanolic extract (MEL) displayed cytotoxicity against HT-29 cells in a dose- and time-dependent manner. In addition, the AEL demonstrated stronger cytotoxicity than MEL.	[17]
17	*Myelochroa irrugans* (Nyl.) Elix and Hale	Chichibu City, Nakatsugawa, Prefecture Saitama, Province Musashi, Honshu, Japan	Methanol	The extract was found to be inactive for MCF-7, but displayed strong cytotoxicity against HepG2 cell lines ($LD_{50} = 22\,\mu g/mL$).	[11]
18	*Parmelia caperata*	Brittany During the summer 2000	*n*-hexane, diethyl ether and methanol	The extracts showed moderate to high cytotoxicity against L1210 (leukemia), 3LL (lung), DU145 (prostate), MCF-7 (breast), K562 (leukemia), and U251 (glioblastoma) with IC_{50} values ranging from 7.9 to 25.3 μg/mL.	[12] [12]
19	*Parmelia omphalodes* (L.) Ach	Candeleda, Castilla y Leon, Spain July 2015	Methanol	The extract showed low cytotoxicity against MCF-7 (breast) and HepG2 (liver) cell lines with LD_{50} values of >100 μg/mL.	[11]
20	*Parmelia perlata*	Brittany During the summer 2000	*n*-hexane, diethyl ether, and methanol	Among the studied extract, the *n*-hexane extract showed the best cytotoxicity against L1210 (leukemia) cells (IC_{50} value of 4.9 μg/mL).	[12]
21	*Platismatia glauca*	Piła commune, Poland October 2018	Dichloromethane, acetone, methanol, methanol-water (1:1), and water extracts	The extracts were screened against T98G and glioblastoma multiformeU-138MG cells. It was found that IC_{50} values range from 31.8 ± 10.4 to 65.8 ± 18.9 μg/mL against T98G and from 65.7 ± 5.0 to 88.2 ± 15.2 μg/mL against U-138 MG.	[18]
22	*P. glauca*	Limousin (near Meymac), France	*n*-hexane, diethyl ether, methanol	The *n*-hexane extract was significantly cytotoxic against glioblastomaU251cell with an IC_{50} value of 5.5 μg/mL.	[12]

23	*Ramalina cuspidata*	Brittany During the summer 2000	*n*-hexane, diethyl ether, and methanol	The *n*-hexane extract exhibited good cytotoxicity against 3LL (lung) cell line with an IC_{50} value of 5.7 μg/mL.	[12]
24	*Umbilicaria crustulosa*	Kopaonik, Serbia September of 2010	Acetone	The extract showed good cytotoxic activity against FemX cell (lung)and LS174 (colon) cell with IC_{50} values of 69.43,97.82 μg/mL, respectively.	[19]
25	*Umbilicaria cylindrica*	Kopaonik, Serbia September of 2010	Acetone	The extract showed good cytotoxic activity against FemX (lung) cell (IC_{50} = 28.45 μg/mL) and LS174 (colon) cell (IC_{50} = 40.55 μg/mL).	[19]
26	*Umbilicaria polyphylla*	Kopaonik, Serbia September of 2010	Acetone	The extract showed good cytotoxic activity against FemX (lung) and LS174 cells with IC_{50} values of 35.05 and 39.45 μg/mL, respectively.	[19]
27	*Usnea aurantiacoatra (Jacq.) Bory*	Chile January 2008	Methanol	The extract exhibited low cytotoxicity against MCF-7 (breast) and HepG2 (liver) cell lines (LD_{50} values of >100 μg/mL).	[11]
28	*Usnea contexta* Motyka	Chile January 2005	Methanol	The extract was found to be inactive against MCF-7 (breast), but exhibited strong cytotoxicity against HepG2 (liver) cell lines (LD_{50} = 54 μg/mL).	[11]
29	*Usnea rubicunda*	UK Summer 2000	*n*-hexane, diethyl ether and methanol	The *n*-hexane extract exhibited potent cytotoxicity against DU145(prostate) cell line (IC_{50} = 8.1 μg/mL).	[12]
30	*Usnea filipendula* Stirt.	Mount Uludag, Bursa, Turkey May 2010	Methanol-water	The extract displayed a dose and time-dependent antiproliferative effect against HCT-15 (colon) and HT-29 (colon) cells, with IC_{50} values of 17.92 and 41.87 μg/mL, respectively. The extract induced apoptosis in both cell lines.	[20]
31	*Usnea longissima*	Osla region in Uttarkashi district of Uttarakhand state of India November 2013	Acetone extract	The crude acetone extract was tested against several human cancer cell lines (HeLa (cervical), A-549 (lung), MCF-7 (breast), DU-145 (prostate), HEK-293 a Human embryonic kidney) to reveal that the extract showed potent activity against A549 (lung) cells.	[21]

(*Continued*)

Table 14.1 (Continued)

Serial number	Species	Location/date collected	Solvent	Anticancer activity	References
32	*Usnea undulata* Stirt.	Lam Dong province, Vietnam December 2017	*n*-hexane, acetone and methanol extracts	The *n*-hexane extract exhibited promising cytotoxic activity against MCF-7 (breast), ($IC_{50} = 5.26 \pm 0.13\,\mu g/mL$) and NCI-H460 (lung) ($IC_{50} = 6.83 \pm 0.64\,\mu g/mL$). On the other hand, the acetone extract displayed moderate activity ($IC_{50} = 66.44 \pm 1.37\,\mu g/mL$ and $92.14 \pm 1.49\,\mu g/mL$, respectively), while the methanol extract showed no activity.	[22]
33	*Usnea barbata* (L.) F. H. Wigg	Călimani Mountains, Romania	Dry acetone extract	The extract was found to inhibit human tongue squamous cell carcinoma (CAL 27) after 24 h of treatment; this response is directly proportional to the tested extract concentrations.	[23]
34	*Usnea barbata*	Traditional herbal medicine pharmacy, Jakarta, Indonesia July, 2014	methanol extract	The cytotoxicity of the extract was evaluated against oral cancer cells (Ca9-22, OECM-1, CAL 27, HSC3, and SCC9). It was suggested that the extract showed potential killing activity for the tested cells through apoptosis and DNA damage, but minimally affected normal oral cell lines (HGF-1).	[24]
35	*Usnea barbata* (L.)Weber ex F.H. Wigg (U. barbata)	Călimani Mountains, Romania March 2020	Canola oil (UBO) and cold-pressed canola seed oil (CNO) extracts	Cytotoxic activity was evaluated using a brine shrimp lethality (BSL) test after *Artemia salina* (A. salina) larvae. The results showed that both UBO and CNO emulsions induced dissimilar and progressive morphological changes in *A. salina* larvae, and UBO activity was higher than CNO.	[25]

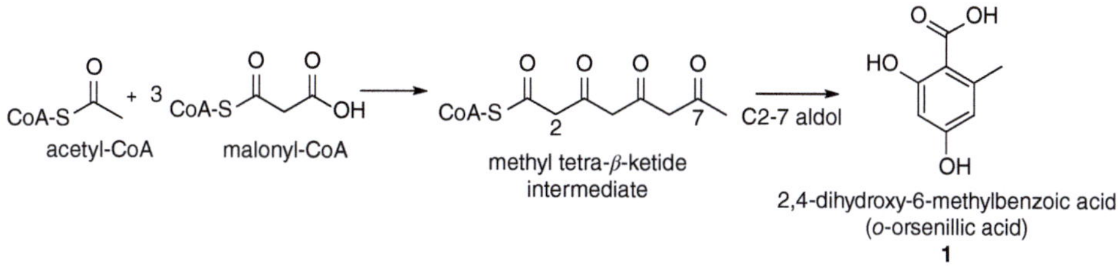

Figure 14.2 Biosynthesis depside monomer *o*-orsenillic acid (**1**).

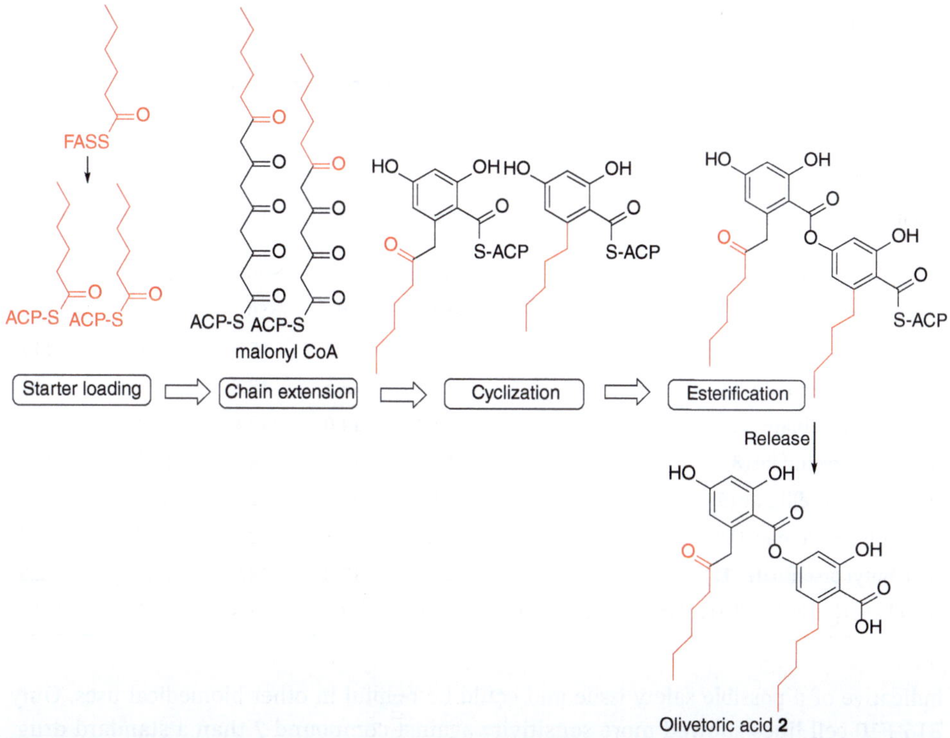

Figure 14.3 Proposed biosynthesis mechanism of oliveric acid **2** [31].

The depside congeners, orsenilates, were also biosynthesized by *Parmelia subrudecta* Nyl. Either the acid or its ester derivate (e.g. orsellinic acid methyl ester **3**) indicated moderate cytotoxicity with IC_{50} values of 50 µg/mL against all L-929 mouse fibroblast cells, K-562 human leukemia cells, and HeLa human cervix carcinoma cells [32]. Further orsenilate derivatives were produced from lecanoric acid **4** (isolated from *Parmotrema tinctorum* [Nyl.] Hale) with the corresponding alcohols to produce methyl, ethyl, *n*-propyl, *n*-butyl, *n*-pentyl, *iso*-propyl, *sec*-butyl, and *tert*-butyl orsellinates **3**, **5–11** (Figure 14.4) [33]. The antitumor activities of these derivatives were evaluated against HEp-2 (larynx carcinoma), Vero (African green monkey kidney), B16-F10 (murine melanoma), MCF7 (breast carcinoma), and 786-0 (kidney carcinoma) with significant to moderate activities (Table 14.2). The study suggested that *n*-butyl ester provided optimum lipophobicity with the highest cytotoxicity to all of the tested cell lines. However, cytotoxicity to normal cells was

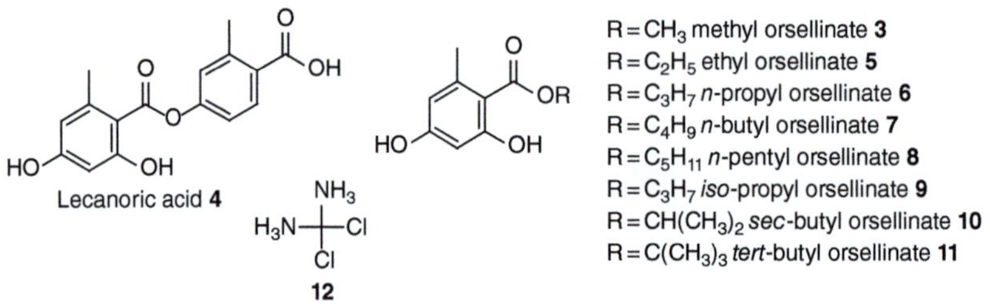

Figure 14.4 Orsenillate derivatives **3–11.** A standard anti-cancer drug (positive control), cisplatin **12**.

Table 14.2 Cytotoxicity activities of lecanoric acids and its orsellinate derivatives.

	IC_{50} (μg/mL)				
Compound	**Hep-2**	**MCF7**	**786-0**	**B16-F10**	**Vero**
Lecanoric acid (**4**)	>50	>50	>50	>50	>50
Methyl orsellinate (**3**)	>50	>50	>50	>50	>50
Ethyl orsellinate (**5**)	31.2	70.3	47.5	64.8	28.1
n-Propyl orsellinate (**6**)	13.5	23.5	18.8	33.4	12.1
n-Butyl orsellinate (**7**)	7.2	14.0	12.6	11.4	18.9
n-Pentyl orsellinate (**8**)	17.3	18.5	14.3	17.9	47.3
iso-Propyl orsellinate (**9**)	34.9	30.2	22.0	26.5	15.0
sec-Butyl orsellinate (**10**)	8.9	16.2	20.6	17.3	26.3
tert-Butyl orsellinate (**11**)	10.2	17.8	14.9	15.4	2.8
Cisplatin ([$Pt(NH_3)_2Cl_2$] (**12**) (positive control)	1.8	3.22	1.4	12.5	2.8

indicative of a possible safety issue and could be helpful in other biomedical uses. Only B16-F10 cell lines showed more sensitivity against compound **7** than a standard drug, cisplatin **12**.

The depside congener (*o*-orsinolic acid unit) can undergo esterification with other units to form didepside and tridepside, namely lecanoric acid (**4**) and gyrophoric acid (**13**). Starting point diversities in the biosynthesis were able to produce different depsides (Figure 14.5). However, lecanoric acid (**4**), isolated from Brazilian folious lichen *Parmelia tinctorum* (Nyl.) Hale, exerted less cytotoxicity against HeLa Human cervix carcinoma HEp-2, breast cancer MCF7, 786-0 kidney carcinoma, murine melanoma B16-F10 and Vero normal cell lines with an IC_{50} value of more than 50 μg/mL [33]. The lecanoric acid obtained from *P. subrudecta* grown on Canary Island in Spain also showed only moderate activity against cancer cell lines (K-562 human leukemia cells, and HeLa human cervix carcinoma) as well as normal L-929 mouse fibroblast cells with GI_{50} value of 50 μg/mL each [32]. In addition, the tridepside, gyrophoric acid (**13**) isolated from *Umbilicaria hirsute* (collected from Harmanec, central Slovakia) was also an ineffective growth inhibitor of cancer cell lines (A2780 [ovary], HeLa [cervic], MCF-7 [breast], SK-BR-3 [breast], HT29 [colon],

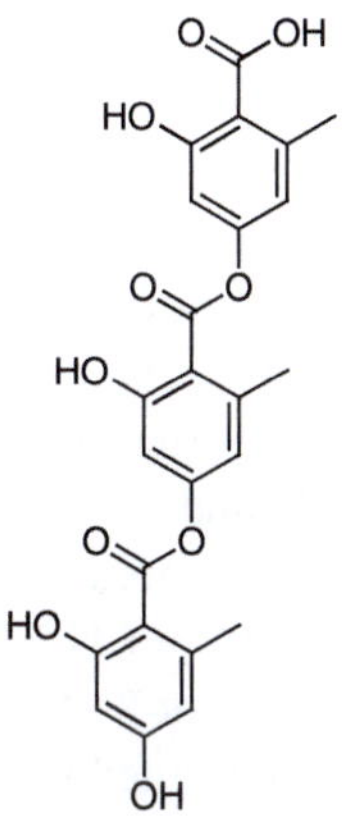
gyrophoric acid
13

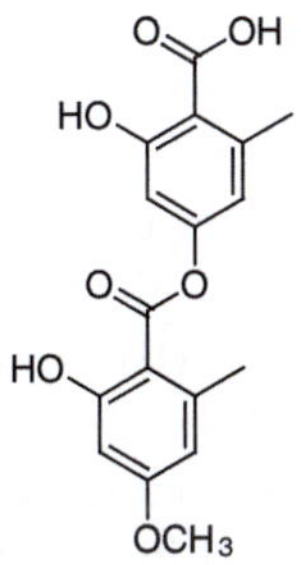
evernic acid
14

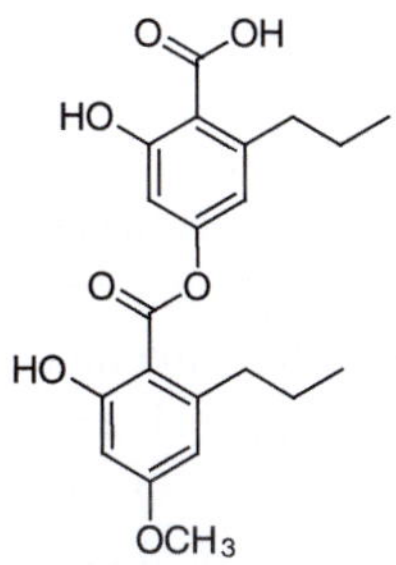
divaricatic acid
15

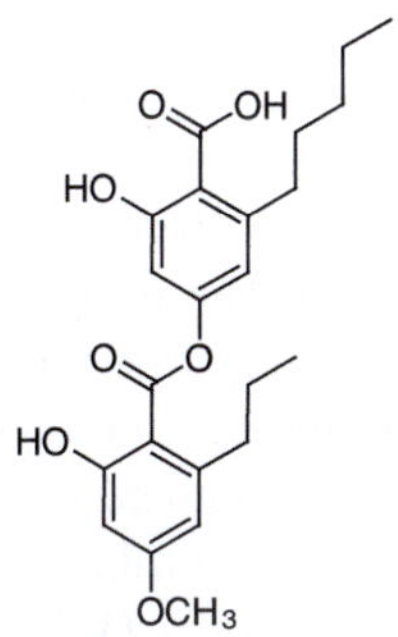
stenosporic acid
16

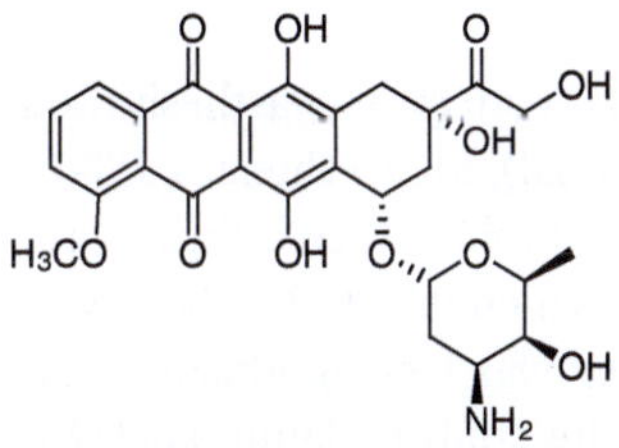
doxorubicin
17

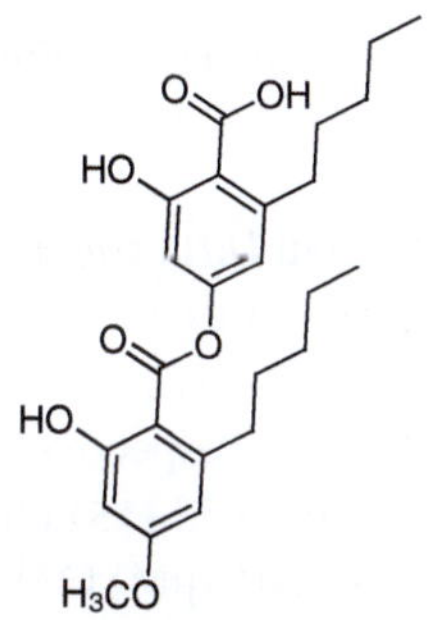
perlatolic acid
18

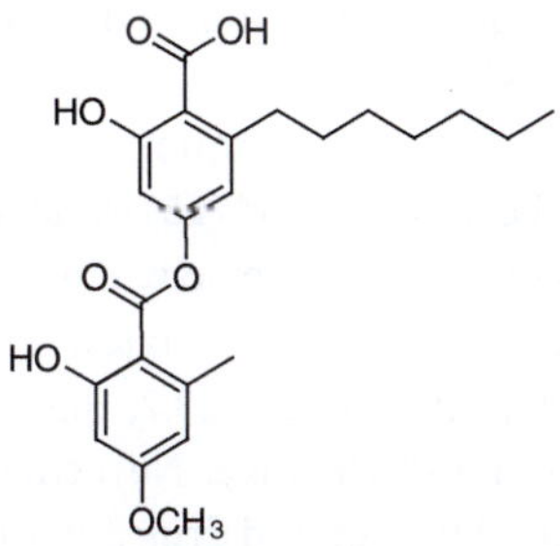
sphaerophorin
19

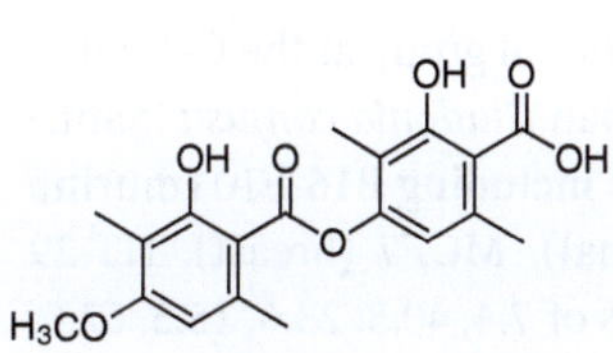
barbatic acid
20

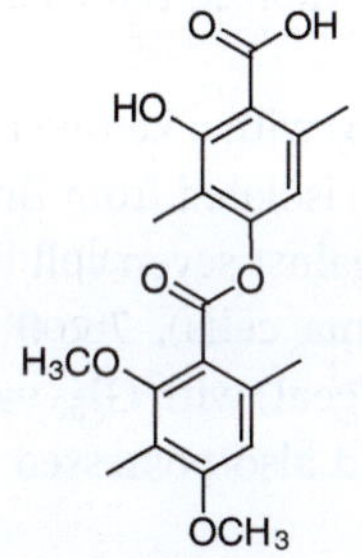
diffractaic acid
21

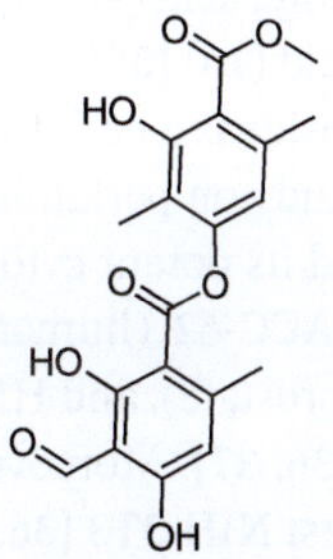
atranorin
22

Figure 14.5 Bioactive depsides exerted by lichen.

HCT-116 p53+/+ [colon] and HCT-116 p53/ [colon]) with the majority IC_{50} values above 200 μM [34]. The study on the molecular mechanism of action showed that lecanoric acid (**4**) slightly reduced Axin2 expression in HCT116 cells [35].

The C-4′ methoxylated derivate of lecanoric acid, evernic acid (**14**), indicated a stronger anticancer activity. From the Indonesian folious lichen, *Physcia millegrana* Degel Nugraha et al. isolated evernic acid (**14**), and a derivative, divaricatic acid (**15**), which the later was the major constituent and presented as readily precipitated solids from the methanol extract with yield of 1.2 mg of dried sample [1]. This study also reported evernic acid (**14**) from *Parmelia aurulenta* along with its derivate, stenosporic acid (**16**). These didepsides were evaluated against three cancer cell lines, including HepG2 liver cancer, A549 lung cancer, and HL-60 leukemia cell lines with IC_{50} values of 68, 180, and 197 μM, for evernic acid (**14**),15, 72, 6 μM for divaricatic acid (**15**), and for stenosporic acid (**16**) were 15, 28, and 6 μM, respectively [1]. The study revealed a consistent cytotoxicity towards all three cell lines tested, with the most promising responses observed for divaricatic (**15**) and stenosporic acid (**16**) in the acute promyelocytic leukemia cell line (HL-60 cells). Compared to evernic acid (**14**), the lengthening of the alkyl chain (R1 = R2 = C_3H_7 vs. R1 = R2 = CH_3) in divaricatic acid (**15**), contributed to increasing toxicity. Similarly, the extension of R1 to the pentyl group stenosporic (**16**) (vs. the propyl group for R1 in evernic acid) resulted in even greater toxicity among the di-depsides. The resultant increased lipophilicity of the molecules may be beneficial to molecular transport across the cell membrane and allow a greater contribution to the observed cytotoxicity [1]. In addition, the free acid groups on the depside derivatives are unlikely to contribute to the anticancer activity as also shown by less efficacy of lecanoric acid (**4**).

Divaricatic acid (**15**) isolated from Brazilian *Dirinaria aspera* (H. Magn.) Awasthi showed significant cytotoxicity effect against six cell lines, i.e. 786-0 (renal), MCF7 (breast), HT-29 (colon), PC-3 (prostate), Hep2 (liver), NIH/3T3 (normal fibroblast) cell lines with GI_{50}valuesof 72.2, 20.2, 16.6, 9.8, 14.2, 37.4 μM, respectively, in which the PC-3 cell line was the most sensitive. Nevertheless, divaricatic acid (**15**) efficacy was less significant compared to standard drug doxorubicin (**17**) (GI_{50} 0.5 μM) [36]. In this study perlatolic acid (**18**) exerted lower activity with a GI_{50} value of 23, 15, 62, 16, 29, 59 μM. These outcomes suggested elongation at C-6’ was crucial to the cytotoxicity of evernic acid (**14**) derivatives since the perlatolic acid (–C_5H_7) (**18**) experienced reduced efficacy by 1.5–4 fold compared to divaricatic acid (**15**) [36].

A unique depside is a perlatolic acid (**18**) with a ketone functional group at the C-3′ alkyl side chain. A study on perlatolic acid (**18**) isolated from Brazilian *Cladonia confusa* (Sant.) Folmm revealed its potent cytotoxicity against seven cell lines including B16-F10 (murine melanoma), UACC-62 (human melanoma cells), 786-0 (renal), MCF7 (breast), HT-29 (colon), PC-3 (prostate), and HEp2 (laryngeal) with GI_{50} values of 7.4, 40.5, 23.6, 15.5, 62.0, 16.1, 29.8 μM [36, 37]. Moreover, the acid also possessed a moderate cytotoxicity against normal fibroblast NIH-3T3 [36, 37].

In addition to this, another evernic acid (**14**) derivative, sphaerophorin (**19**) (isolated from *Sphaerophorus globosus* and the genus Psoroma, i.e. *Psoroma reticulatum*, *Psoroma pulchrum*, *Psoroma palladium*, from Antarctic Chile) with a heptyl side chain at C-6 and a methyl substituent at C-6′ significantly inhibited M14 melanoma cell at 12.5–50 μM concentrations and induced cell death, as demonstrated by a DNA fragmentation experiment, caspase-3 activity, and ROS generation increase [38]. Barbatic acid (**20**) (isolated from Vietnamese *Usnea aciculifera* Vain.), the C-3 and C-3′ methylated derivative of lecanoric acid (**4**) showed moderate cytotoxicity against MCF-7 breast cancer cell, HeLa cervix cancer cell and NCI-H460 hypotriploid cancer cell lines with percentage growth inhibition values of 50.8, 52.7, and 15.2% at the concentration of 100 μg/mL [39]. Another lecanoric acid (**4**) derivative, difractaic acid (**21**), in which the methyl substituents are at the C-3 and C-3′positionsand the methoxy at C-2′, was isolated from *Parmotrema dilatatum* (Vain.)

Hale originated from Aquidauana county-Brazil. Difractaic acid (**21**) inhibited 50% cell growth of seven human cancer cell lines, i.e. B16-F10 (murine melanoma), UACC-62 (human melanoma cells), 786-0 (renal), MCF7 (breast), HT-29 (colon), PC-3 (prostate), and HEp2 (laryngeal) with GI_{50} values of 67.8, 66.0, 81.3, 97.9, 74.1, 58.6, 98.9 μM, respectively [36, 37]. This study also showed that a higher concentration was required to cause cell death while the same cell lines indicated sensitivity to doxorubicin (**17**) (a standard drug) with respected GI_{50} values of 3.2, 30.7, 0.4, 0.09, 0.1, 0.5, 1.0, 0.9 μM, respectively [36, 37]. Diffractaic acid (**21**), isolated from Vietnamese *U. aciculifera* Vain., showed moderate sensitivity against NCI–H460 human lung cancer with percentage growth inhibition of 79.9% at a concentration of 100 μg/mL [39]. The depside also possessed moderate cytotoxicity against HeLa cervix adenocarcinoma and HCT-116 colon carcinoma with IC_{50} values of 64.6 and 42.2 μM [40] but neither possessed significant antioxidant activity nor affected ROS production in the Hela cell experiment [40]. On the other hand, toxicity toward normal cells evaluation indicated a mild cytotoxicity against mouse non-tumoral fibroblast cell line (NIH-3T3) with an IC_{50} value of >100 μM, which was detrimental to further studies [40]. Further evaluation using human blood tissues showed that diffractaic acid (**21**) caused no chromosome aberration and micronucleus production implying the depside was non-genotoxic [41].

Another lichen didepside with comparable cytotoxicity with lecanoric acid (**4**) or gyrophoric acid (**13**) is atranorin (**22**), where the structure possesses modifications from the basic depside template, lecanoric acid, with methyl and formyl substituents at the C-3 and C-3′ positions. Atranorin (**22**) from folious lichen *P. dilatatum* (Vain.) Hale originated from Aquidauana county-Brazil, was reported to have moderate-low efficacy against seven cell lines including B16-F10 (murine melanoma), UACC-62 (human melanoma cells), 786-0 (renal), MCF7 (breast), HT-29 (colon), PC-3 (prostate), and HEp2 (laryngeal) and normal fibroblast cell NIH/T3T with GI_{50} values of >667.8, 393.3, 438, 754, 993, 526, 1093, 745 μM, respectively [36, 37]. Other cytotoxicity experiments on cell lines including A2780 ovarian cancer cell, HeLa cervix cancer cell, MCF-7 breast cancer, SK-BR-3 breast cancer cell, HT-29 colorectal adenocarcinoma cell, HCT-116 (p53+/+) colon cancer cell, HCT-116 (p53) colon cancer cells, HL-60 leukemia cancer cell, and Jurkat T cell leukemia confirmed its moderate to low activity with IC_{50} value of 197, >200, >200, >200, >200, >200, 197.5, 93.5, 181.6 μM, respectively [34].

The molecular mechanism of atranorin (**22**) in inducting apoptosis was evaluated on A2780 human ovarian cancer cell and HT-29 human colorectal adenocarcinoma cell lines in which the didepside induced apoptosis through massive disruption of mitochondrial potential, caspase-3 activation and phosphatidylserine externalization. In addition, the lichen metabolite also contributed to the increased intracellular level of both reactive oxygen (ROS) and nitrogen species production (RNS), which are responsible for its cytotoxic activity [42]. Further studies on protein expression associated with apoptosis (PARP, p53, Bcl-2/Bcl-xL, Bax, p38, pp38) concluded that atranorin (**22**) is able to activate programmed cell death in the cell lines model [42].

14.3.2 Depsidone and Pseudodepsidone

Depsidones are common lichen compounds composed of the depside congener, which undergoes esterification and further cyclization forming fused rings. Similar to depside, biosyntheses start with starter loading and chain extension followed by cyclization and esterification prior to oxidative cyclization to form the depsidone fused ring (Figure 14.6).

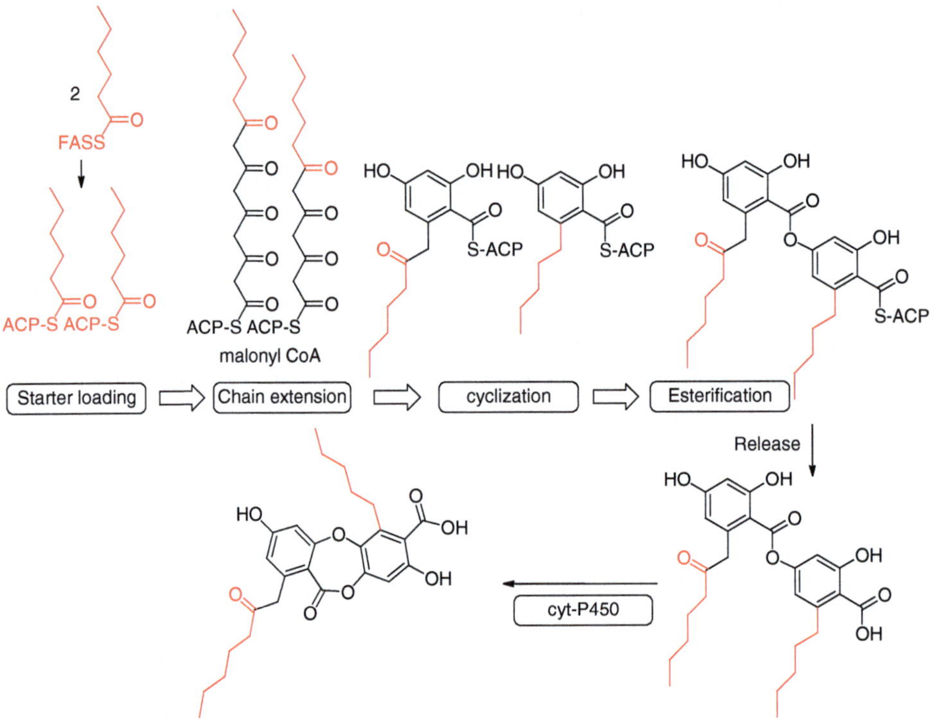

Figure 14.6 Putative biosynthesis of physodic acid [31].

Some notable depsidones (Figure 14.7) were reported as promising anticancer candidates. Physodic acid **23**, a cyclization product of olivetoric acid isolated from *Pseudoevernia furfuraceae* (collected at Kopaonik, Serbia), possessed high cytotoxic activity against human melanoma FemX and human colon carcinoma LS174 cell lines with IC_{50} values of 19.52 and 17.89 μg/mL, respectively [43]. This compound was also able to reduce the cancer cell progression represented by the increase of sub-G1 DNA, which is related to cell death [43]. Despite the moderate toxicity of physodic acid **23** against Hela cancer cells (IC_{50} 171 μg/mL), a study on *Hypogymnia physodes* (L.) Nyl. (collected from Donje Vlase, Serbia) revealed that 3-hydroxy physodic acid (**24**) displayed better cytotoxicity with an IC_{50} value of 97 μg/mL [44]. Another depsidone with a ketone-containing side chain was lobaric acid (**25**) produced by Chilean *Stereocaulon alpinum* Laurer ex Funck, and exerted insignificant cytotoxicity against MCF-7 (breast adenocarcinoma), HeLa (cervix adenocarcinoma), HCT-116 (colon carcinoma), and NIH-3T3 (normal cell, mouse embryonic fibroblast) cell lines with IC_{50} values of 93.2, 78.0, >100 and >100 μM, respectively [40]. A molecular cytotoxicity study of lobaric acid (**25**) showed its capacity to reduce DNA synthesis based on thymidine uptake measurement on three malignant cell lines (T-47D and ZR-75-1 from breast carcinomas and K-562 from erythro-leukemia) with significant malignant cell death at concentrations above 30 μg/mL [45].

Brandão et al. successfully isolated protocetraric acid (**26**) from the Brazilian *Usnea* species, which showed high cytotoxic efficacy against UACC-62 (human melanoma cells)

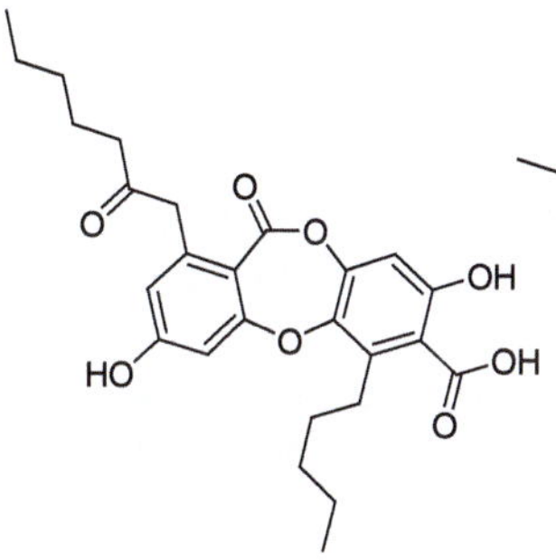

physodic acid
23

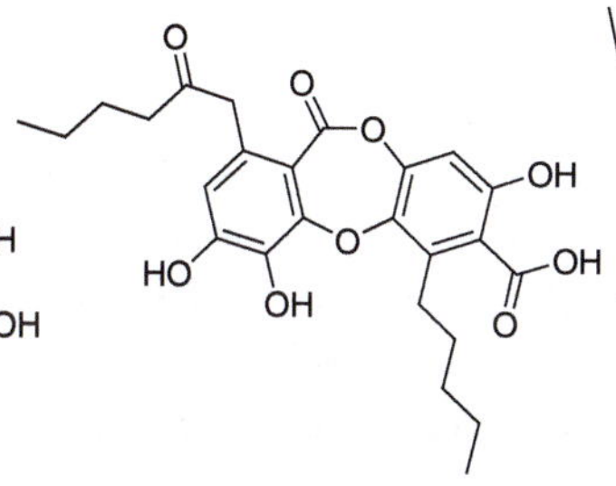

3-hydroxyphysodic acid
24

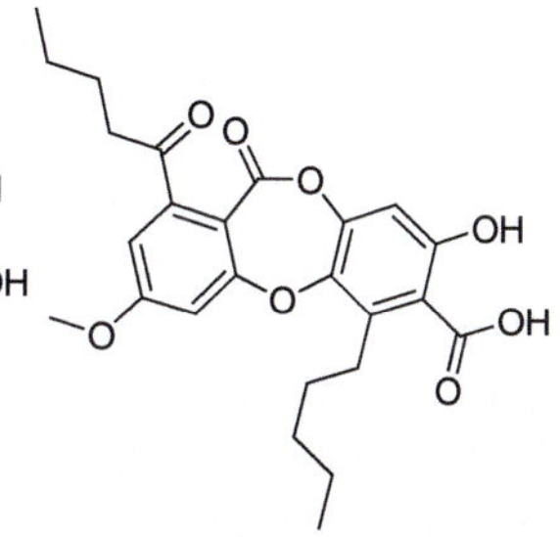

lobaric acid
25

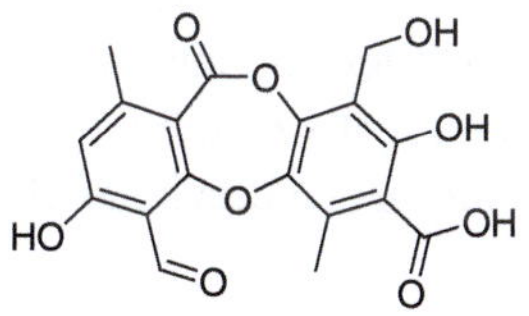

protocetraric acid
26

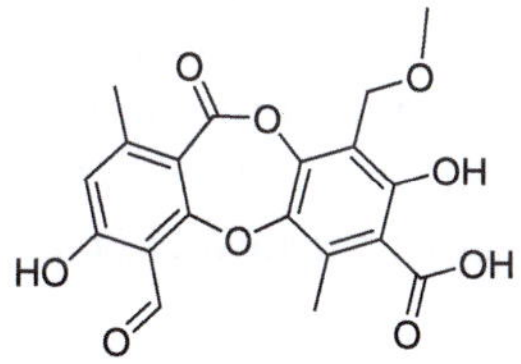

9′-(*O*-methyl)protocetraric acid
27

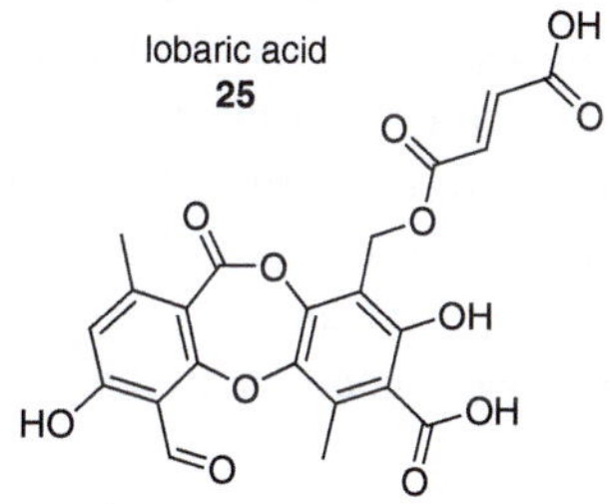

fumarprotocetraric acid
28

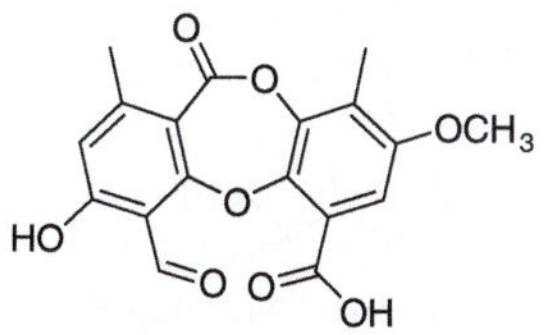

psoromic acid
29

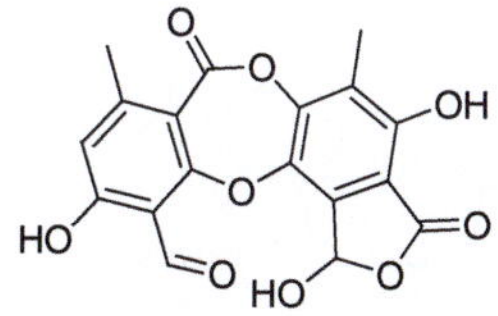

norstictic acid
30

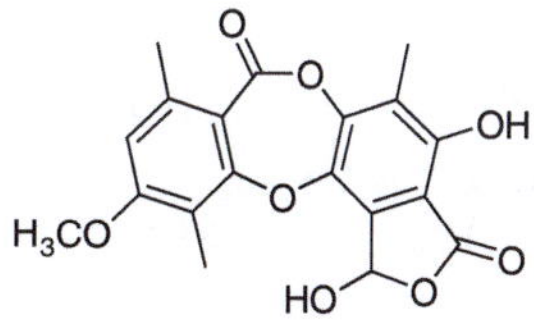

hypostictic acid
31

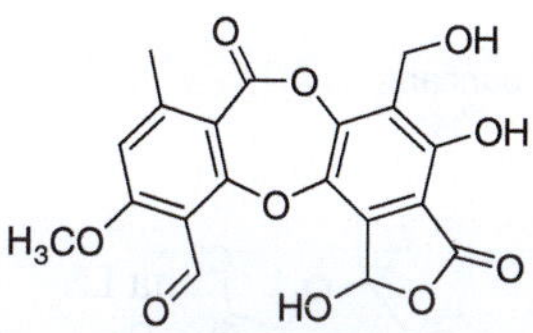

salazinic acid
32

Figure 14.7 Notable depsidone molecules produced by lichen, *Pseudoevernia furfuraceae, Hypogymnia physodes, Stereocaulon alpinum, Usnea* sp., *Parmotrema tsavoense, Pseudoparmelia sphaerospora,* and *Parmotrema cetratum.*

and B16-F10 (murine melanoma cells) with GI_{50} values of 1.4 and 64.0 μM, respectively. This activity was comparable to a standard drug, doxorubicin (**17**) (GI_{50} 0.81 and 0.43 μM, respectively) [37]. Nevertheless, 9′-*O*-methylprotocetraric acid (**27**) and fumarprotocetraric acid (**28**) obtained from either Vietnamese *Parmotrema tsavoense* (Krog and Swinscow) *Cladonia convolute* (Lam.) Anders (Cressensac, France) possessed insignificant cytotoxicity against HepG2 (liver), MCF-7 (breast), NCI-H460 (lung), and L1210 (murine leukemia) cell lines, and hence indicated that alteration of the hydroxyl group at C-9′ was disadvantages for the cytotoxicity of protocetraric acid (**26**) [46].

Compared to protocetraric acid, other aldehyde-containing depsidones, i.e. psoromic acid (**29**) and norstictic acid (**30**), were less active against UACC-62 (human melanoma), cells with GI_{50} values of 102.2 and 88.4 μM. In addition, the depsidones were not sensitive to B16-F10 (murine melanoma) cells [37].

Alexandrino et al. successfully isolated depsidone, hypostictic acid (**31**) and salazinic acid (**32**) from Piraputanga-Brazil origins *Pseudoparmelia sphaerospora* (Nyl.) Hale and *Parmotrema cetratum* (Ach.) Hale, respectively. *In vitro* studies displayed the significant antiproliferative activity of hypostictic acid against K562 (leukemia)s, B16-F10 (melanoma), and 786-0 (renal) cells with GI_{50} values of 2.20, 13.78, 14.24 μM, respectively. **32** possessed less efficacy against similar cell lines with GI_{50} values of 64.36, 67.91, and 78.64 μM, respectively [47].

Interestingly, the lichen clade was also capable of producing halogenated secondary metabolites with most of them being chlorinated compounds, as is common in mycobiome [48]. Halogenated depsidones (Figure 14.8), vicanicin (**33**) and pannarin (**34**), were reported to possess anticancer potency. A bioprospecting study on Chilean lichen *Psoroma pallidum* Nyl. and *Psoroma pulchrum* Malme resulted in the isolation of vicanicin, which possessed moderate to insignificant activity against HCT-116 colon, HeLa cervix, MCF-7

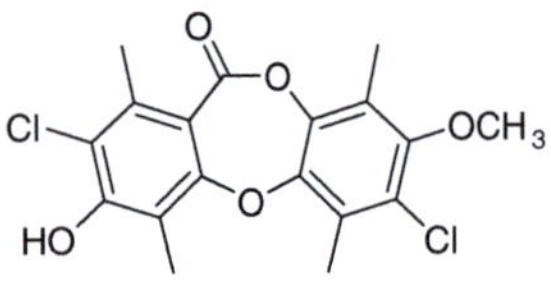

vicanicin
33

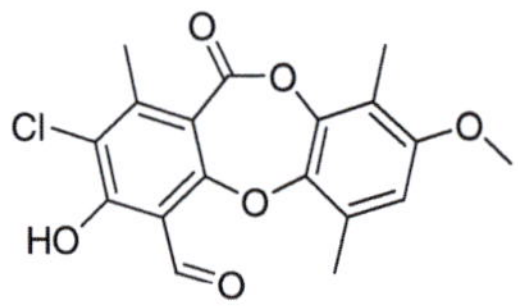

pannarin
34

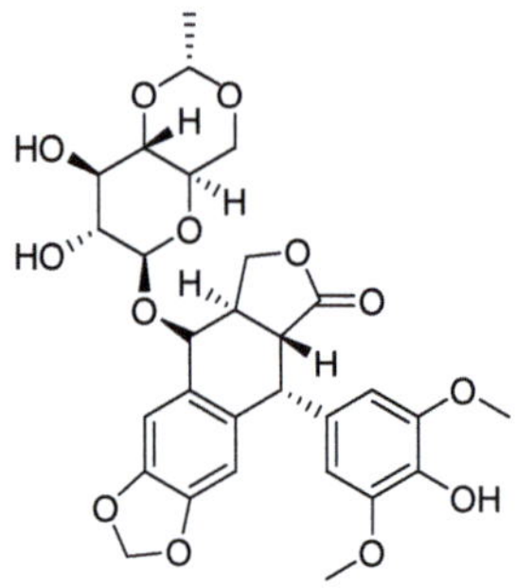

etoposide
35

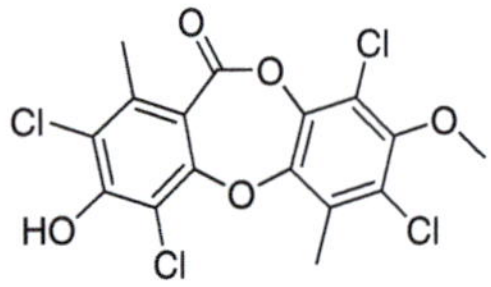

diploicin
36

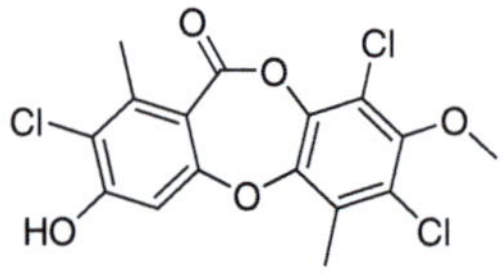

dechlorodiploicin
37

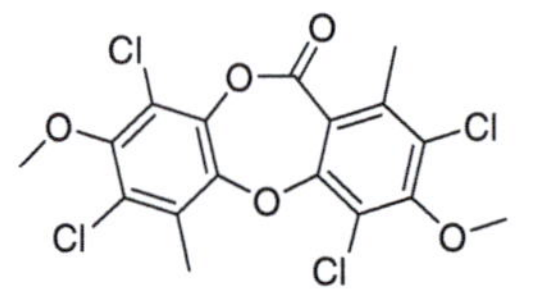

4-*O*-methyldiploicin
38

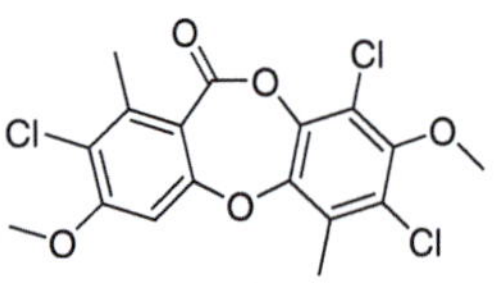

4-*O*-methyldechlorodiploicin
39

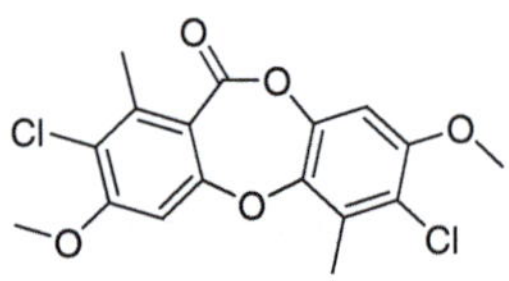

scensidin
40

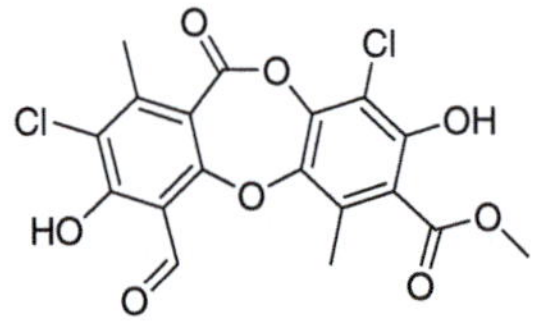

physciosporin
41

Figure 14.8 Biologically active halogenated depsidone molecules produced by lichen *Psoroma pallidum* Nyl., *Psoroma pulchrum* Malme, *Diploicia canescens* (Dicks.) A., *Pseudocyphellaria coriacea* (Hook f. and Taylor) D.J. Galloway and P. James *Pseudocyphellaria coriacea* (Hook f. and Taylor) D.J. Galloway and P. James.

breast cancer cell lines and normal fibroblast NIH-3T3 with IC_{50} values of 40.5, 67.0, > 100, > 100 μM, respectively [40]. Some of the results were comparable to etoposide (**35**) (IC_{50} values of 40.3, 45.2, 67.6, 2.3 μM) [40]. Another study of vicanicin (**33**) isolated from Chilean *Psoroma dimorphum* showed the depsidone activating apoptotic process on DU-145 and LNCAP human prostate cancer cell lines with a dose–response relationship in the range of 6.25–50 μM [49]. This dose–response manner was also observed for another chlorinated depsidone, pannarin (**34**), isolated from Chilean *P. palladium* and *P. pulchrum* [38]. **34** was able to reduce M14 cell vitality to as low as 17% at a concentration of 50 μM [38]. Further study revealed that **34** also induced apoptosis in DU-145 prostate cancer cell experiment [50].

Phytochemical investigation of *Diploicia canescens* (Dicks.) A. revealed halogenated depsidones, i.e. diploicin (**36**), dechlorodiploicin (**37**), 4-*O*-methyldiploicin (**38**), 4-*O*-methyldechlorodiploicin (**39**) and scensidin (**40**). Cytotoxicity evaluation conducted on **33** and **34** indicated significant efficacy with IC_{50} values of 4.0 and 1.7 μM against B16 murine melanoma and IC_{50} values of >10, 4.3 μM against normal HaCaT human keratinocyte cells, respectively (etoposide IC_{50} values 0.28 and 0.55 μM, respectively) [51].

Lichen bioprospecting on 13 Chilean lichen species resulted in the discovery of physciosporin (**41**) from folious *Pseudocyphellaria coriacea* (Hook f. and Taylor) D.J. Galloway and P. James which displayed significant inhibition of lung cancer cell A549 motility. Further studies revealed its mechanism was through inhibition of epithelial-mesenchymal marker [52].

14.3.3 Dibenzofuran

Dibenzofuran is a mancude ring system composed of two benzene rings fused to a central furan ring with 2,3- and 4,5-positions of the furan ring [53]. One of the naturally occurring dibenzofuran derivates is usnic acid, which is biosynthetically started from condensation of acetyl-CoA and malonyl-CoA to form a methyl tetra-β-ketide intermediate followed by one carbon addition to produce methylphloroacetophenone. This congener undergoes a coupling reaction through a radical reaction in which molecular stabilization leads to the formation of usnic acid **42** (Figure 14.9) [54].

Many dibenzofuran derivates with cytotoxic effects were isolated from various lichen families (Figure 14.10). The well-known dibenzofuran representing the importance of lichen as a source of bioactive compounds is unsic acid (**42**), which was common in *Cladoniaceae* and *Parmeliaceae* genera. **42** isolated from Chilean *Cladonia lepidophora* Ahti and Kashiw was toxic against three cancer cell lines, MCF-7 (breast adenocarcinoma), HeLa (cervix adenocarcinoma), and HCT-116 (colon carcinoma) with IC_{50} values of 17.7, 23.7, 75.7 μM, respectively, but inactive against NIH-3T3 (mouse embryonic fibroblast) normal cell line (IC_{50} >100 μM) [40]. The *levo* isomer of (−)-usnic acid (**43**) isolated from *Cladonia convoluta* (Lam.) Anders (France origin) displayed moderate efficacy against L1210, 3LL, DU145, MCF7, K-562, and U251 cell lines with respected IC_{50} values of 6, 12.1, 15.8, 17.8, 8.2, and 6.8 μg/mL. The study also reported that this compound induces L1210 murine leukemia cells in a dose–response manner [46b]. Significant cytotoxicity of usnic acid (**42**) (isolated from Serbian *U. barbata* (L.) Weber ex F.H. Wigg) was also shown on FemX (lung) and LS174 (colon) cell lines with IC_{50} values of 12.72 and 15.66 μg/mL,

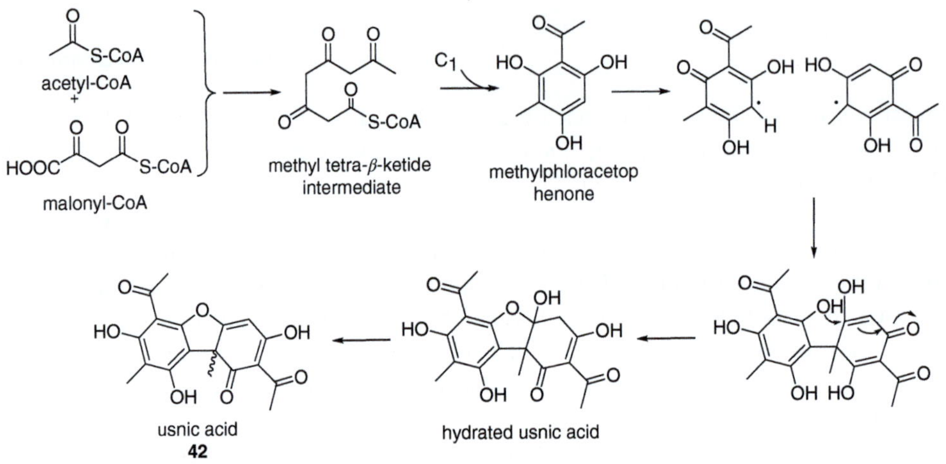

Figure 14.9 Putative biosynthesis mechanism of usnic acid [54].

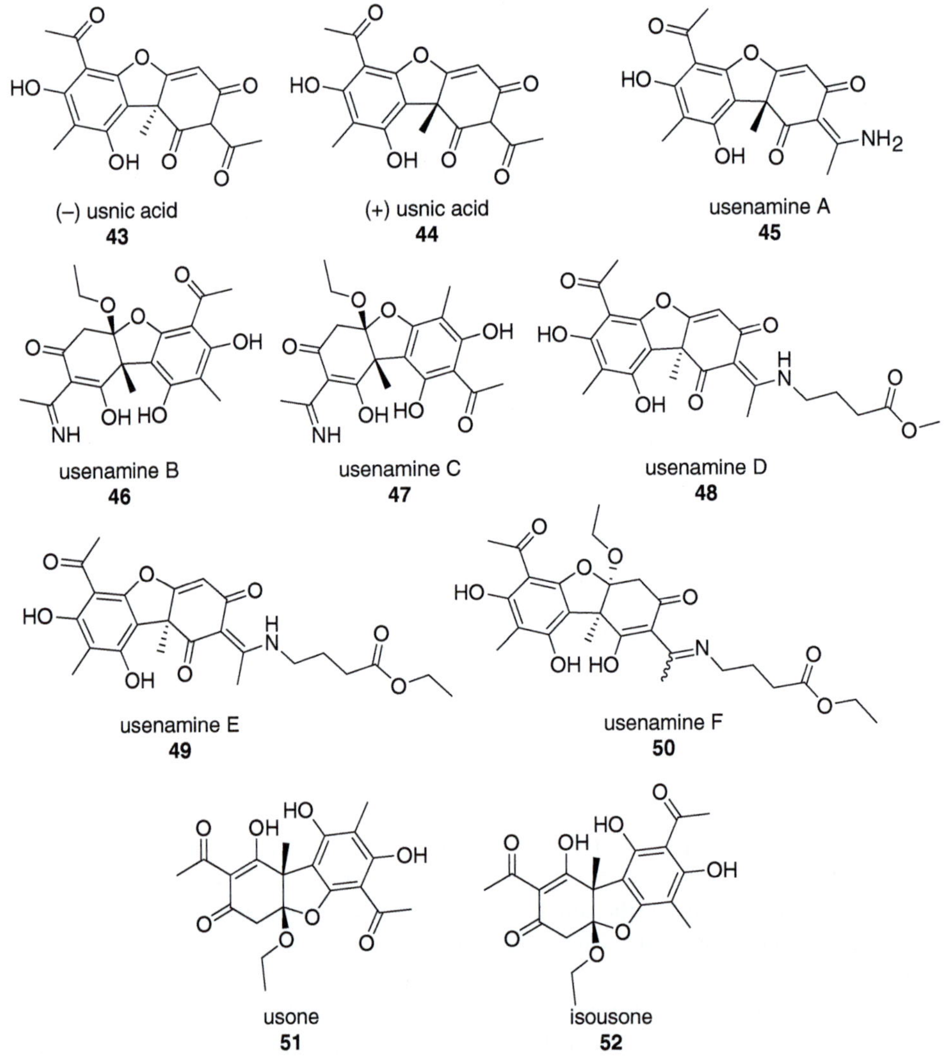

Figure 14.10 Cytotoxic dibenzofuran of lichens *Cladonia arbuscula, Cladonia convolute, Cladonia lepidophora, Parmelia tinctorum, Usnea barbata,* and *Usnea longissima.*

respectively [55]. A cytotoxicity experiment of (+)-usnic acid (**44**) (isolated from *Cladonia arbuscula* [Wallr.] Rabenh of Iceland origin) using T-47D (breast) and Capan-2 (pancreas) cells demonstrated the capability of the acid to significantly inhibit DNA synthesis in which flow cytometric data analysis confirmed S-phase inhibition as well as cell size reduction [56]. Nevertheless, **42** possessed moderate activity against UACC-62 (human melanoma cells), B16-F10 (murine melanoma) cells, and 3T3 (fibroblast cells) with GI_{50} values of 91.5, 137.7, and 120.0 μM, respectively [37]. The (+)-usnic acid (**44**) (isolated from *P. tinctorum* Despr. ex Nyl. of Nepal origin) indicated significant cytotoxicity against human keratinocyte cell line HaCaT with an IC_{50} value of 1.7 μM [57].

Despite the high possibility of **42** being developed as an anticancer agent, the process was deactivated by its liver toxicity. The acid was able to decouple oxidative phosphorylation, induce adenosine triphosphate (ATP) depletion, decrease glutathione (GSH), and induce oxidative stress markedly leading to lipid peroxidation and organelle stress. Animal model experiments using rats displayed massive vacuolization of rat liver tissues associated with cellular swellings [58]. Several usnic acid derivates were isolated from *Usnea longissima* Ach. collected from Chengdu, Sichuan Province, People's Republic of China, including usenamine A-F (**45–50**), usone (**51**), and isousone (**52**) along with (+)-usnic acid (**44**) [59]. Usenamine A (**45**), usenamine B (**46**), and isousone (**52**) were cytotoxic against HepG2 cell lines with IC_{50} values of 6.0, 50.2, and 53.3 μM, respectively, which were comparable to IC_{50} values of methotrexate, the positive control, 15.8 μM [59]. Yu et al. suggested that the usnic acid (**42**) enamines with substituent alteration at C-4a could improve solubility leading to an increase in bioavailability and hence demonstrating better bioactivity as well as reducing severe side effects.

14.3.4 Anthraquinone

Quinone class compounds are also found in lichenized fungi, arising from polyketide formation from acetyl-CoA, before cyclization forming quinone molecular backbones such as emodin (**53**) from *Pyrenula japonica* Kurokawa mycobiont and chrysophanol (**54**) from whole thallus of *Asahinea chrysantha* (Tuck.) W.L. Culb. and C.F. Culb [60] (Figure 14.11).

Phytochemical study on *Xanthoria parietina* (L.) Beltr. of Slovakia origins produced a quinone-type compound from the acetone extract, parietin (**55**). Up to nine human cancer

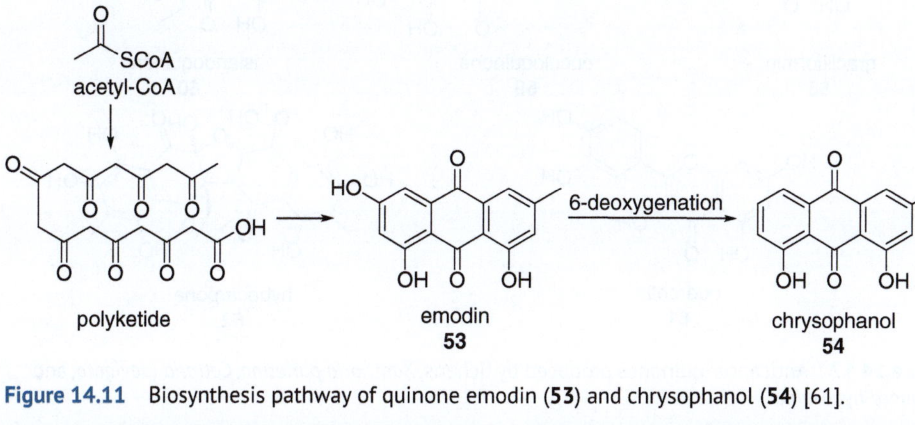

Figure 14.11 Biosynthesis pathway of quinone emodin (**53**) and chrysophanol (**54**) [61].

cell lines (A2780 [ovary], HeLa [cervix], MCF-7 [breast], SK-BR-3 [breast], HT-29 [colon], HCT-116 p53+/+ [colon], HCT-116 p53−/− [colon], HL-60 [leukemia] and Jurkat [leukemia]) were sensitive to the antiproliferative/cytotoxic effects of parietin (**55**) (Figure 14.10). The effect of **55** proved to be significant in some cell lines (e.g. A2780, Jurkat, or HT-29) already at 50 μM. However, noticeable differences among cell lines were also observed. Whereas Jurkat and A2780 cells were found to be markedly sensitive, others (e.g. HL60, HCT-116 p53+/+, or MCF-7) were rather resistant [34].

Other quinones were also produced by *Cetraria laevigata* Rass. of Russian origins, namely skyrin (**56**), 3-ethyl-2,7-dihydroxynaphthazarin (**57**), graciliformin (**58**), cuculoquinone (**59**), and islandoquinone (**60**) (Figure 14.10). Despite their high antioxidant activities, which are more active than quercetin (**61**), these quinones possess weak cytotoxic activity against human cancer cell line A549 [62].

Interestingly, a potent cytotoxin was reported from mycobiont broth culture of *Lecanora hybocarpa* (Tuck.) Brodo collected from the US. The pentacyclic naphthazarin-derived dimer, hybocarpone (**62**), was able to induce murine P815 mastocytoma cell death at low concentration with an IC_{50} value of 0.27 μM (Figure 14.12) [63].

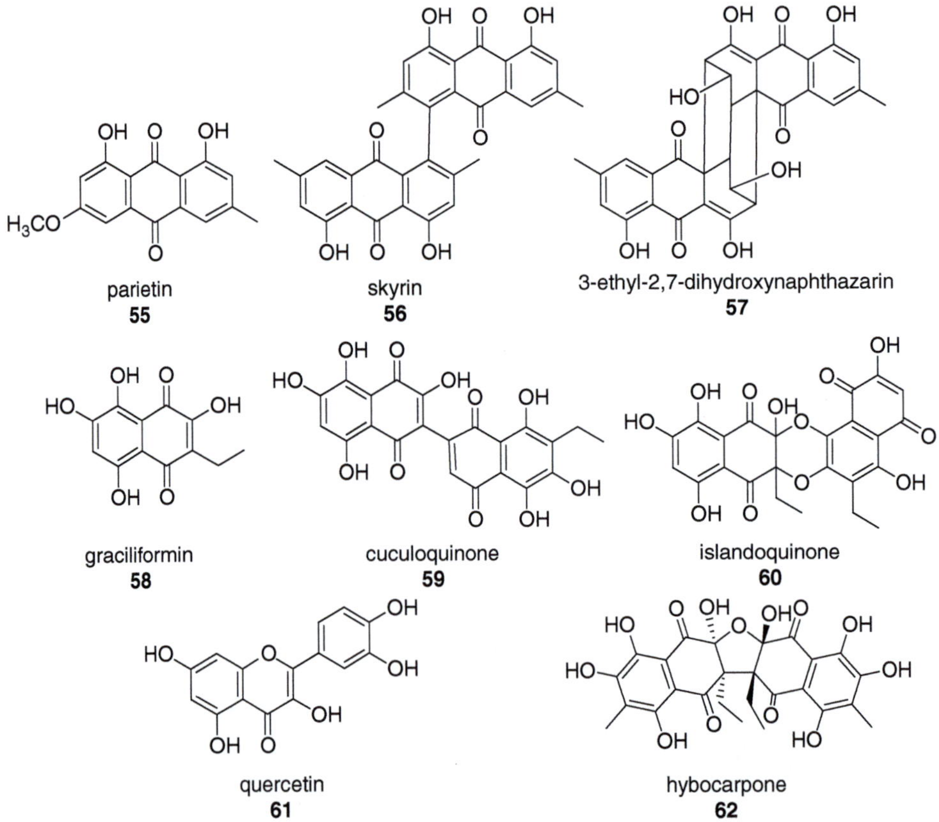

Figure 14.12 Anticancer quinones produced by lichens, *Xanthoria parietina*, *Cetraria laevigata*, and *Lecanora hybocarpa*.

14.3.5 Xanthone Derivates

Xanthones are commonly produced by mycobiome, including lichenized fungi in which the molecules contain a dibenzo-γ-pirone scaffold. Xanthone biosynthesis in lichen proceeds via polyketide synthesis, which can evolve in two pathways, lichenxantyhone-type lichen xanthone (A) and thiomielin-type lichen xanthone pathways (B) (Figure 14.13) [61]. In pathway A, acetyl-CoA forms a polyketide intermediate, which finally produces 1,3,6-trihydroxy-8-methylxanthone **63**). In pathway B, the polyketide intermediate formed a quinone (emodin, **53**), chrysophanol (**54**), and subsequent oxidation produces an intermediate *o*-carboxybenzophenone (**64**) or islandicin (**65**) prior to ravanelin (**66**) formation.

Among hundreds of xanthones isolated and identified from lichen mycobiont sources, only a limited number have been evaluated for their bioactivities, including their anticancer potency [64]. Brazilian folious lichen *Parmotrema lichexanthonicum* Eliasaro and Adler, were investigated to obtain lichexanthone (**67**), which was inactive against UACC-62 and B16-F10 cells [37]. Subnudatone A (**68**) obtained from *Pseudopyrenula subnudata* Müll.Arg. exerted moderate cytotoxic activity against MCF-7 (breast) and K562 (myelogenous leukemia) tumor cell lines (IC_{50} values of 23.5 and 51.9 μM, respectively) (Figure 14.14) [65].

On the other hand, eumitrin A1 (**69**), a C—C bond formed dixanthone, isolated from *Usnea blepharea* Motyka obtained from Sulawesi—Indonesia, was highly cytotoxic against Murine Leukemia P388 cells with an IC_{50} value of 4.5 μg/mL [66]. In addition, eumitrins C-E (**70–72**) were isolated from Vietnamese *Usnea baileyi* (Stirt.) Zahlbr. with null to mild cytotoxicity against seven tested cell lines, Huh7 (liver), Caco 2 (colon), MDA-MB-231 (breast), HCT-116 (colon), PC-3 (prostate), NCI-H2 (lung), and diploid skin fibroblasts as normal cell lines for control (Figure 14.11) [65].

Three heterodimeric tetrahydroxanthones, usneaxanthones A, C, and D (**73–75**) isolated from *U. aciculifera* Vain (Parmeliaceae) were evaluated on HT–29 human colorectal cancer cells resulting in potent cytotoxicity of usneaxanthone D (**75**) with an IC_{50} value of 2.41 μM [67]. In addition, usneaxanthones E-H (**76–79**) isolated from *U. aciculifera* Vain (Parmeliaceae) were evaluated on four human cancer cell lines, including HCT116 colorectal cancer, MCF-7 breast cancer, A549 lung cancer, and OVCAR-3 ovarian cancer showing their good cytotoxicity against the first three tested cancer cell lines with the best IC_{50} value of 3.37 μM. All compounds showed potent cytotoxicity against HCT116 colon cancer with IC_{50} values ranging from 3.37 to 4.53 μM [68].

14.3.6 Diketopiperazines

Another unique molecule type produced by lichens is the diketopiperazines, the simplest naturally occurring cyclic peptide in the form of tricyclic fused rings consisting of two proline subunits, in which its biosynthesis relied on non-ribosomal peptide synthetases (NRPs) and tRNA-dependent cyclodipeptide synthases (CDPs) [69].

Phytochemical studies on *Usnea* sp. collected from Sri Lanka resulted in crystalline diketopiperazine, ambewelamide A (**80**), and B (**81**). Ambewelamide A (**80**) exerted potent *in vitro* cytotoxicity against P388 murine leukemia cell line with an IC_{50} value of 8.6 ng/mL.

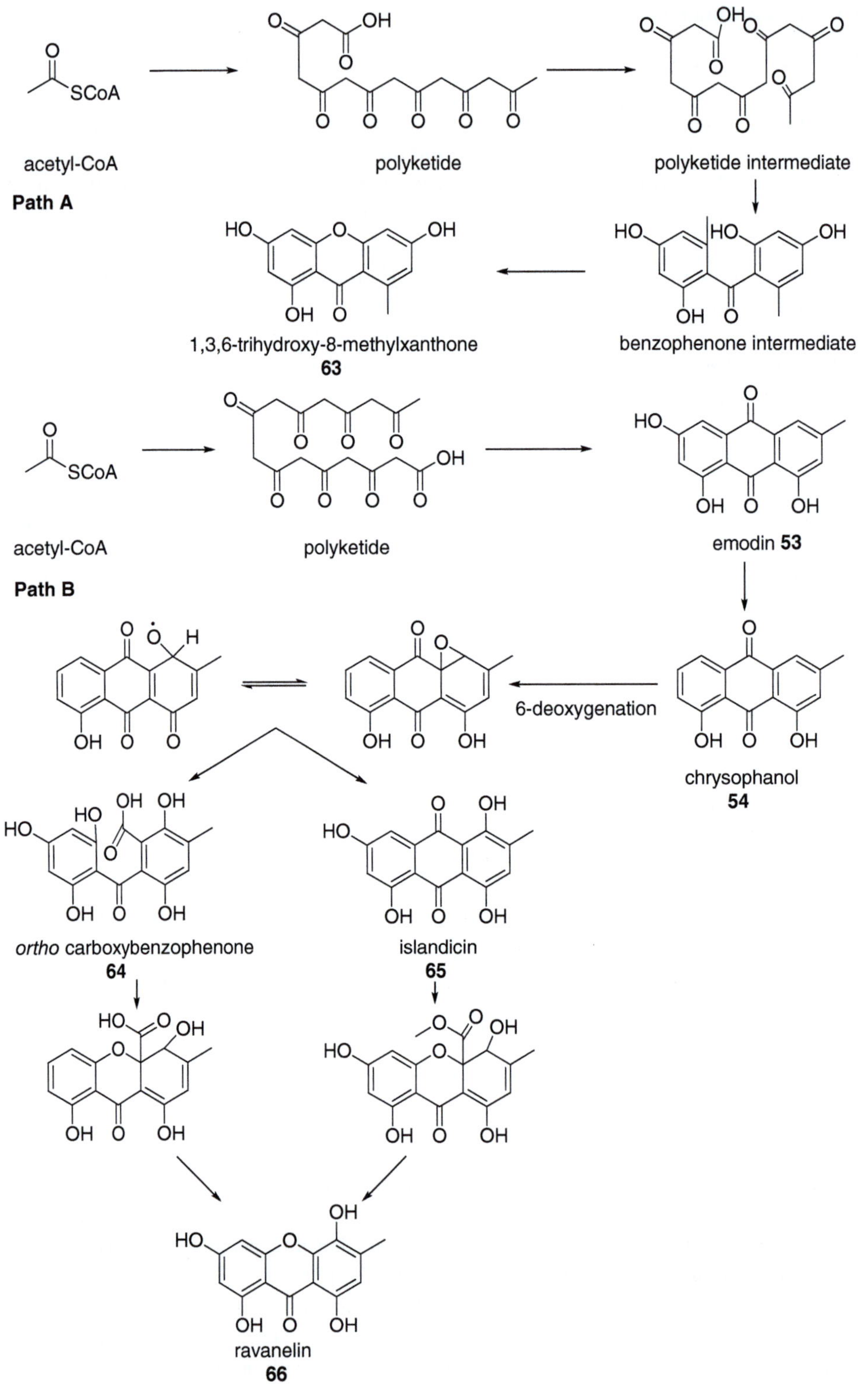

Figure 14.13 Lichenxantyhone-type lichen xanthone and thiomielin-type lichen xanthone pathways.

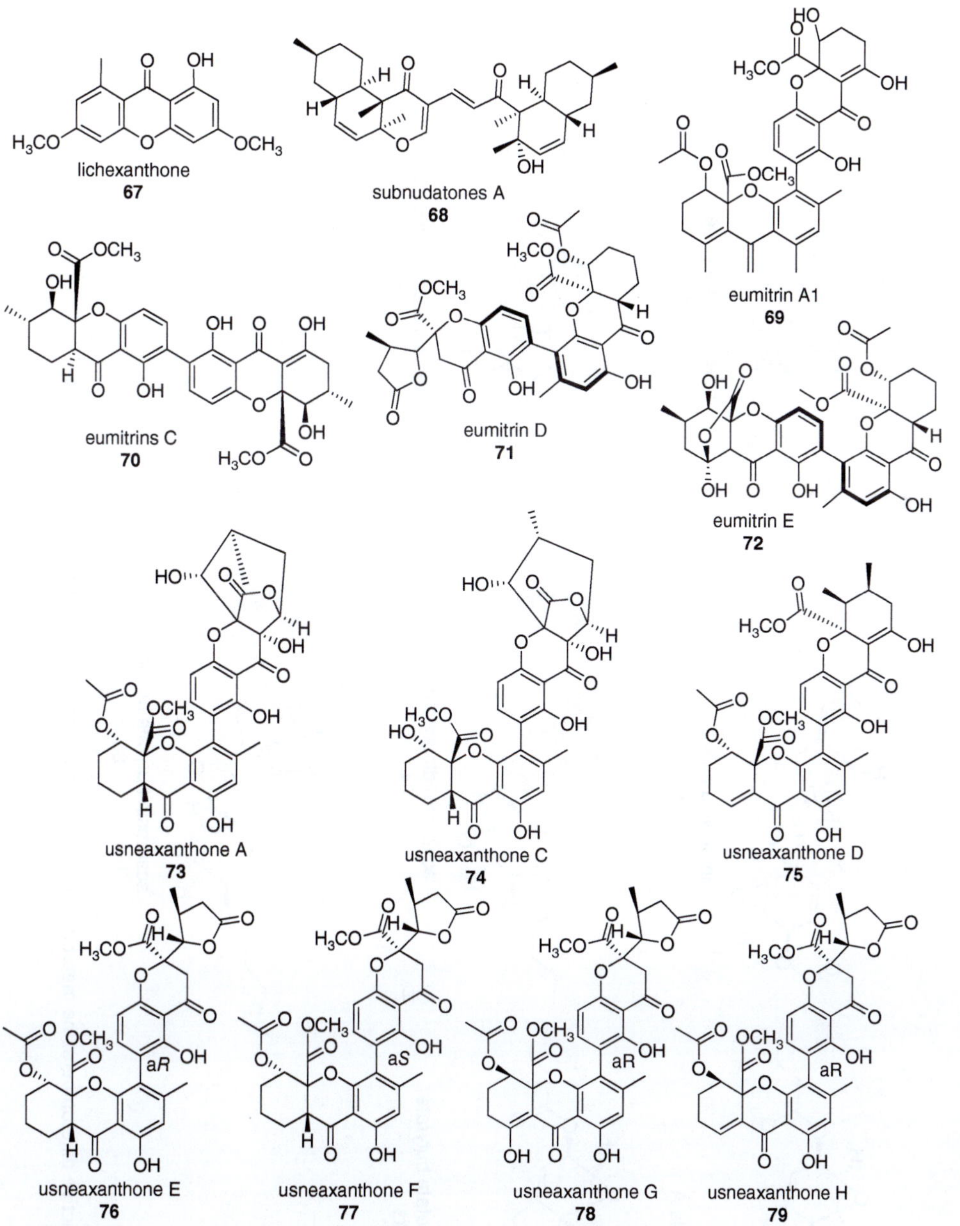

Figure 14.14 Cytotoxic xanthones produced by lichens, *Parmotrema lichexanthonicum*, *Pseudopyrenula subnudata*, *Usnea aciculifera*, *Usnea baileyi*, and *Usnea blepharea*.

It was suggested that functionality combination of ring A/E epoxides and a diketopiperazine epidisulfide bridge might be required for cytotoxicity (Figure 14.15) [70].

The epidithiopiperazinediones were also discovered from *Xanthoparmelia scabrosa* (Taylor) Hale grown on a coastal cliff face southeast of Batemans Bay, New South Wales, Australia. The investigation revealed scabrosin diacetate (**82**), scabrosin acetate butyrate (**83**), scabrosin dibutyrate (**84**), scabrosin acetate hexanoate (**85**), and scabrosin butyrate

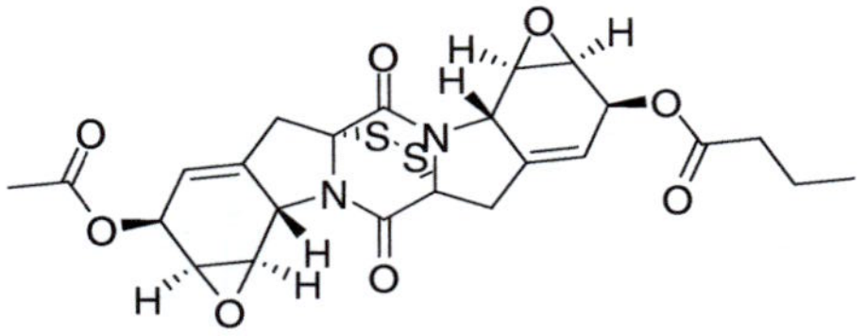

ambewelamide A
80

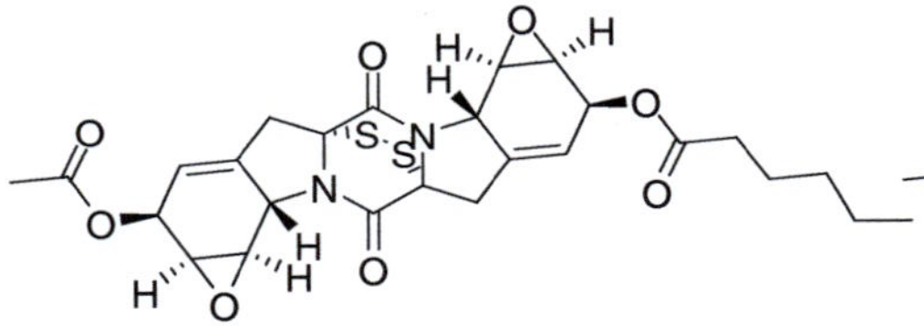

ambewelamide B
81

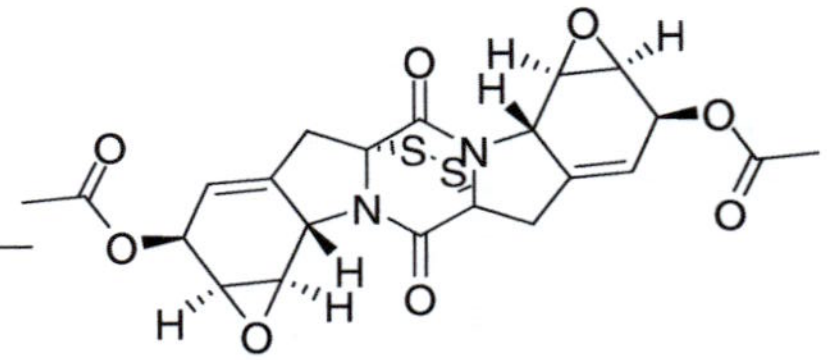

scabrosin diacetate
82

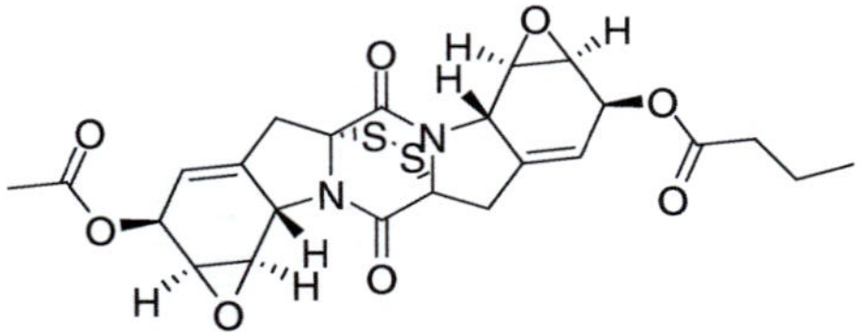

scabrosin acetate butyrate
83

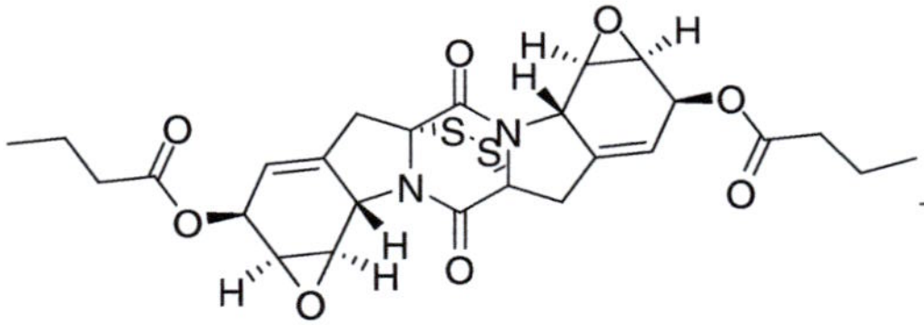

scabrosin dibutyrate
84

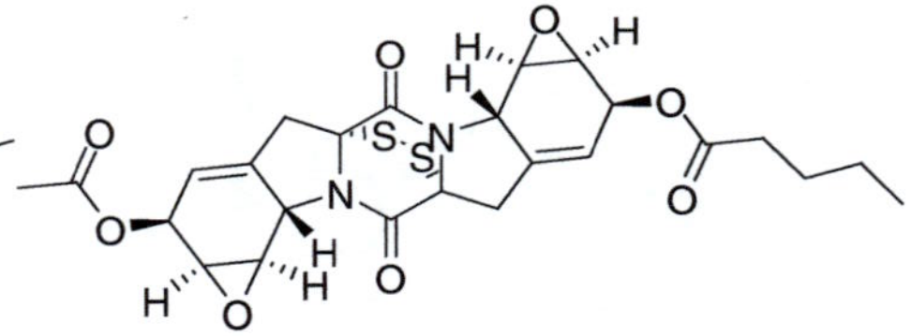

scabrosin acetate hexanoate
85

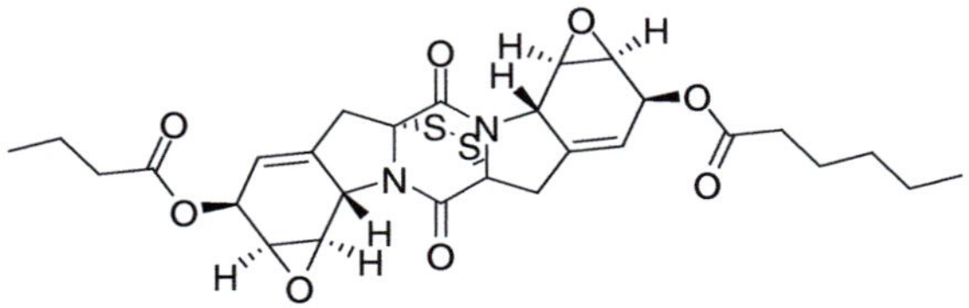

scabrosin butyrate hexanoate
86

Figure 14.15 Cytotoxic active diketopiperazine metabolites from lichens, *Usnea* sp., and *Xanthoparmelia scabrosa*.

hexanoate (**86**). Cytotoxic evaluation of scabrosin diacetate and scabrosin acetate butyrate against P815 mastocytomia cancer cell line showed a significant efficacy with IC_{50} values of 0.56 and 0.50 μM, respectively, comparable to a standard inhibitor gliotoxin (IC_{50} 2.9 μM). The analog studies suggested that variation of the side chain lipophilicity of the esters is negligible (Figure 14.12) [71].

14.3.7 Terpenoids

Terpenoids are one of the most chemically, structurally, and functionally diverse metabolites produced by organisms. The molecules are derived from two 5-carbon isopentenyl diphosphate building blocks [60a]. In the case of lichen, few terpenoids were successfully isolated compared to depside congener type compounds with some of these terpenes being important in anticancer studies (Figure 14.16). Investigation on *Myelochroa aurulenta* (Tuck.) Elix and Hale of Kyoto-Japan origin discovered the naturally occurring pentacyclic triterpenoid acid, 16-*O*-acetyl-leucotylic acid **87** along with its deacetylation derivate, leucotylic acid **88**, which is close to the lupane skeleton. Further studies reported 16-*O*-acetyl-leucotylic acid (**87**) to possess significant cytotoxicity against HL-60 human leukemia cells with an EC_{50} value of 21 μM which is higher than that of the commercially available betulin (**89**) and betulinic acid (**90**) with EC_{50} values >50 μM. The deacetylated compound showed reducing efficacy (EC_{50} 72 μM) despite its increased solubility [72]. Another study by Liu et al. isolated two lupane-type triterpenoids, retigeric acid A (**91**) and B (**92**), from *Lobaria kurokawae* Yoshim [73]. Retigeric acid A (**91**) exerted cytotoxicity against RPE1 (human retinal pigment epithelial-1), MCF-7 (breast), and 3AO (ovary) with IC_{50} values of 64.71, 47.84, and 61.47 μM, respectively, whereas retigeric acid B (**92**) indicated stronger cytotoxicity against eight cell lines RPE1, MCF-7, 3AO, 3AO/CDDP, KB (oral), KB/VCR (oral), U2OS (osteosarcoma), and Saos2 (osteosarcoma) cells with IC_{50} values of 67.71, 52.40, 27.57, 86.14, 28.26, 45.97, 18.41, and 13.31 μM, respectively [73]. Further studies revealed that retigeric acid B induced LNCap apoptosis in a dose-dependent manner with an IC_{50} value of 7.97 μM and blocked PC-3 cell cycle progression through cell cycle regulators' alteration [73].

Another lichen terpenoid was 2-*O*-caffeoyl alphitolic acid (2OCAA, **93**) isolated from *U. barbata* (L.) Weber ex F.H. Wigg, which was tested on the human colorectal cancer cell line HCT116. The results showed that **93** exhibited concentration-dependent anticancer activities (IC_{50} 12.5–50 μM) by suppressing antiapoptotic genes, such as MCL-1, and inducing apoptotic genes, such as BAX, TP53, and CDKN1A (p21). Moreover, **93** inhibited the migration and invasion of colorectal cancer cells in a concentration-dependent manner [74].

Lichen bioprospecting in Vietnam conducted by Tuan et al. discovered nine lupane triterpenoids from *Cryptothecia faveomaculata* Makhija and Patw of Vinh City origin. The terpenoids (i.e. cryptothecin A (**94**), ergosterol (**95**), ergosterol peroxide (**96**), 20(*S*)-hydroxydammar-24,25-epoxy-3b-ol (**97**), ursolic acid (**98**), 20,29,30-trisnorlupane-3b, 19a-diol, **99**) were individually screened using five cell lines for their cytotoxic potency (Table 14.3). Among the tested compounds, cryptothecin A was concluded to have the best activities to overall cancer cell lines with MCF7 (breast) showed the most sensitivity. Nevertheless, efficacy was still far lower than that of the standard drug, ellipticine [75].

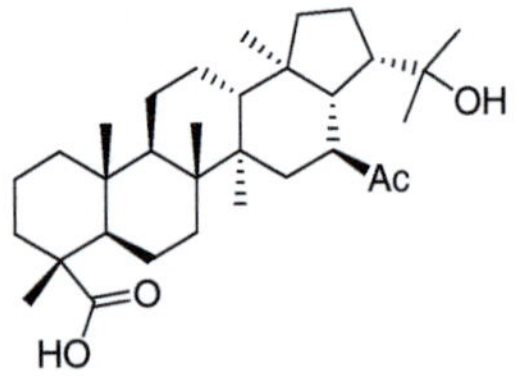

16-*O*-acetyl-leucotylic acid
87

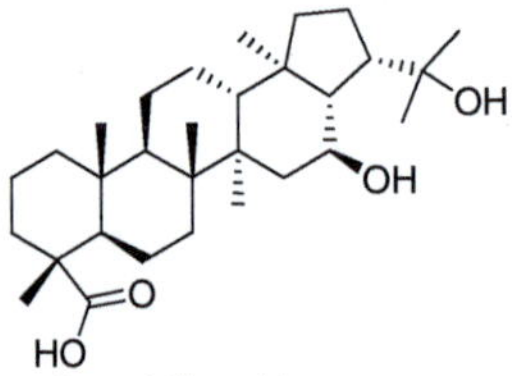

leucotylic acid
88

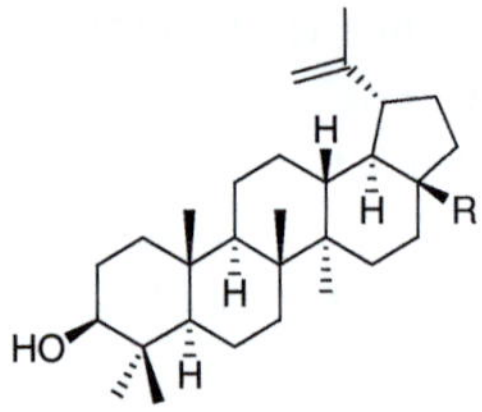

betulin R=CH_2OH **89**
betulinic acid R=COOH **90**

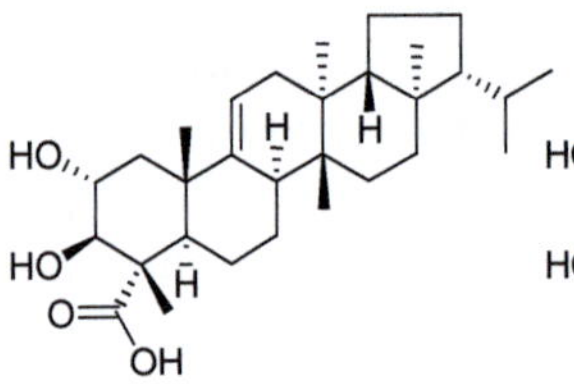

retigeric acid A
91

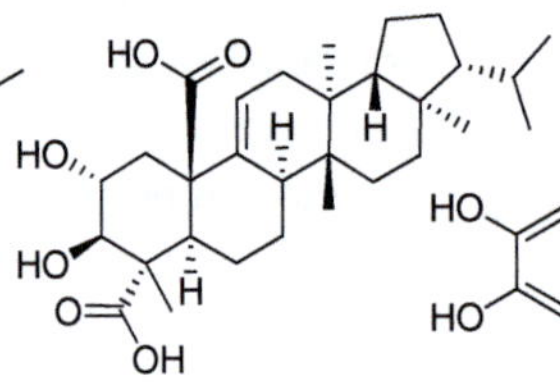

retigeric acid B
92

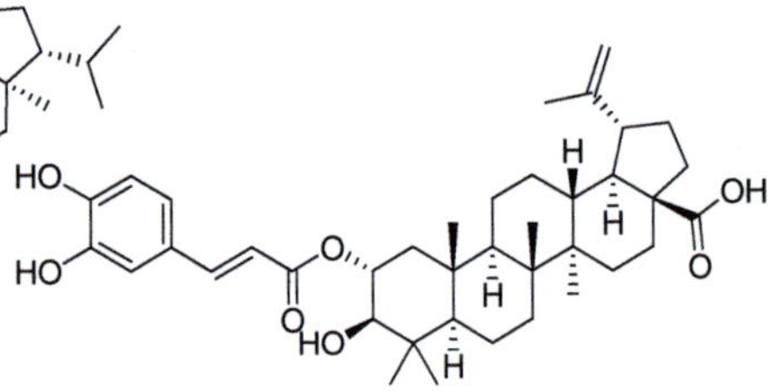

2-*O*-caffeoyl alphitolic acid (OCAA)
93

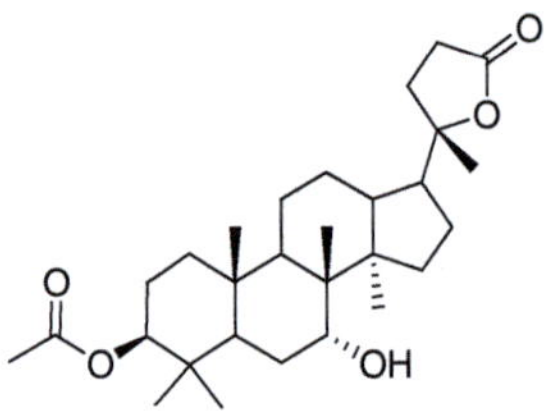

cryptothecin A
94

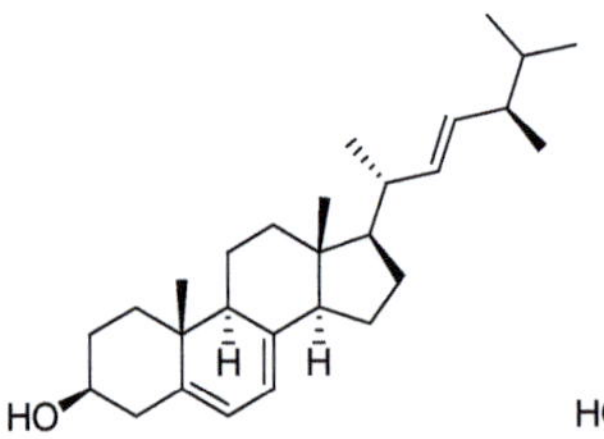

ergosterol
95

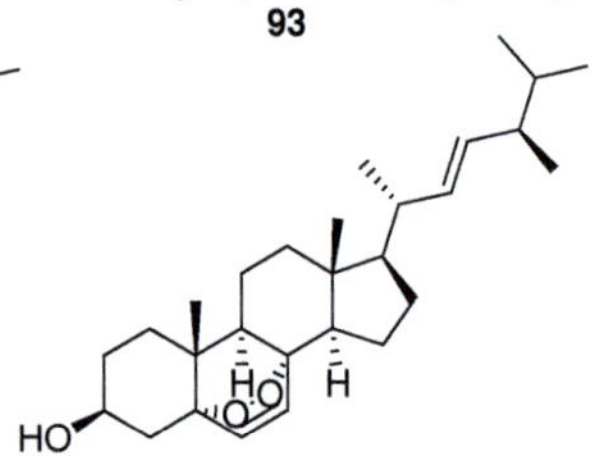

ergosterol peroxide
96

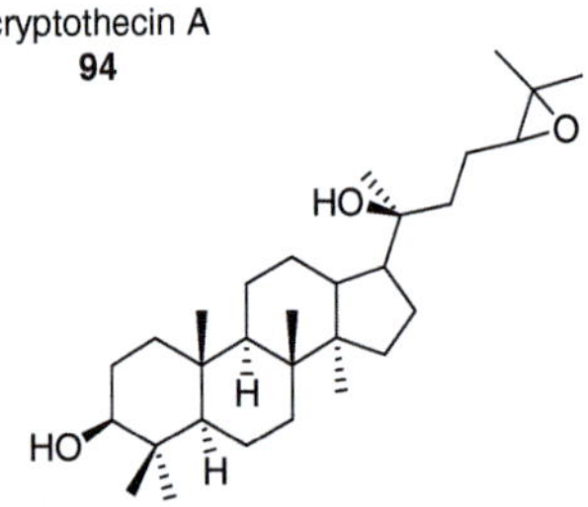

20(*S*)-hydroxydammar-24,25-epoxy-3b-ol
97

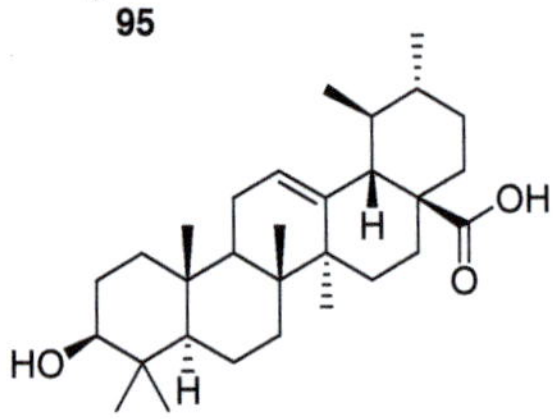

ursolic acid
98

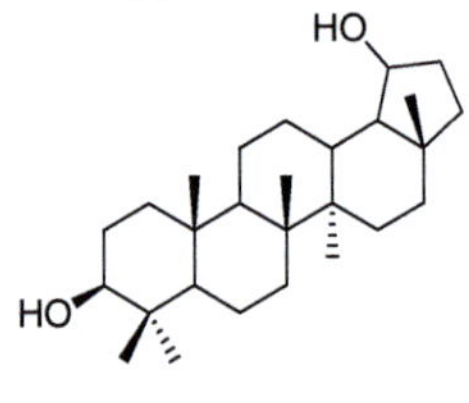

20,29,30-trisnorlupane-3b,19a-diol
99

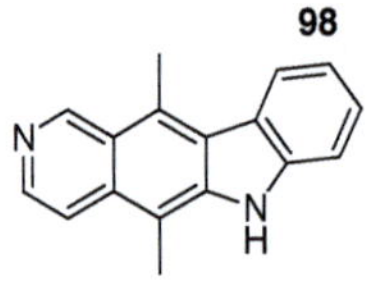

ellipticine
(intercalator)
100

Figure 14.16 Cytotoxic active terpenoids derived from lichens, *Cryptothecia faveomaculata*, *Lobaria kurokawae*, *Myelochroa aurulenta*, and *Usnea barbata*.

Table 14.3 Cytotoxic terpenoids of *Cryptothecia faveomaculata*.

	IC$_{50}$ (μM)				
Triterpenoid of *C. faveomaculata*	**SK-LU-1**	**HepG2**	**Hep3B**	**SW480**	**MCF7**
Cryptothecin A (**94**)	45.4	17.5	39.3	41.1	27.5
Ergosterol (**95**)	>50	21.5	21.7	>100	43.6
Ergosterol peroxide (**96**)	>50	46.9	35.2	>100	>100
20(*S*)-hydroxydammar-24,25-epoxy-3b-ol (**97**)	40.2	55.7	53.1	58.2	35.2
Ursolic acid (**98**)	53.4	32.8	42.4	27.0	30.7
20,29,30-trisnorlupane-3b,19a-diol (**99**)	46.2	63.7	38.1	48.2	42.1
Ellipticine (**100**)	0.4	0.4	0.4	0.5	0.4

Note: MCF-7 (human breast carcinoma), SW480 (human colon adenocarcinoma), SK-LU-1 (human lung carcinoma), HepG2 (human hepatocellular carcinoma), and Hep3B (human hepatocellular carcinoma) cells.

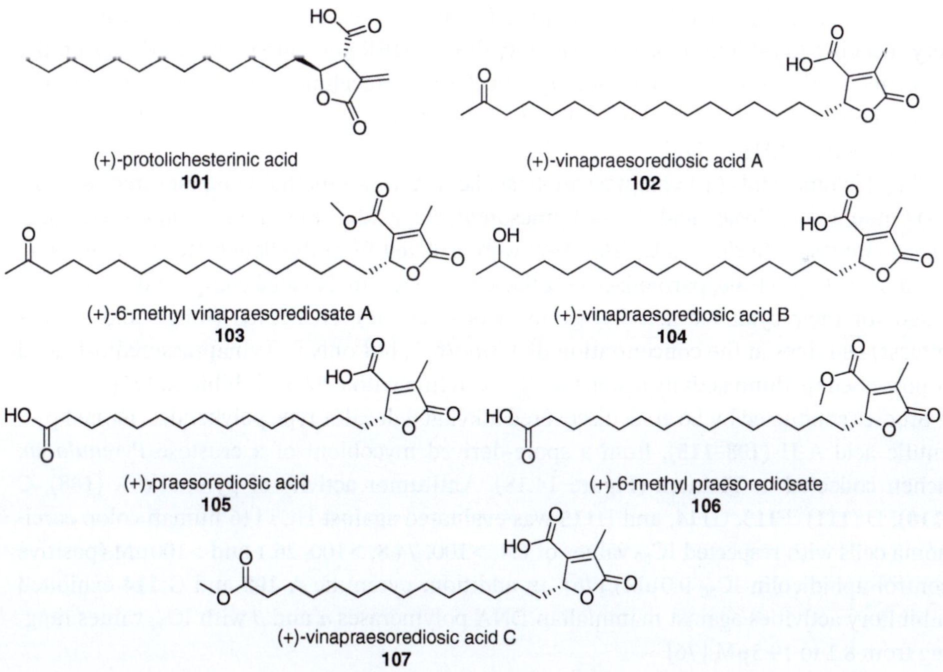

Figure 14.17 Lactonic acid derivatives of lichens with cytotoxic activities, *Cetraria islandica, Cornicularia aculeata, Rhizoplaca melanophthalma, Parmelia nepalensis,* and *Parmotrema praesorediosum.*

14.3.8 γ-Lactonic Acid and Other Important Derivatives

There are several reports on cytotoxic lactonic acids from lichens (Figure 14.17), including protolichesterinic acid (**101**), which was reported as a constituent of *C. islandica* L. (Ach.) of Iceland origin, *Cornicularia aculeata* (Schreb.) Ach. and *Rhizoplaca melanophthalma* of Chile origin, as well as *Parmelia nepalensis* Taylor of Nepal origin [45], [40].

Protolichesterinic acid (**101**) exerted moderate cytotoxicity against MCF-7 (breast adenocarcinoma), HeLa (cervix adenocarcinoma), HCT-116 (colon carcinoma), and NIH-3T3 (normal cell, mouse embryonic fibroblast) cell lines with IC_{50} values of 34.3, 46.7, >100, and 78.5 μM, respectively [40]. In addition, the human keratinocyte cell line HaCaT was not sensitive to the acid at the 5 μM concentration [57].

Nevertheless, protolichesterinic acid (**101**) (exerted by *R. melanophthalma* (DC.) Leuckert) could activate an apoptotic process in DU-145 and LNCaP prostate cell lines with a clear dose–response relationship in the concentration range 6.25–50 μM [49]. In another cell line study (using T-47D and ZR-75-1 from breast carcinomas and K-562 from erythroleukemia), protolichesterinic acid (**101**) caused a significant cell death at a dose of 20 μg/mL and a significant reduction in DNA synthesis, as measured by thymidine uptake with the dose inducing 50% of maximum inhibition (ED_{50}) between 1.1 and 24–6 pg/mL. Moreover, the proliferative response of mitogen-stimulated lymphocytes was inhibited (ED_{50} 8.4 pg/mL). In contrast, the same dose did not cause any changes in the normal fibroblast cells [45]. In SK-BR-3 cells (breast cancer), treatment with (+)-protolichesterinic acid (**101**) obtained from *C. islandica* L. (Ach.) caused fatty acid synthetase (FASN) increase and human epidermal growth factor receptor 2 (HER2) decrease which implied a compensatory response to inhibition of FASN activity during HER2 downregulation. Moreover, the experiment also showed an inhibitory effect on extracellular signal-regulated protein kinase (ERK1/2) and Protein kinase B (AKT) signaling pathways, which is probably due to reduced levels of HER2 [76].

Six γ-lactonic acids (+)-vinapraesorediosic acid A, (+)-6-methyl vinapraesorediosate A, (+)-vinapraesorediosic acid B, (+)-praesorediosic acid, (+)-6-methyl praesorediosate, (+)-vinapraesorediosic acid C, (**102–107**) were isolated from the lichen *Parmotrema praesorediosum* (Nyl.) Hale, parmeliaceae collected in Vietnam. Isolated compounds were evaluated for their cytotoxic activity against HeLa (cervix), NCI-H460 (lung), and MCF-7 (breast) cell lines at the concentration of 100 μg/mL, but only (+)-vinapraesorediosic acid A possessed medium activity against MCF-7 cell line with a 52.5% inhibition [77].

Studies conducted by Le et al. discovered alkylated decalin-type polyketides, named pyrenulic acid A-H (**108–115**), from a spore-derived mycobiont of a crustose *Pyrenula sp.* lichen collected in Vietnam (Figure 14.18). Antitumor activity of pyrenulic A (**108**), C (**110**), D (**111**), F**113**, G**114**, and H**115** was evaluated against HCT116 human colon carcinoma cells with respected IC_{50} values of 6.4, >100, 74.8, >100, 26.1 and >100 μM (positive control aphidicolin IC_{50} 9.0 μM) [76]. In addition, pyrenulic A **108** and G **114** exhibited inhibitory activities against mammalian DNA polymerases α and β with IC_{50} values ranging from 8.1 to 19.5 μM [76].

14.4 Anticancer of Lichen in Animal Model

There are several *in vivo* data revealing the potential anticancer properties of lichens, which have been supported by various experiments using animal xenograft/allograft models [78]. It is noteworthy that the molecular mechanisms of lichen metabolites inhibiting cancer cells have not yet been fully understood, with some important questions remaining open. Nevertheless, several studies have indicated plausible anticancer activities via cell cycle

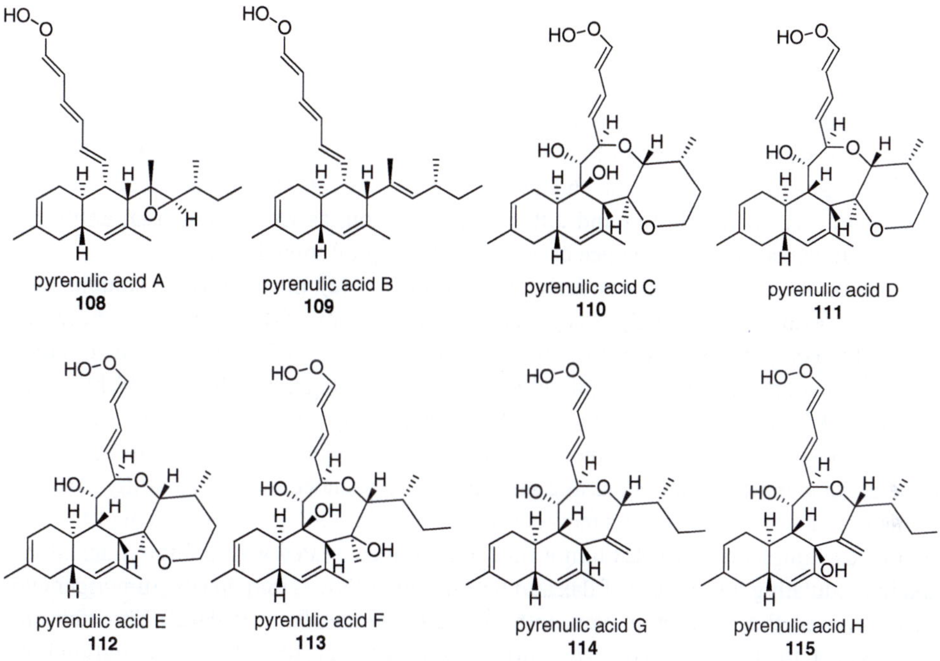

Figure 14.18 Alkylated decalin-type polyketides from *Pyrenula sp.* lichen with cytotoxic activity.

arrest, suppressing growth factor signaling, modulating anticancer immunity, preventing tumor-promoting inflammation, preventing tumor invasion and metastasis, inhibiting telomerase activity, inhibiting angiogenesis, and initiating apoptosis [79].

The frequently reported lichen metabolites, i.e. usnic acid (**42**), atranorin (**22**), barbatic acid (**20**), diffractaic acid (**21**), hypostictic acid (**31**), salazinic acid (**32**), and physciosporin (**41**), and their promising *in vivo* anticancer activities are discussed in this section.

Usnic acid (2,6-diacetyl-7,9-dihydroxy-8,9b-dimethyldibenzo [*b*,*d*]furan-1,3(2*H*,9b*H*)-dione; UA, **42**), a dibenzofuran derivative, is one of the most investigated lichen metabolites, and is generally found in several genera, including *Alectoria*, *Usnea*, *Lecanora*, *Cladonia*, *Evernia*, and *Ramalina* [80]. The first natural UA was isolated from the species *U. barbata* in 1880, which has become a driving force for the search for this molecule in other lichen species, including *Usnea deffractas* Vain., *Lecanora pseudogangalevides*, *Usnea laevis* (Eschw.) Nyl., *Roccella montagnei* Bél., *Sticta weigelii* (Ach.) Vain., *Cladonia substellata* Vainio, and *U. longissima* Ach [81].

42 has been found exhibiting different biological activities, such as anticancer, antimicrobial, antiparasitic, and antiviral activities [81]. It exists in nature as either (−) or (+) isomers as well as a racemic mixture, and both isomers have shown promising anticancer properties [82]. A study by Song et al. showed that **42** actively suppressed *in vivo* angiogenesis in a chick embryo chorioallantoic membrane and in a vascular endothelial growth factor-induced mouse corneal angiogenesis model [83]. Zuo et al. demonstrated that **42** strongly inhibited tumor growth in MCF-7 (breast) cancer-bearing mice in a dose-dependent manner with no significant toxicity observed [84]. **42** has also been found to inhibit tumor growth on H22-bearing mice. When combined with bleomycin, **42** exhibited a more effective activity in suppressing the tumor growth, arresting the cell cycle at the G0/G1 phase, and promoting tumor cellular apoptosis by initiating cleaved caspase-3 and caspase-8 activities [85]. An effort to improve *in vivo* anticancer properties of **42** by synthesizing several **42** analogs has been made by Ebrahim and colleagues. The authors revealed that several **42** analogs displayed potent antiproliferative activities *in vitro*. In particular, one analog-bearing *p*-OCF_3-benzylidene unit possessed a potent mammalian target of rapamycin (mTOR) inhibition and autophagy induction, as well as potential antitumor property in two athymic nude mice breast cancer xenograft models [86].

Furthermore, to enhance the anticancer effects of UA (**42**), the more water-soluble form, namely potassium usnate (KU), was prepared by salinization. KU exhibited enhanced bioavailability compared to UA in the tumor, plasma, and liver of a CT26syngeneic mouse tumor xenograft model after oral administration, as suggested by LC-MS/MS data. Moreover, KU showed significant antitumor properties by inhibiting liver metastasis in a mouse model of orthotopic colorectal cancer. KU therapy also suppressed the epithelial-mesenchymal marker expressions and several genes associated with cell motility [87].

In addition, UA treatment was markedly effective in inhibiting tumor growth of the subcutaneously implanted BGC823 tumor model than mice treated with 5-fluorouracil alone, and in modulating the amount of Bax and Bcl2 in tumor tissues [88]. *In vivo*, tumorigenicity assay revealed that in combination with *Flavocetraria cucullate* (Bellardi) Kärnefelt and Thell extract, the UA displayed *in vivo* anticancer activities at sub-lethal concentration [89].

Although **42** is one of the most attractive lichen metabolites, it is suggested that **42** has significant toxic effects, which can cause severe hepatotoxicity, allergic reactions, and teratogenic effects. Studies have shown that **42** can significantly bind to proteins as well as possesses a high bioavailability, which may contribute to the hepatotoxicity effects in mammals [81, 90]. However, **42** still arises greater scientific interest and has become a focus of clinical investigations compared to other lichen metabolites, a fact confirmed by the progressive growth in the number of publications in the past years. Accordingly, research has driven the introduction of innovative approaches to minimize side effects while maintaining therapeutic efficacy, such as encapsulation in controlled release systems [81].

In addition to UA (**42**), further lichen metabolites were also studied *in vivo* using tumor animal models. In this context, Solar et al. examined the anticancer effects of atranorin (ATR, **22**) in a mouse model of 4T1-induced tumor. It was found that ATR administration led to noticeably longer survival time of the animals, in line with higher numbers of apoptotic 4T1 cells and reduced tumor size. In addition, this study revealed the protective properties of ATR against oxidative stress in the livers of the tumor mouse model [91]. Furthermore, Zhou et al. demonstrated that ATR reduced tumor volume and weight, as well as reduced Ki-67 marker of proliferation in a mouse model of lung carcinoma. ATR was also found to suppress the β-catenin-mediated TOPFLASH and KAI1 C-terminal interacting tetraspanin (KITENIN)-mediated AP-1 activities [92]. More recently, the ART-superparamagnetic iron oxide nanoparticles (SPION) complex was studied on a mouse model of gastric cancer. The experiment revealed that the complexes inhibited the *in vitro* and *in vivo* activity of gastric cancer stem cells [93].

It also suggested that barbatic acid (BAR, **20**) isolated from lichens possessed potential anticancer activities. For example, **20** were obtained from the lichen *Cladia aggregata* (Sw.) Nyl. actively inhibited tumor growth (reduced tumor weight) in albino Swiss mice [94]. Karagoz and colleagues showed that diffractaic acid (DA, **21**) from lichen metabolites displayed great antitumor potential in Swiss albino mice with Ehrlich ascites carcinoma [95]. Moreover, hypostictic acid (HA, **31**) and salazinic acid (SA, **32**) isolated from lichen species *P. sphaerospora* (Nyl.) Hale. And *P. cetratum* (Ach.) Hale. (in open forests, near Piraputanga, Aquidauana Country, Mato Grosso do Sul State, Brazil), respectively, were found to exhibit anticancer activities in the BALB/c mouse model of melanoma (B16-F10). These isolated metabolites inhibited tumor growth with respect to weight and volume; treatment with HA and SA inhibited 72% and 88% of tumor volume ($p < 0.05$), respectively [47].

Another lichen metabolite, physciosporin (PHY, **41**) has shown promising anticancer effects in *in vivo* experiments. As revealed by Tas et al., the administration of PHY showed considerable bioavailability as well as great potential in decreasing the growth of colorectal cancer xenografts in BALB/c mice without promoting toxic effects [96].

The *in vivo* efficacy of lichen extracts isolated from different lichen species in the Yercaud hills of Tamil Nadu, India was evaluated in the cancer-induced albino Wistar rats by Poornima. This study demonstrated that bioactive constituents from *Parmotrema reticulatum* (Taylor) M. Choisy, *Parmotrema hababianum* (Gyeln.) Hale. and *Roccella montagnei* Bel. displayed a potent antitumor activity, as suggested by cell-line studies and animal model studies [97].

14.5 Conclusion and Future Perspective

Lichen have been part of human civilization for arrays of purposes including medicine in which their empirical information as medicine is inherited throughout generations. The need for evidence-based uses of herbal medicine led to bioprospecting protocols being applied. A number of investigations on lichen shown they have the potential as a source of anticancer agents. The crude *n*-hexane extract of *P. perlata* showed significant cytotoxicity against L1210 cells. Similarly, the *n*-hexane extract of *Platismatia glauca* exhibited good cytotoxicity against the U251 cell line. Moreover, diketopiperazines on scabrosin diacetate (**82**), scabrosin acetate butyrate (**83**) from Australian *X. scabrosa* (Taylor) Hale produce nano molar sensitivity against P815 mastocytomia cancer cells. Nevertheless, one of the common hurdles in anticancer drug discovery is cytotoxicity against normal cells, in which synthetic medicinal approaches are needed for molecular modification to reduce toxicity while increasing efficacy.

References

1 Nugraha, A.S., Laksono, T.A., Firli, L.N. et al. (2020). Anti-cancer evaluation of depsides isolated from indonesian folious lichens: *Physcia millegrana, Parmelia dilatata* and *Parmelia aurulenta*. *Biomolecules* 10: 1–10.

2 Nugraha, A.S., Pratoko, D.K., Damayanti, Y.D. et al. (2019). Antibacterial and anticancer activities of nine lichens of Indonesian Java island. *Journal of Biologically Active Products from Nature* 9: 39–46.

3 Nugraha, A.S., Wangchuk, T., Willis, A.C. et al. (2019). Phytochemical and pharmacological studies on four Indonesian epiphytic medicinal plants: *Drynaria rigidula, Hydnophytum formicarum, Usnea misaminensis* and *Calymperes schmidtii*. *Natural Product Communications* 14: 1–6.

4 Guarrera, P.M., Lucchese, F., and Medori, S. (2008). Ethnophytotherapeutical research in the high molise region (central-southern Italy). *Journal of Ethnobiology and Ethnomedicine* 4: 7.

5 Afolayan, A.J., Grierson, D.S., Kambizi, L. et al. (2002). *In vitro* antifungal activity of some south African medicinal plants. *South African Journal of Botany* 68: 72–76.

6 Rego-Cunha, N., Silva, N., Honda, E. et al. (2017). Lichens used in the traditional medicine by the Pankararu indigenous community, Pernambuco-Brazil. *Global Journal of Science Frontier Research: C Biological Science* 17: 15–22.

7 CHMP (2023). *European Union Herbal Monographs: Overview of Recommendations for the Uses of Herbal Medicinal Products in the Paediatric Population*. Amsterdam: European Medicines Agency.

8 Sung, H.A.-O., Ferlay, J., Siegel, R.A.-O. et al. (2021). Global cancer statistics 2020: Globocan estimates of incidence and mortality worldwide for 36 cancers in 185 countries. *CA: A Cancer Journal for Clinicians* 71: 209–249.

9 Newman, D.J. and Cragg, G.M. (2020). Natural products as sources of new drugs over the nearly four decades from 01/1981 to 09/2019. *Journal of Natural Products* 83: 770–803.

10 Maurya, I.K., Singh, S., Tewari, R. et al. (2018). Antimicrobial activity of *Bulbothrix setschwanensis* (zahlbr.) Hale lichen by cell wall disruption of *Staphylococcus aureus* and *Cryptococcus neoformans*. *Microbial Pathogenesis* 115: 12–18.

11 Fernández-Moriano, C., González-Burgos, E., Divakar, P.K. et al. (2016). Evaluation of the antioxidant capacities and cytotoxic effects of ten parmeliaceae lichen species. *Evidence-Based Complementary and Alternative Medicine* 2016: 3169751.

12 Bézivin, C., Tomasi, S., Lohézic-Le Dévéhat, F., and Boustie, J. (2003). Cytotoxic activity of some lichen extracts on murine and human cancer cell lines. *Phytomedicine* 10: 499–503.

13 Kosanić, M., Ristić, S., Stanojković, T. et al. (2018). Extracts of five cladonia lichens as sources of biologically active compounds. *Farmácia* 66: 644–651.

14 Ranković, B.R., Kosanić, M.M., and Stanojković, T.P. (2011). Antioxidant, antimicrobial and anticancer activity of the lichens *Cladonia furcata*, *Lecanora atra* and *Lecanora muralis*. *BMC Complementary and Alternative Medicine* 11: 97.

15 Triggiani, D., Ceccarelli, D., Tiezzi, A. et al. (2009). Antiproliferative activity of lichen extracts on murine myeloma cells. *Biologia* 64: 59–62.

16 Ari, F., Celikler, S., Oran, S. et al. (2014). Genotoxic, cytotoxic, and apoptotic effects of *Hypogymnia physodes* (l.) nyl. on breast cancer cell. *Environmental Toxicology* 29: 804–813.

17 Ren, M.R., Hur, J.-S., Kim, J.Y. et al. (2009). Anti-proliferative effects of *Lethariella zahlbruckneri* extracts in human HT-29 human colon cancer cells. *Food and Chemical Toxicology* 47: 2157–2162.

18 Studzińska-Sroka, E., Majchrzak-Celińska, A., Bańdurska, M. et al. (2022). Is caperatic acid the only compound responsible for activity of lichen *Platismatia glauca* within the nervous system? *Antioxidants* 11: 1–30.

19 Kosanić, M., Ranković, B., and Stanojković, T. (2012). Antioxidant, antimicrobial, and anticancer activity of 3 umbilicaria species. *Journal of Food Science* 77: T20–T25.

20 Disoma, C., Erkisa, M., Oran, S. et al. (2018). *Usnea filipendula* induces apoptosis in human colon cancer cell lines, *Usnea filipendula* induces apoptosis in human colon cancer cell lines. *Indian Journal of Pharmaceutical Sciences* 80: 1078–1085.

21 Reddy, S.D., Siva, B., Kumar, K. et al. (2019). Comprehensive analysis of secondary metabolites in *Usnea longissima* (lichenized ascomycetes, parmeliaceae) using uplc-esi-qtof-ms/ms and pro-apoptotic activity of barbatic acid. *Molecules* 24: 2270.

22 Nguyen, T.T.T., Anh, D.H., Thuc, H.H., and Nguyen, T.T. (2020). Investigation of chemical constituents and cytotoxic activity of the lichen *Usnea undulata*. *Vietnam Journal of Chemistry* 58: 63–66.

23 Popovici, V., Bucur, L.A., Schroder, V. et al. (2020). Evaluation of the cytotoxic activity of the *Usnea barbata* (l.) f. H. Wigg dry extract. *Molecules* 25: 1865.

24 Tang, J.-Y., Wu, K.-H., Wang, Y.-Y. et al. (2020). Methanol extract of *Usnea barbata* induces cell killing, apoptosis, and DNA damage against oral cancer cells through oxidative stress. *Antioxidants* 9: 694.

25 Popovici, V., Bucur, L., Gird, C.E. et al. (2022). Antioxidant, cytotoxic, and rheological properties of canola oil extract of *Usnea barbata* (l.) weber ex f.H. Wigg from Calimani mountains, Romania. *Plants* 11: 854.

26 Gómez-Serranillos, M.P., Fernández-Moriano, C., González-Burgos, E. et al. (2014). Parmeliaceae family: Phytochemistry, pharmacological potential and phylogenetic features. *RSC Advances* 4: 59017–59047.

27 González-Burgos, E., Fernández-Moriano, C., and Gómez-Serranillos, M.P. (2019). Current knowledge on parmelia genus: Ecological interest, phytochemistry, biological activities and therapeutic potential. *Phytochemistry* 165: 112051.

28 Oh, S.O., Wang, X.Y., Wang, L.S. et al. (2014). A note on the lichen genus ramalina (ramalinaceae, ascomycota) in the Hengduan mountains in China. *Mycobiology* 42: 229–240.

29 Taura, F., Iijima, M., Yamanaka, E. et al. (2016). A novel class of plant type III polyketide synthase involved in orsellinic acid biosynthesis from *Rhododendron dauricum*. *Frontiers in Plant Science* 7: 1452.

30 Staunton, J. and Weissman, K.J. (2001). Polyketide biosynthesis: a millennium review. *Natural Product Reports* 18: 380–416.

31 Singh, G., Armaleo, D., Dal Grande, F., and Schmitt, I. (2021). Depside and depsidone synthesis in lichenized fungi comes into focus through a genome-wide comparison of the olivetoric acid and physodic acid chemotypes of *Pseudevernia furfuracea*. *Biomolecules* 11: 1445.

32 Ivanova, V., Backor, M., Dahse, H.-M., and Graefe, U. (2010). Molecular structural studies of lichen substances with antimicrobial, antiproliferative, and cytotoxic effects from *Parmelia subrudecta*. *Preparative Biochemistry and Biotechnology* 40: 377–388.

33 Bogo, D., Matos, M.D.F.C., Honda, N.K. et al. (2010). *In vitro* antitumour activity of orsellinates. *Zeitschrift für Naturforschung C* 65: 43–48.

34 Bačkorová, M., Bačkor, M., Mikeš, J. et al. (2011). Variable responses of different human cancer cells to the lichen compounds parietin, atranorin, usnic acid and gyrophoric acid. *Toxicology in Vitro* 25: 37–44.

35 Paluszczak, J., Kleszcz, R., Studzinska-Sroka, E., and Krajka-Kuzniak, V. (2018). Lichen-derived caperatic acid and physodic acid inhibit Wnt signaling in colorectal cancer cells. *Molecular and Cellular Biochemistry* 441: 109–124.

36 Bogo, D., Honda, N.K., Alcantara, G.B. et al. (2020). Cytotoxic activity of compounds from lichens of the cerrado biome in Brazil. *Orbital: The Electronic Journal of Chemistry* 12: 7–16.

37 Brandão, L.F.G., Alcantara, G.B., Matos, M.D.F.C. et al. (2013). Cytotoxic evaluation of phenolic compounds from lichens against melanoma cells. *Chemical and Pharmaceutical Bulletin* 61: 176–183.

38 Russo, A., Piovano, M., Lombardo, L. et al. (2008). Lichen metabolites prevent UV light and nitric oxide-mediated plasmid DNA damage and induce apoptosis in human melanoma cells. *Life Sciences* 83: 468–474.

39 Truong, T.L., Nga, V.T., Huy, D.T. et al. (2014). A new depside from *usnea aciculifera* growing in Vietnam. *Natural Product Communications* 9: 1179–1180.

40 Brisdelli, F., Perilli, M., Sellitri, D. et al. (2013). Cytotoxic activity and antioxidant capacity of purified lichen metabolites: an *in vitro* study. *Phytotherapy Research* 27: 431–437.

41 Demir, L., Togar, B., Turkez, H. et al. (2015). The investigation of cytogenetic and oxidative effects of diffractaic acid on human lymphocyte cultures. *Brazilian Archives of Biology and Technology* 58: 75–81.

42 Bačkorová, M., Jendzelovsky, R., Kello, M. et al. (2012). Lichen secondary metabolites are responsible for induction of apoptosis in HT-29 and A2780 human cancer cell lines. *Toxicology in Vitro* 26: 462–468.

43 Kosanić, M., Manojlović, N., Janković, S. et al. (2013). *Evernia prunastri* and *Pseudoevernia furfuraceae* lichens and their major metabolites as antioxidant, antimicrobial and anticancer agents. *Food and Chemical Toxicology* 53: 112–118.

44 Stojanović, I.Ž., Najman, S., Jovanović, O. et al. (2014). Effects of depsidones from *Hypogymnia physodes* on HeLa cell viability and growth. *Folia Biologica* 60: 89–94.

45 Ögmundsdóttir, H.M., Zoëga, G.M., Gissurarson, S.R., and Ingólfsdóttir, K. (1998). Natural products: anti-proliferative effects of lichen-derived inhibitors of 5-lipoxygenase on malignant cell-lines and mitogen-stimulated lymphocytes. *The Journal of Pharmacy and Pharmacology* 50: 107–115.

46 (a) Duong, T.-H., Chavasiri, W., Boustie, J., and Nguyen, K.-P.-P. (2015). New meta-depsidones and diphenyl ethers from the lichen *Parmotrema tsavoense* (krog & swinscow) krog & swinscow, parmeliaceae. *Tetrahedron* 71: 9684–9691. (b) Bézivin, C., Tomasi, S., Rouaud, I. et al. (2004). Cytotoxic activity of compounds from the lichen: *Cladonia convolute*. *Planta Medica* 70: 874–877.

47 Alexandrino, C.A.F., Honda, N.K., Matos, M.D.F.C. et al. (2019). Antitumor effect of depsidones from lichens on tumor cell lines and experimental murine melanoma. *Revista Brasileira de Farmacognosia* 29: 449–456.

48 Cochereau, B., Meslet-Cladière, L., Pouchus, Y.F. et al. (2022). Halogenation in fungi: What do we know and what remains to be discovered? *Molecules* 27: 1-45.

49 Russo, A., Caggia, S., Piovano, M. et al. (2012). Effect of vicanicin and protolichesterinic acid on human prostate cancer cells: Role of Hsp70 protein. *Chemico-Biological Interactions* 195: 1–10.

50 Russo, A., Piovano, M., Lombardo, L. et al. (2006). Pannarin inhibits cell growth and induces cell death in human prostate carcinoma DU-145 cells. *Anti-Cancer Drugs* 17: 1163–1169.

51 Millot, M., Tomasi, S., Studzinska, E. et al. (2009). Cytotoxic constituents of the lichen *Diploicia canescens*. *Journal of Natural Products* 72: 2177–2180.

52 Yang, Y., Park, S.-Y., Thanh, N. et al. (2015). Lichen secondary metabolite, physciosporin, inhibits lung cancer cell motility. *PLoS One* 10: e0137889.

53 Ji Ram, V., Sethi, A., Nath, M., and Pratap, R. (2019). Lichen secondary metabolite, physciosporin, inhibits lung cancer cell motility. In: *The Chemistry of Heterocycles* (ed. V.J. Ram, A. Sethi, M. Nath, and R. Pratap), 149–478. Elsevier.

54 Taguchi, H., Sankawa, U., and Shibata, S. (1969). Biosynthesis of natural products. VI. Biosynthesis of usnic acid in lichens. *Chemical and Pharmaceutical Bulletin* 17: 2054–2060.

55 Ranković, B., Kosanić, M., Stanojković, T. et al. (2012). Biological activities of *Toninia candida* and *Usnea barbata* together with their norstictic acid and usnic acid constituents. *International Journal of Molecular Sciences* 13: 14707–14722.

56 Einarsdóttir, E., Groeneweg, J., Björnsdóttir, G.G. et al. (2010). Cellular mechanisms of the anticancer effects of the lichen compound usnic acid. *Planta Medica* 76: 969–974.

57 Kumar KC, S. and Müller, K. (1999). Lichen metabolites. 2. Antiproliferative and cytotoxic activity of gyrophoric, usnic, and diffractaic acid on human keratinocyte growth. *Journal of Natural Products* 62: 821–823.

58 Kwong, S.P. and Wang, C. (2020). Review: usnic acid-induced hepatotoxicity and cell death. *Environmental Toxicology and Pharmacology* 80: 103493.

59 Yu, X., Guo, Q., Su, G. et al. (2016). Usnic acid derivatives with cytotoxic and antifungal activities from the lichen *Usnea longissimi*. *Journal of Natural Products* 79: 1373–1380.

60 (a) Takenaka, Y., Tanahashi, T., Nagakura, N., and Hamada, N. (2000). Production of xanthones with free radical scavenging properties, emodin and sclerotiorin by the cultured lichen mycobionts of *Pyrenula japonica*. *Zeitschrift für Naturforschung C: A Journal of Biosciences* 55: 910–914. (b) Prateeksha, Yusuf, M.A., Singh, B.N. et al. (2019). Chrysophanol: a natural anthraquinone with multifaceted biotherapeutic potential. *Biomolecules* 9: 1–24.

61 Badiali, C., Petruccelli, V., Brasili, E., and Pasqua, G. (2023). Xanthones: biosynthesis and trafficking in plants, fungi and lichens. *Plants* 12: 1–34.

62 Prokopiev, I., Sleptsov, I., Serebryakov, E., and Sharoyko, V. (2023). Antioxidant and cytotoxic activities of quinones from *Cetraria laevigata*. *Natural Product Research* 38: 1–5.

63 Ernst-Russell, M.A., Elix, J.A., Chai, C.L.L. et al. (1999). Hybocarpone, a novel cytotoxic naphthazarin derivative from mycobiont cultures of the lichen *Lecanora hybocarpa*. *Tetrahedron Letters* 40: 6321–6324.

64 Resende, D.A.-O., Pereira-Terra, P., Inácio, Â.S. et al. (2018). Lichen xanthones as models for new antifungal agents. *Molecules* 23: 2617.

65 Duong, T.-H., Nguyen, H.-H., Le, T.-T. et al. (2020). Subnudatones a and b, new trans-decalin polyketides from the cultured lichen mycobionts of *Pseudopyrenula subnudata*. *Fitoterapia* 142: 104512.

66 M. Maulidiyah, Cahyana, A.H., Suwarso, W.P., and Nurdin, M. (2015). Isolation and structure elucidation of eumitrin a1 from lichen *Usnea blepharea* Motyka and its cytotoxic activity. *International Journal of PharmTech Research* 8: 782–789.

67 Tuong, T.L., Aree, T., Do, L.T.M. et al. (2019). Dimeric tetrahydroxanthones from the lichen *Usnea aciculifera*. *Fitoterapia* 137: 104194.

68 Tuong, T.L., Do, L.T.M., Aree, T. et al. (2020). Tetrahydroxanthone-chromanone heterodimers from lichen *Usnea aciculifera* and their cytotoxic activity against human cancer cell lines. *Fitoterapia* 147: 104732.

69 Mishra, A.K., Choi, J., Choi, S.-J., and Baek, K.-H. (2017). Cyclodipeptides: an overview of their biosynthesis and biological activity. *Molecules* 22: 1–13.

70 Williams, D.E., Bombuwala, K., Lobkovsky, E. et al. (1998). Ambewelamides a and b, antineoplastic epidithiapiperazinediones isolated from the lichen *Usnea* sp. *Tetrahedron Letters* 39: 9579–9582.

71 Ernst-Russell, M.A., Chai, C.L.L., Hurne, A.M. et al. (1999). Structure revision and cytotoxic activity of the scabrosin esters, epidithiopiperazinediones from the lichen *Xanthoparmelia scabrosa*. *Australian Journal of Chemistry* 52: 279–283.

72 Tokiwano, T., Satoh, H., Obara, T. et al. (2009). A lichen substance as an antiproliferative compound against HL-60 human leukemia cells: 16-*o*-acetyl-leucotylic acid isolated from *Myelochroa aurulenta*. *Bioscience, Biotechnology, and Biochemistry* 73: 2525–2527.

73 Liu, H., Liu, Y.-Q., Liu, Y.-Q. et al. (2010). A novel anticancer agent, retigeric acid b, displays proliferation inhibition, S phase arrest and apoptosis activation in human prostate cancer cells. *Chemico-Biological Interactions* 188: 598–606.

74 Chae, H.-J., Kim, G.-J., Deshar, B. et al. (2021). Anticancer activity of 2-*o*-caffeoyl alphitolic acid extracted from the lichen, *Usnea barbata* 2017-kl-10. *Molecules* 26: 3937.

75 Ngoc Tuan, N., Kuo, P.-C., Trung Hieu, T. et al. (2021). A new triterpenoid and other compounds from lichens *Cryptothecia faveomaculata* Makhija & Patw. *Natural Product Research* 35: 1349–1356.

76 Bessadóttir, M., Skúladóttir, E.Á., Gowan, S. et al. (2014). Effects of anti-proliferative lichen metabolite, protolichesterinic acid on fatty acid synthase, cell signalling and drug response in breast cancer cells. *Phytomedicine* 21: 1717–1724.

77 Huynh, B.L., Duong, T.H., Do, T.M.L. et al. (2016). New γ-lactone carboxylic acids from the lichen *Parmotrema praesorediosum* (nyl.) Hale, parmeliaceae. *Records of Natural Products* 10: 332–340.

78 Solarova, Z., Liskova, A., Samec, M. et al. (2020). Anticancer potential of lichens' secondary metabolites. *Biomolecules* 10: 87.

79 Adenubi, O.T., Famuyide, I.M., McGaw, L.J., and Eloff, J.N. (2022). Lichens: an update on their ethnopharmacological uses and potential as sources of drug leads. *Journal of Ethnopharmacology* 298: 115657.

80 Ingolfsdottir, K. (2002). Usnic acid. *Phytochemistry* 61: 729–736.

81 Macedo, D.C.S., Almeida, F.J.F., Wanderley, M.S.O. et al. (2020). Usnic acid: from an ancient lichen derivative to promising biological and nanotechnology applications. *Phytochemistry Reviews* 20: 609–630.

82 Pyrczak-Felczykowska, A., Narlawar, R., Pawlik, A. et al. (2019). Synthesis of usnic acid derivatives and evaluation of their antiproliferative activity against cancer cells. *Journal of Natural Products* 82: 1768–1778.

83 Song, Y., Dai, F., Zhai, D. et al. (2012). Usnic acid inhibits breast tumor angiogenesis and growth by suppressing VEGFR2-mediated AKT and ERK1/2 signaling pathways. *Angiogenesis* 15: 421–432.

84 Zuo, S.-T., Wang, L.-P., Zhang, Y. et al. (2015). Usnic acid induces apoptosis via an ros-dependent mitochondrial pathway in human breast cancer cells *in vitro* and *in vivo*. *RSC Advances* 5: 153–162.

85 Su, Z.Q., Liu, Y.H., Guo, H.Z. et al. (2017). Effect-enhancing and toxicity-reducing activity of usnic acid in ascitic tumor-bearing mice treated with bleomycin. *International Immunopharmacology* 46: 146–155.

86 Ebrahim, H.Y., Akl, M.R., Elsayed, H.E. et al. (2017). Usnic acid benzylidene analogues as potent mechanistic target of rapamycin inhibitors for the control of breast malignancies. *Journal of Natural Products* 80: 932–952.

87 Yang, Y., Bae, W.K., Lee, J.Y. et al. (2018). Potassium usnate, a water-soluble usnic acid salt, shows enhanced bioavailability and inhibits invasion and metastasis in colorectal cancer. *Scientific Reports* 8: 16234.

88 Geng, X., Zhang, X., Zhou, B. et al. (2018). Usnic acid induces cycle arrest, apoptosis, and autophagy in gastric cancer cells *in vitro* and *in vivo*. *Medical Science Monitor* 24: 556–566.

89 Nguyen, T.T., Yoon, S., Yang, Y. et al. (2014). Lichen secondary metabolites in *Flavocetraria cucullate* exhibit anti-cancer effects on human cancer cells through the induction of apoptosis and suppression of tumorigenic potentials. *PLoS One* 9: e111575.

90 Wang, H., Xuan, M., Huang, C., and Wang, C. (2022). Advances in research on bioactivity, toxicity, metabolism, and pharmacokinetics of usnic acid *in vitro* and *in vivo*. *Genes* 27: 7469.

91 Solar, P., Hrckova, G., Koptasikova, L. et al. (2016). Murine breast carcinoma 4T1 cells are more sensitive to atranorin than normal epithelial NMuMG cells in vitro: Anticancer and hepatoprotective effects of atranorin *in vivo*. *Chemico-Biological Interactions* 250: 27–37.

92 Zhou, R., Yang, Y., Park, S.Y. et al. (2017). The lichen secondary metabolite atranorin suppresses lung cancer cell motility and tumorigenesis. *Scientific Reports* 7: 8136.

93 Ni, Z., Nie, X., Zhang, H. et al. (2022). Atranorin driven by nano materials SPION lead to ferroptosis of gastric cancer stem cells by weakening the mRNA 5-hydroxymethylcytidine modification of the Xc-/GPX4 axis and its expression. *International Journal of Medical Sciences* 19: 1680–1694.

94 Barroso Martins, M.C., Alves Rocha, T., Santos Silva, T.D. et al. (2016). *In vitro* and *in vivo* antineoplastic activity of barbatic acid. *International Archives of Medicine* 9: 1–9.

95 Karagoz, I.D., Ozaslan, M., Guler, I. et al. (2014). *In vivo* antitumoral effect of diffractaic acid from lichen metabolites on swiss albino mice with Ehrlich ascites carcinoma: An experimental study. *International Journal of Pharmacology* 10: 307–314.

96 Tas, I., Han, J., Park, S.Y. et al. (2019). Physciosporin suppresses the proliferation, motility and tumourigenesis of colorectal cancer cells. *Phytomedicine* 56: 10–20.

97 Poornima, S. (2016). Evaluation of anti-cancer properties of lichens using *Albino wistar* rats as an animal model. *Cancer Research Journal* 4: 84.

15

Ethnobotanical and Pharmacological Properties of *Parmelia, Cetraria, Cladonia,* and *Usnea*

Varsha Sharma[1] *and Alka Rajput*[2]

[1] *Department of Zoology, School of Basic, Applied and Biosciences, RIMT University, Mandi-Gobindgarh, India*
[2] *Department of Botany, School of Applied, Basic and Biosciences, RIMT University, Mandi-Gobindgarh, India*

15.1 Introduction

Lichens are sui generis organisms that exert extensive biological actions in conjunction with anti-inflammatory, antibiotic, antiviral, analgesic, antipyretic, and cytotoxic effects. Lichens are often used to treat respiratory, digestive, obstetric, and gynecological concerns. The doctrinal use of lichens can traced back to the 18th dynasty (1700–1800 BC) once *Evernia furfuracea* (L.) Mann (Parmeliaceae) was first time used as a drug [1]. Some lichens were beneficial in restoring hair loss, coughs, jaundice, and rabies [2]. They exist on every continent and have a history of being used as animal feed, a source of dyes, medicine, and food [3]. Lichens have been used to treat leprosy, kidney diseases, arthritis, alopecia, pharyngitis, rabies, constipation, worm-related infection, and infestation. Few countries have developed commercial pharmacological active products from lichen, such as usnic acid and atranorin [4].

Worldwide cultures use lichens in traditional medicine, mainly in temperate and arctic regions. They have been used for their primary or secondary metabolites [5]. In different zones of the world, lichens are generally important genera in traditional medicines, out of which *Usnea* is the most widely used genus [6]. It is not only traditionally used medicinal lichen in Europe, but plentiful other lichen genera have meticulous value in different parts of the world. A few of the important genera are given in Table 15.1.

The current challenge in front of pharmaceuticals is the authentication and development of new pharmacologically active compounds from traditionally important phytomedicines [7]. From a therapeutic point of view, lichens have a unique position in the pharmaceutical industry due to their metabolic activities. Different biologically active compounds of lichen thalli have been effectively used to treat overweight, diabetes, malignancy, and Alzheimer's disease [8]. Lichen and their secondary metabolites consist mainly of depsides, depsidones, dibenzofurans, terpene, and xanthones derivatives that account for their analgesic, antipyretic, antiproliferative, anti-inflammatory, antiviral, anti-tumor, and

Chemistry, Biology and Pharmacology of Lichen, First Edition. Edited by Ashoke Kumar Das, Ajay Sharma, Deepika Kathuria, Mohammad Javed Ansari, and Garima Bhardwaj.

Table 15.1 Traditionally important lichen genera and their geographic location.

Serial number	Lichen genera	Geographic location
01	*Cetraria*	Europe
02	*Evernia* and *Pseudevernia*	Europe and North Africa
03	*Cladonia* and *Cladina*	North America, Europe, and Asia
04	*Letharia*	North America
05	*Lobaria* and *Peltigera*	North America, Europe, and Asia
06	*Lethariella*	China
07	*Parmotrema* and *Hypotrachyna*	India
08	*Ramalina*	North America, Europe, and Asia
09	*Umbilicaria*	North America and Asia
10	*Thamnolia*	Asia
11	*Usnea*	Worldwide (except Australia)
12	*Xanthoparmelia*	North America and Africa

antiprotozoal activities [9]. In addition, depside compound has been associated with antimicrobial and anticancer properties and could be of significant pharmaceutical value as an anticancer drug [10].

The secondary metabolites of lichens have been investigated for their hepato-protecting activities. Diffractaic acid has been reported as a liver protective agent against toxicity induced by CCl_4[11]. Another one is atranorin (a secondary metabolite of *Stereocaulon caespitosum*), which has been proclaim for tumor-suppressive activity in hepatocellular carcinoma. This metabolite accomplished particular activity by regulating the cell cycle, cell death, and metastatic activity [12]. The secondary metabolites of lichens have been reported for hepatoprotective efficacy (Table 15.2) and thus possess a crucial role in pharmaceuticals [17].

Table 15.2 Hepatoprotective effects of various species of lichens.

Serial number	Lichen	Hepatoprotective efficacy
01	*Cladina rangiferina*	Jaundice [13]
02	*Cladonia pleurota*	Hepato-cooling agent [13]
03	*Evernia divaricata*	Hepatitis [13]
04	*Parmelia perlata*	Hepatic dysfunction [14]
05	*Parmelia saxatilis*	Jaundice [13]
06	*Parmelioid lichens*	Cleanser for liver [15]
07	*Usnea certain*	Hepatitis [13]
08	*Usnea ghattensis*	Anti-oxidative potential for liver [16]

The lichens are widely used in medicine due to their active biological substances, mainly with antimicrobial potential. These active substances are considered in the chemotaxonomy of lichen and as natural antibiotics [18]. Due to inbuilt medicinal potential, lichens have been used as an economically viable substance from ancient times. Three biosynthetic pathways, the shikimic acid, the mevalonic acid, and the polymalonate pathway, may be used by lichens for the production of their active metabolites [19]. In several ways, lichens are being used by different communities of the world. In Nepal, among the Brahman, Dalit, Lama, Rai, Chhetri, Gurung, Limbu, Tamang, and Sherpa communities, there is the ethnic use of seven species of lichens belonging to four families (Usneaceae, Parmeliaceae, Ramalinaceae, and Physciaceaem) and six genera (*Usnea*, *Everniastrum*, *Parmotrema*, *Ramalina*, *Thamnolia*, and *Heterodermia*) [20]. The Hills of Uttar Pradesh, a part of the Indian Himalayan region, is the main area for the collection of lichens. The superior grade of trade lichens widely includes *Parmotrema* and *Everniastrum* (Parmelia) species are collect from the Chirpine, Himalayan Oak, Indian Bayberry, and Rhododendron [21]. Lichens are used not only as medicines but also as a source of dyes, food, fodder, perfumes, and spices [22]. It is estimated that around 8% of the terrestrial ecosystem is occupied by lichens, and more than 20,000 lichen species are distributed worldwide. The biological activities of different lichen species and their active compounds are found unexplored to date [23].

15.2 Ethnobotanical and Pharmacological Properties of the Genus *Parmelia*

Parmelia is medium to giant foliose lichen and has a global distribution. It extends from the Arctic to the Antarctic continent but chiefly concentrates in temperate regions. The Parmeliaceae family mainly includes around 80 genera and over 2000 species [24]. Earlier to 1810, five different culturally significant genera of Parmeliaceae, including *Alectoria*, *Evernia*, *Cetraria*, *Parmelia*, and *Usnea*, were described. In 1903, *Evernia* was split into *Letharia* and *Pseudevernia*, whereas *Parmotrema* and *Hypogymnia* had been split from *Parmelia*. These genera have not been recognized by all authors. The taxonomy of Parmeliaceae existed in nature in 1965 through the subsequent splitting of the genus *Cetraria* into different genera. In 1974, the genus *Parmelia* splitting was started and completed by the early 1990s by including more than 80 genera in the family [25]. Currently, genus *Parmelia* and *Cetraria* consist of 32 and 22 genera, respectively. Lichens are classified into three types according to their growth forms: fruticose (*Usnea*), Crustose (*Rhizocarpon*), and Foliose (*Umbilicaria*) lichens. However, some are between the above types, for example, *Cladonia*.

Medicinal values of different lichens are associated with folklore. In distinctive cultures, a folk taxonomy of living organisms is developed and goes along with clans. It has been observed that folk taxonomies are peculiar to a culture and, more often than not, a reflection of particular environments and values. Although, many cultures have to show detailed folk taxonomy for the identification of lichens. Pharmacological and doctrinal activities of *Parmelia* species are recapitulated in Tables 15.3 and 15.4, respectively.

Table 15.3 Active phytoconstituents and pharmacological activities of *Parmelia* species.

Lichen species	Active phytoconstituents	Pharmacological activities
Parmelia arseneana	Usnic acid, atranorin, stictic acid, norstictic acid	Antimicrobial, Cytotoxic, Antioxidant
Parmelia sulcata	Salazinic acid	Antimicrobial, Cytotoxic, Antioxidant
Parmelia pulla	Glomelliferic acid, divaricatic acid, stenosporic acid, gyrophoric acid	Anti-genotoxic
Parmelia saxatilis	Salazinic acid, usnic acid	Anti-inflammatory, Antioxidant, Antimicrobial, Cytotoxic
Parmelia crinita	Stictic acid, atranorin	Antioxidant, Antimicrobial
Parmelia reticulata	Protolichesterinic acid, atranorin	Antimicrobial
Parmelia erumpens	2-hydroxy-4-methoxy-3,6-dimethylbenzoic acid, atranorin, usnic acid	Cytotoxic, Antimicrobial
Parmelia nepalensis	Gyrophoric acid, usnic acid, diffractaic acid, methyl β-orcinolcarboxylate, ethylhematommate, atranorin, protolichesterinic acid	Antiproliferative
Parmelia subrudecta	Lecanoric acid, orsellinic acid methyl ester, orcinol, usnic acid	Antimicrobial, Cytotoxic,Antiproliferative
Parmelia tinctorum	Atranorin, diffractaic acid, gyrophoric acid, methyl β-orcinolcarboxylate, ethyl hematommate, protolichesterinic acid, usnic acid	Uncoupler of oxidative Phosphorylation, Radioprotective, Antiproliferative
Parmelia caperata	Protocetraric acid	Cytotoxic, Antimicrobial, Antioxidant
Parmelia perlata	Usnic acid, atranorin	Antimicrobial, Antioxida, Antidiabetic

Table 15.4 Doctrinal use of parmelioid lichens.

Country	Doctrinal use
Canoparmeliae caperata	
India	Respiratory congestion [26]
Flavoparmelia caperata	
Mexico	Burns [27]
China	Heat removal
Hypotrachyna cirrhata [syn. *P. kamtschadalis*]	

Table 15.4 (Continued)

Country	Doctrinal use
India	Benefits same as *Usnea baileyi*. Smoke relieves headaches Powder used for cephalic snuff [28] Fever and cold (Kumaon) [26]
Nepal	Along with unidentified parmelioid lichen paste of the whole plant is applied on cuts and wounds [29]
New Guinea	In chew form, "*Evernia* sp." used on skin eruptions [30]
Hypotrachyna nepalensis [syn. *P. nepalensis*]	
India	Benefits similar to *Usnea baileyi* (mix in tobacco). In charila is used as an infrequent adulterant [21]
India	Respiratory problems [31]
Nepal	Application similar as *Ramalina* spp. as antiseptic, act on burn, and wounds. Used as a powder in tincture of iodine after application of leaf juice of *Eupatorium adenophorum* or *Artemisia dubia*[32]
Nepal	In sore throat, toothache, and pain [33]
Parmelia omphalodes	
Britain	In Scotland, apply to prevent inflamed feet while traveling [34] In soup form to strengthen invalids in Ireland. Poultice for sores, cuts, and burns in Scotland and Ireland [35] Perhaps interchangeable with *P. saxatilis*.
Parmelia saxatilis	
Bhutan	In traditional medicine, Sowa Rigpa heals white leprosy, skin diseases, and sore feet [36]
China	Vomiting blood, chronic dermatitis, jaundice, blurred vision, uterus bleeding, and oral ulcers in children. The decoction is used with wine, and powder is applied to the affected area [13]
Britain	Interchangeable with *P. omphalodes* as crottle.
Shetland Is., Britain	In 1966, mixed with tobacco and smoked [37]
Parmelia sulcata	
Alberta, Canada	To prevent pain and discomfort in the gums of teething babies [38]
BC, Canada	As birth control. Not differentiated from *Lobaria pulmonaria*[39]
Parmelinella *salacinifera*	
Brazil	A component of flor de pedra [40]
Parmelinella *abessinicum* [syn. ***Parmelia*** *abessinica*]	
India	Taken as medicine [41]
Parmotrema *andinum* [syn. ***Parmelia*** *hyporysalea*]	
India	Occasionally used as an adulterant in charila [42]
Parmotrema nilgherrense [syn. ***Parmelia*** *nilgherrensis*]	

(*Continued*)

Table 15.4 (Continued)

Country	Doctrinal use
India	According to Ayurvedic texts, it was considered an important drug. In India, *Parmotrema nilgherrense* (high-quality charila) coupled with the lichen mixture *of P. perlatum, Parmotrema perforatum, Parmotrema chinense, Hypotrachyna nepalensis* and *H. cirrhata*, with some adulterant *Ramalina inflata, R. farinacea, Usnea longissima, U. baileyi, U. subsordida, Parmelia hyporysalea, Leptogium* spp., and *Anaptychia* spp. are used. It is beneficial in indigestion, stomach disorders, flatulence, loss of appetite, diarrhea, painful urination, kidney stones, hemorrhoids, congestion, shortness of breath, bronchitis, enlarged spleen, sore throats, fevers, toothaches, excessive salivation, musculoskeletal pain, headaches, rheumatism, broken bones, in swelling, leprosy, soothing irritated skin, scabies, involuntary semen emission, painful menstruation, lack of menstruation, and prenatal and postnatal care. In addition, it has an aphrodisiac, astringent, antibiotic, antiseptic, diuretic, sedative, and a demulcent to reduce inflammation. In powdered form, it is used for wounds. Smoke is used as incense in medicated oil as a cephalic snuff to relieve headaches. It is used as a poultice to lumbar and renal regions for urine induction and as a remedy for headaches [42, 43]. A
Nepal	In Nepal, the paste is applied externally for kidney stones, and soup is used as an aphrodisiac. In China, the paste is applied externally for kidney stones. Soup is beneficial in male infertility. In Malaysia, soup is used as an aphrodisiac and for seminal weakness. In Afghanistan, soup is for chest disorders and paste in wound healing. In Saudi Arabia, it is a painkiller and cephalic snuff for headaches [44].
India	In cough, throat (Kumaon), and unspecified respiratory problems (Sikkim) [31]
Nepal	For pain, toothache and sore throat [33]
***Parmotrema** perforatum* [syn. ***Parmelia** perforata*]	
India	Charila [28] and as diuretic [45]
Parmotrema reticulatum [syn. ***Parmelia** reticulata*]	
India	For cough and throat (Kumaon) and unspecified respiratory problems (Sikkim) [26, 31]
Chihuahua, Mexico	Tea is prepared especially in afternoon and left overnight before being drunk to relieve discomfort related to venereal disease [27]
Parmotrema saccatilobum	
Papua New Guinea	Oil extract of fresh or dried lichen is applied to the joint. Beneficial in muscle pain or inflammation [46]
Parmotrema sancti-angelii syn. *Parmelia sancti-angelii*	
India	In skin disease (sem.), salve was used. In this, 30–50 g of jhavila was burned and ash mixed with mustard or linseed oil.
Parmotrema subtinctorium [syn. *Parmelia subtinctoria*]	
China	Beneficial in localized swelling, pain, and bleeding from external injury [13]
Parmotrema tinctorum [syn. *Parmelia tinctoria*]	
China	Used for sores and swelling, blurred vision, chronic dermatitis, uterus bleeding, and external injuries. In this, either drink decoction or apply powdered form on the affected area [13]

Table 15.4 (Continued)

Country	Doctrinal use
Parmotrema wrightii	
Brazil	One of the components in a medicinal mixture with *Heterodermia galactophylla* and *Parmelinella salacinifera*. In the utility of pregnant women neither *P. salacinifera* nor *H. galactophylla* is added. In case of children *H. Galactophylla* is not recommended. Aqueous extract is basically beneficial in digestive problems (diarrhea and vomiting). In smokers, it is used to treat epilepsy and cultural diseases [40]
Parmotrema zollingeri syn. *Parmelia zollingeri*	
Philippines	Burn kalas and let the child smell the fumes to treat high fever and convulsions [47]
Punctelia borreri	
China	Used in the treatment of bleeding from the uterus and any external injuries, chronic dermatitis, blurred vision, and sores and swelling. Drunk the decoction or applied powder form to the affected area [13]
Xanthoparmelia spp.	
(AZ and NM, USA) New Mexico	New Mexico: According to Elmore tschétláatas remedy for impetigo [48] céλá·d or nihaλá·d: chewed for swollen gums, canker, and decayed teeth (sometimes include *Peltigera* sp.) [49] Arizona: Whiting mentioned an unidentified rock lichen called owa'si or owa'huru'suki, and have benefits for sore mouth and toothache [50]
Xanthoparmelia conspersa	
South Africa	In syphilis eruption, powder is applied externally on the eruption (perhaps after scarified) and sometimes used internally (sources disagree). For the treatment of suspected and known snakebites, in this one tablespoon of lichen made in cold water is prescribed to drink, scarify the bite, and then powder form is sprinkled on it [51]
Ontario, Canada	Use in fever-associated raw throat and inflamed gums. One cup of cold water mixed with the contents collected from the bark of the tree of *Fraxinus nigra* and *Coptis trifolia*. For treatment, one teaspoon is taken in the mouth until warmed and finally swallowed. Repeat this process for the entire cup [52]
Arizona, USA	Traditional benefits were described by Curtin[53]. Lipp identified it as *X. conspersa* [54], while Hawksworth disagreed with this identification [37]. It is usually mixed with tobacco and then smoked to "make young men crazy." The powdered form sprinkled on sores or cuts, but not bound, as it would cause blisters. It is widely used to heal rattlesnake bites.
Xanthoparmelia hottentotta [syn. *Xanthoparmelia hottentotta*]	
Kuiseb, Namibia	Decoction is used for the treatment of cough and pain in the stomach and chest [55]
Xanthoparmelia tinctina	
China	Used for blurred vision, chronic dermatitis, sores,swelling, bleeding from the uterus, and external injuries. For cure, decoction is drunk or applied decoction or powdered form to the affected area [13]

15.3 Ethnobotanical and Pharmacological Properties of the Genus *Cetraria*

Cetraria comes from the Latin word cetra, a small and light leather shield of Roman fiction [56]. The *Cetraria* genus was first time described by Erik Acharius (known as the father of lichenology) in 1803, when his work Methodus qua omnes detectos lichenes ad genera secundum organa carpomorpha was first time published. *Cetraria* genus was foliose form having membranous, cartilaginous thallus with curled and irregular lobes. The presence of certain morphological features such as growth form and color of the thallus, structures like soredia, isidia, pycnidia and cilia, and pseudocyphellae on the upper or lower surfaces, or both are used for the identification of *Cetraria* species [57]. From a pharmacological point of view, *Cetraria islandica* has been considered the most beneficial and investigated species. In addition to therapeutic activity, *C. islandica* has been used in folk medicine to mitigate the symptoms related to digestion and respiration [58]. The pharmacological and doctrinal activities of *Cetraria* species are summarized in Tables 15.5 and 15.6, respectively.

15.4 Ethnobotanical and Pharmacological Properties of the Genus *Cladonia*

*Cladonia*isconsidered a sub-cosmopolitan genus, and its photobiont belongs to Asterochloris (a green algae genus). *Cladonia* includes those species commonly known as reindeer moss, British soldiers, and cuplichen. They are fruticose lichens that consist of two types of the thallus, one is horizontal thallus (primary, seldomcrustose or squamulose), and another is vertical thallus (secondary, podetia). The active phytoconstituents and doctrinal activities of the*Cladonia* species are summarized in Tables 15.7 and 15.8, respectively.

Table 15.5 Active phytoconstituents and pharmacological activity of *Cetraria* species.

Lichen species	Active phytoconstituents	Pharmacological activities
Cetraria aculeate	Lichesterinic acid, protolichesterinic acid	Antioxidant, Cytotoxic, Antibacterial, Antifungal, Antigenotoxicity
Cetraria annae	Isonephrosterinic acid, usnic acid	Antioxidant, Cytotoxic
Cetraria islandica	Fumarprotocetraric acid, protolichesterinic acid, protocetraric acid, usnic acid	Antioxidant, Cytotoxic, Antibacterial, Antifungal, Antidiabetic, Anti-inflammatory, Antitrypanosomal, Cellular differentiation, Depigmenting, Genotoxicity, Immunomodulating, Neuroprotective

Table 15.6 Doctrinal use of *Cetraria* lichens.

Country	Doctrinal use
Cetraria islandica	
Europe	According to European pharmacopoeias, it has been used as medicinal lichen from the 1500s to till date. In Europe and Greenland, it is widely used for digestive and pulmonary problems. To remove bitterness solution is boiled in water or in weak alkaline solution once or twice times. Use as a decoction or sometimes infusion in milk or water. It is administrated as a drink or a jelly. Sugar, lemon, or other flavorings substances are also mixed in final product. Use as a powder for salves like strengthening tonic and for the treatment of asthma, chronic congestion, whooping cough, pulmonary tuberculosis, indigestion, edema, wounds, scurvy, dysentery, chronic diarrhea, kidney stones, and uterine cysts. It had laxative, sedative, and aphrodisiac properties [78, 79, 154, 157–162]
Spain	Use as a tea for inflammation, asthma, tuberculosis, congestion, and high blood pressure [107, 163, 164]
France	Lung-associated problems and an emollient are cured by decoction [165]
Northern Italylic: Kyé, RoccaforteMondovì; Waldensian; Sondrio	In cough, suppressant (Occitans), drunk as a tea. Decoction: used in intestinal problems, bronchitis, and cough; sometimes served with milk; in overdose body becomes weak. It is also usedto treat stomach disorders in cattle and as an expectorant (Waldensian). It is drunk as digestive. In bronchitis and pneumonia, fed to livestock (Sondrio). Use in congestion and tuberculosis (Venezia Giulia). Related application in Trentino [166–170]
Austria	Use as a medicinal drink and mixed with *Plantago lanceolata*[171]
Austria	Immunity booster; as restorer after sickness, weight loss, or delivery and improve skin. Use in weight gain and milk production. Use as fodder for cattle, pigs, horses, and sheep. Additionally, have applications in hoof quality, udder health, and coat gloss. In pregnant animals, it is used to facilitate delivery, prevent birth-related complications, and prevent fever [172]
Lukomir	A decoction is drunk for pulmonary problems and improves blood. Decoction is also used with *Pimpinella anisumis* for croakness and cough. In cold and cough, urinary tract inflammation, and anemia. The powdered lichen in cold water is used. Taken for chest or heart pain [173–176]
Macedonia	Indeterminate medicinal properties [177]
Bulgaria	Coughs [178]
Lebanon	Alcohol-soaked lichen is rubbed externally to treat neuralgia and rheumatism [179]

(*Continued*)

Table 15.6 (Continued)

Country	Doctrinal use
Ukraine	Milk-soaked lichen is taken internally for cough [180]
Lithuania	Tea form, alcohol extract, or decoction is usually drunk for respiratory tract disorders. In addition, decoction is externally applied for burns and skin diseases [181]
Estonia	Tea is used in cancer [182], bronchitis, cough, colds, respiratory problems, and fever [183]
Sweden Islandslav	Warm or cold infusion or decoction in milk or water is drunk to treat whooping cough, asthma, tuberculosis, colds, chest ailments, congestion, diabetes, and nephritis. Sometimes honey or chocolate is used [109]
Iceland	Listed as an important medicinal plant [184]
Siberia	Coughs [62]
Tibet	As famous Tibetan medicine [128]
China	Decoction is used to improve digestion and strengthen the stomach [13]
China (Arabic)	One of famous medicine in Hui Hui Yao Fang [98]
NWT, Canada	In tuberculosis, water-based decoction is used. The water is boiled for 0.5–1 h until the liquid is red. For treatment, the one-third cup is usually drunk three times daily [185]
Cetraria pseudolivetorum	
China	Use for bleeding from the uterus and any trauma. It is used to cure sores, blurred vision, and chronic psoriasis. For treatment, either decoction is drunk or decoction/powdered is applied to the affected area [13]
Flavocetraria cucullata	
Pallars (Spain)	It is used as a tea to cure symptoms associated with asthma [164]
Flavocetraria nivalis	
Europe	It is not commonly used lichen in Europe as *C. islandica* buthas similar medical properties [78]
Qollahuayas, Bolivia	Use in the form of tea for the treatment of heart attack and motion sickness [186]
Masonhalea richardsonii [syn. *Cornicularia richardsonii*]	
Alaska, BC	Use in lungs-associated inflammation [75]
Nephromopsis pallescens [syn. *Cetrariopsis pallescens*]	
China	It has an unspecified medicinal value but is eaten by local people [13]
Vulpicida canadensis	
BC, Canada	Natural remedy for cough and cold. The tea is made in 1 liter of water by adding a couple of handfuls of dagha. If chewed freshly, it is helpful for lungs associated aliments [187]

Table 15.7 Active phytoconstituents of *Cladonia* species.

Lichen species	Common name	Active phytoconstituents
Cladonia arbuscula	Reindeer lichens	Usnic acid, isousnic acid
Cladoniaborealis	Boreal cup lichen	Usnic acid, barbatic acid, 4-O-demethylbarbatic acids
Cladoniacariosa	Split-leg lichen	Atranorin, fumarprotocetraric acid
Cladonia chlorophaea	Mealy pixie cup	Fumarprotocetraric acid
Cladonia coccifera	Madam's cup	Usnic acid, zeorin, isousnic acid, porphyrilic acid
Cladoniadeformis	Lesser sulphur-cup lichen	Usnic acid, zeorin, isousnic acid
Cladonia fimbriata	Trumpet lichen	Fumarprotocetraric acid, atranorin
Cladonia gracilis	Smooth horn lichen	Fumarprotocetraric acid
Cladonia macrophyllodes	Large-leaved cladonia	Fumarprotocetraric acid, atranorin
Cladonia phyllophora	Felt horn lichen	Fumarprotocetraric acid, atranorin
Cladonia pleurota	Red-fruited pixie cup	Usnic acid, zeorin, porphyrilic acid
Cladonia pyxidata	Pebbled pixie lichen	Fumarprotocetraric acid
Cladonia rangiferina	Grey reindeer lichen. actual reindeer lichen	Atranorin, fumarprotocetraric acid
Cladonia uncialis	Thorn lichen	Usnic acid, squamatic acid
Cladonia verticillata	Ladder lichen	Fumarprotocetraric acid

Table 15.8 Doctrinal useof *Cladonia* lichens.

Country	Doctrinal use
Cladina spp.	
Alaska, USA	The decoction is used for the treatment of diarrhea [59]
AK, USA	Plant-based "liquor" prepare and drink to treat colds [60]
Alaska, USA	Tea is used for chest pains. Chewed by hunters during climbing hills chew the lichen for the maintenance of wind [61]
Siberia	Use in scurvy [62]
Scandinavia	Decoction is used as unspecified medicine [62, 63]
Cladina arbuscula	
China	Use for pulmonary tuberculosis, fever, dizziness, hypertension, and skin infections.
Cladina mitis	
India	Medicine used for hepatic and respiratory problems [31]
Cladina laevigata	
Tierra del Fuego, Argentina	Personal hygiene: apply on body before drying with *Protousnea magellanica* [64]

(*Continued*)

Table 15.8 (Continued)

Country	Doctrinal use
Cladina rangiferina	
MN and WI, USA	Boil in water and use for the bath of newborn baby [65]
Quebec, Canada	Use for diabetes-associated inflammation [66]
Finland	Boil in water and drink to treat cough and tuberculosis [67]
China	Remedy for blurred vision, cloudy cornea, headaches, cuts, fever, and phlegm due to dry throat, coughing up blood, jaundice, difficulty in urination, irritable depression urinary tract infection, and rheumatism. For treatment, either drink or apply decoction to the affected area [13]
India	Use in respiratory health [31]
India	In kidney stones, half a teaspoon of dried lichen is added to one cup of boiled water and drank empty stomach in the morning for one month or until problem is not cured [68]
Cladina stellaris [*Cladina alpestris*]	
SK, Canada	In case of intestinal worm infection, decoction or powder lichen add in water and drink [69]
Nunavut, Canada	The broth form use to cure sickness and eye infections [70]
Russian Far East	Powder form used to treat infections and wounds [71]
China	Use for hypertension, tuberculosis, headaches, nose bleeds, eye problems, menstrual problems, and vaginal discharge. Drink decoction [13, 72]
Cladina subtenuis	
NC, USA	Use to relieve pain caused by insect stings. For treatment, either chew or mix with tobacco or put on the area affected by insect sting [73]
Cladonia sp.	
Oaxaca, Mexico	In the treatment of "culture-bound syndromes" [74]
Cladina amaurocraea	
China	Use in dizziness and headaches [13]
Cladina bellidiflora	
Alaska, USA	For the treatment of eye disease, mix with mother's milk [75]
BC, Canada	In sore eyes, red ends are dipped in mother's milk and applied [76]
Cladina cervicornis	
China	Use for the treatment of cuts, scalds, and coughing up blood. [13]
Cladina chlorophaea	
BC, Canada	In slow-healing sores, decoction is used to wash sores [77]
Welsh	Application same as *C. pyxidata* for whooping cough. It is used for the same purpose in Waterford (Ireland) [35]
Cladina coccifera [syn. *Scyphophorus cocciferus*]	
Europe, early modern era	The decoction used in children to treat whooping cough and fever, like *C. pyxidata*[78–81]

Table 15.8 (Continued)

Country	Doctrinal use
Cladina fenestralis	
Sichuan, China	Use as medicinal tea [13]
Cladina fruticulosa	
China	To treat bacterial infections of the skin [13]
Cladina gracilis	
China	Use for pink eye nose bleeding, impetigo, dizziness and difficult or painful urination. In treatment, either drink or apply decoction/powder lichen to the affected area [13, 72]
Cladina macroceras	
China	The decoction drink to remove toxic substances, relieve from block urination and bring down swelling [13]
Cladina pleurota	
China	Use to dissolve phlegm, eliminate dampness, to clear heat, and cool liver [13]
Cladina scabriuscula	
Papua New Guinea	Heated form taken orally for the treatment of vaginal discharge or bleeding [82]

15.5 Ethnobotanical and Pharmacological Properties of the Genus *Usnea*

Usnea (beard lichen) is a genus of yellow fruticose lichens that grows as a mini leafless shrub. It has long stems and disk-shaped holdfasts resembling tangled threads. They are dispersed in the Arctic and the tropics and collected as fodder or eaten by wild animals. These beard lichens were widely used as a remedy for catarrh, whooping cough, dropsy, and epilepsy, and they have also been extensively used as a tonic, a diuretic, and an astringent [83]. *Usnea* sp. as a drug was first described in a Chinese herbal classic (Shen Nong's Herbal Classic). In addition, *Usnea* extract has been documented in TCM around 100 years BC. According to TCM recommendation, 6–9 g of dried *Usnea* sp. has been used as "Song Lo," an antimicrobial tea used for snakebites, malaria, and liver tonic [84]. According to Okuyama and co-workers, *Usnea* exhibited several healing properties [85]. It has been used in the Unani system (Perso-Arabic system of medicine). In different Unani formulations, it has been described as an antidote, analgesic, stomachic, and astringent [84]. *Usnealongissima* has been used orally to stimulate menstruation or vaginally to stimulate abortion [86]. To treat cholera air (air-pollution-related disease), *Usnea* sp. has been used by Persian scholars (Rhazes and Avicenna). Ancient Greeks usedit as a decoction for the treatment of insomnia, jaundice, growing hair, dandruff, diarrhea, internal and external bleeding, cold, whooping cough, nausea, uterine problems, dried skin, and smallpox [86]. Some ethnic research conducted on *Usne alongissima* has explained it as a spice, stuffing material, and poultice ingredient for bone setting by Indian tribes [87]. *Usne aamblyoclada* (known as

"Barba de piedra" or "yerba de la piedra") is known as an expectorant or used against influenza in Argentina, whereas *Usne asubflorida* is used as a painkiller [88]. The active phytoconstituents and doctrinal activities of *Usnea* species are summarized in Tables 15.9 and 15.10, respectively.

Table 15.9 Active phytoconstituents of *Usnea* species.

Lichen species	Active phytoconstituents
Usnea articulata	Atranorin, barbatic acid, cryptostictinolide, fumarprotocetraric acid, norstictic acid, peristictic acid, stictic acid, usnic acid
Usnea aciculifera	Constictic acid, stictic acid, usnic acid
Usnea barbata	Cinnamic acid, chlorogenic acid, caffeic acid, ellagic acid, gallic acid, norstictic acid, p-coumaric acid, usnic acid
Usnea compressa	Salazenic acid, usnic acid
Usnea florida	Evernic acid, 3-hydroxyphysodic acids, physodic acid, thamnolic acid
Usnea himalayana	Norstictic acid, usnic acid
Usnea longissima	Barbatic acid, diffractaic acid, evernic acid, Isousone, fumarprotocetraric acid, usnic acid, squamatic acid, usenamines, usone
Usnea orientalis	Salazenic acid, usnic acid
Usnea taylori	Fumarprotocetraric acid, usnic acid
Usnea undulata	Usnic acid
Usnea subfloridana	Norstictic acid, usnic acid

Table 15.10 Doctrinal useof *Usnea* species.

Country	Doctrinal use
Usnea spp. [includes Neuropogon, Eumitria, and Dolichousnea]	
Kenya	Crush, water boil, and sieve extracts are used for fever, malaria, loss of appetite, typhoid, backache, stomach ache, and heartburn [89, 90]
Tanzania	One of the ingredients in herbal tea is used to relieve altitude sickness [91]
Iran	To improve breath [92]
India	Used as an astringent, antidote, and analgesic. Use it to improve digestion, appetite, heart troubles, and reduce inflammation. It heals wounds and is helpful for lactating women. It stimulates menstruation or induces abortion. Also identified as *U. longissima*[93–95]
Bhutan	Present in Sowa Rigpa to heal lungs, nerves, liver, and any poison-associated diseases [36]
Nepal	For the treatment of fever, and throat pain, fired powder of jhyau in water was used [32]
Tibet	Tibetan medicine: water-soaked form used to wash feet for treatment of ulcers (Lhoba). Also used for pneumonia, tracheitis, toxic fever, and mastitis [96]

Table 15.10 (Continued)

Country	Doctrinal use
Ancient China	As a Chinese herbal, it was used against anger, indignation, and xieqi ("evil energy"). It relieves head feng ("wind") and vacuity sweating. It relieves genital bans, pain, and swelling in females. For medicinal application, collected in the fifth lunar month and used after shade drying process [91, 97]
Yuan Dynasty, China	As Islamic medicines [98]
New Ireland	Helpful in induce menstruation [99]
Thailand	It was used to prevent infection and to aid parturition after childbirth [91]
New Zealand	In venereal disease, steeped in water and place on the affected parts [100]. In various skin afflictions, dry powder was rubbed on the skin [101, 102]. The crush form was bandaged onto the wound to stop bleeding [103].To keep baby warm, along with moss it was used as a diaper and sanitary napkin [102]
Europe	The fruticose arboreal lichen (splanchnon) consisted of *Evernia prunastrior* and *Pseudevernia furfuracea* had important medicinal benefits in Ancient Greeks. In the early 1500s, this lichen was entered into European pharmacopoeias and incorporated all fruticose arboreal lichens. It was only *Usnea* species in the late 1700s. The decoction of Usnea species was used as a styptic, an anti-inflammatory, a skin moisturizer for drying skin lesions, and whooping cough, insomnia, diarrhea, smallpox, nausea, uterine medicine, and umbilical hernias [104]. It was also used in scalp-related diseases and dandruff [35]
Egypt (Cairo)	Use in cold, cough, heart and eye diseases. Also used to improve eyesight and as an anesthetic agent [105]
Austria	As a medicinal plant [106]
Spain	In respiratory ailments [107]
Italy	To prevent any kind of blisters, Shepherds were kept in their footwear [91]
Scandinavia	The decoction is used for bathing chapped skin on babies and feet in adults. In addition, *Peltigera aphthosa*, *Alectoria* species and *L. pulmonaria*were also used[67]
Scandinavia	Powder applied on sores and external wounds appeared by long journeys. Benefits of scabies and ringworm infection [63, 104]
Finland	It has been taken orally for toothache and sore throat. Fresh lichen applies on the athlete's foot, infected wounds, and skin eruptions. Alectoria species was also used [108]
Sweden	To cure foot blisters [109]
BC, Canada	Similar benefits as Nuxalk (*Alectoria sarmentosa*) [110]
SK, Canada	To prevent nose bleeding fresh lichen was inserted into the nostril [69]
AB, Canada	In sore or infected eyes decoction was use to wash eyes. In this, possibly *U. hirta*[38]
OR, USA	Same as *Umatilla* and application of *Alectoria sarmentosa*[111]
Utah, USA	An infusion or poultice was basically applied on frozen body parts [49]

(Continued)

Table 15.10 (Continued)

Country	Doctrinal use
Loja, Ecuador	Application in children for inflated and sore stomachs. It was boiled in water and adds honey before drinking. *Eucalyptus* or pine plant must not be a favorable collecting site [112]
Peru, South America	In cold, steam bath is done along with other plants. In nervous (a psychological condition) and susto (when someone's anima dissociates from their body), a plant is boiled in water and then drunk in small glasses until problems are not cured [107]
Brazil	In genitourinary diseases along with *Camellia sinensis* drink as an infusion. A saxicolous species of *Usnea* is used in the case of intense yellow color [40]
Usnea sect. *Neuropogon* spp. (Nees and Flot.) Mont. [syn. *Neuropogon* spp.]	
Argentina/Chile	It had medicinal benefits in cough, unspecified gastrointestinal, cardiovascular, respiratory, genitourinary, obstetric, and gynecological afflictions, as well as in many cultural syndromes [113]
Usnea aciculifera	
China	Use to cure urinary retention, swelling, edema of heart and kidney, painful urination, and bladder infection [13]
Usnea angulata [syn. *Usena sulcata*]	
Chorote (Dry Chaco, Argentina)	For the treatment of sweating in shoes and fungal infection. Decoction drink for common cough, pertussis, and stomach ache [114]
Criollos (Dry Chaco, Argentina)	In inflammation, poultice is prepared in goat fat and applied. In infection, decoction in water used to drink once for three consecutive days. In pasmo (possibly tetanus) and asoleadura ("exposure to the sun"): lichen kept under the feet and on head until symptoms exist [115]
Usnea articulata [syn. *Usnea atlantica, Usnea flavescens*]	
Tanzania	In stomach ache, a handful of hewas chew freshly, and juice swallowed that is bitter in taste. It relieves the pain [116]
Samoa	In wounds and shin bruises [117]
Canary Islands barbas	Act as disinfectant [118]
Usnea baileyi	
India	An adulterant in charila. It was mixed with *Valeriana jatamansi* as tobacco flavoring agent, along with *U. subsordida*, *U. longissima*, *H. cirrhata*, *H. nepalensis*, and *Ramalina inflate*[21]
Sikkim, India	Use in respiratory problems [31]
Usnea barbata	
South Africa	In mammary infections of cattle, decoction was used to wash udder (many times in a day). Used to treat wounds and indigestion in humans. For this, decoction or tincture was taken orally several times a day [119, 120]
Nepal	Endangered and banned for their raw export [121]
Philippines	Chop and mix in coconut oil and applied on the wound. The decoction was drunk in case of abdominal pain [47]
West Malaysia	Application in cold and strengthening effect after childbirth [122]

Table 15.10 (Continued)

Country	Doctrinal use
Indonesia	Medicinal benefits as an aphrodisiac, used in rheumatism, syphilis, and postpartum conditions [123]
Europe (early modern era)	In early modern-era pharmacopoeias, it was probably synonymous with *Usnea* species. Use for nausea, insomnia, and uterine affliction. Beneficial effects in whooping cough, hair growth, jaundice, and internal, and external bleeding [80, 124]
Spain	Act as a drying agent used as antiseptic for foot irritationsand cracks [125]
Brazil	In female sterility, a liquid given to particular sterile women [126]
Usnea ceratina	
China	Use for pulmonary tuberculosis, coughs, headache, post-infection, inflamed lymph, hepatitis, mastitis, inflamed lungs, and snake bites [13]
Usnea dasopoga [syn. *Usnea filipendula, Usnea dasypoga*]	
Java	Had some unspecified medicinal properties [127]
Tibet	Basically used in traditional medicine [128]
Russian Far East	To treat wounds and some infections powder is used [71]
Usnea densirostra	
Argentina	Used externally as an antiseptic, astringent, and anti-inflammatory agent. Additionally also used *U. durietzii*[129–131]
Uruguay	Unspecified medicinal benefits [132]
Usnea diffracta [syn. *Dolichousnea diffracta*]	
China	Use for headache, dim vision, cough, phlegm, malaria, tuberculosis of the neck or lungs, traumatic injuries, bone fracture, pus oozing from breasts or sores, bleeding from external injury, vaginal discharge, menstrual disorders, bleeding from the uterus, swelling of female genitalia, burns, and scalds, snake bite, wind damp impediment, vomiting blood, urinary tract afflictions, blood in feces, and ascarid or schistosome infections. It was collected throughout the year and then dried. For the application, apply or drink decoction to the affected area [13, 72]
China	Use as food, herbal tea, and medicinal bath for treatment of rheumatism [133, 134]
Tibet	To treat fever associated with lungs, liver, and channels, as well as poisoning-related fever [135]
Korea	Used to induce menstruation (Pusan) and to treat tuberculosis of neck region (Gongju) [99]
Usnea durietzii [syn. *Neuropogon durietzii*]	
Argentina	In San Luis, it was the same as Argentine and uses *U. densirostra*. Mendocina recorded as an abortive toxin [136]
Usnea florida	
China	To cure bone aching, painful urination, skin diseases, bleeding or infection from external injuries, heart palpitations, coughs, tuberculosis of the lungs or neck, and edema [13]

(*Continued*)

Table 15.10 (Continued)

Country	Doctrinal use
Europe	The decoction used for cold and cough [79]
Bolivia	Unspecified medicinal properties [111]
Chile	Infusion used for diarrhea[137]
Usnea himalayana	
Japan	Used as burned "lichen cigarette" [138]
Usnea himantodes [syn., *Usnea Africana, Usnea Usnea gigas, Eumitria himantodes*]	
Tanzania	Same as *U. articulata*[116]
Usnea hirta	
Europe (early modern)	Application in wound and hair loss [79]
Usnea laevis	
Colombia	The lichen powder is applied in the armpits and feet [88]
Venezuela	Used for fungal infections, dermatosis, pneumonia, and tuberculosis [139, 140]
Usnea longissima [syn. *Dolichousnea longissima*]	
Turkey	Use in tuberculosis, cancer, and ulcers [141, 142]
India	As ushna (a Unani medicine) [93, 94]
India	Same medicinal benefits of *U. baileyi*. Mixed in tobacco; sometimes used as adulterant in charila [21]
India	In bone fractures, wash, air-dried form mixed with other ingredients and soaked overnight in salt water [143]
Nepal	Act as ka (astringent), kha (bitter) and sil (cool). Used as incense and had application against wounds and fever from poisoning[144, 145].
Mongolia	Used for bone fractures, injuries of legs and back, diarrhea, headache, clearing heat, women-related diseases, removing toxins, stomach, lung, and intestine issues, sore muscles and tendons [146]
China	Same use in China as *U. diffracta*[13, 72]. Other names include (sōngluó, nǚluó, jīnqiáncǎo, guāngōngxū, shíxū, "stone beard", hǎifēngténg, xuěfēngténg, jīnsīténg, sōngmáo, xiānréntóufǎ,"fairy hair", shísīxiàn, "stonethread", sōngshàngjìshēng, shùguà, shānguàmiàn, shùfāqī, guòshānlóng, fēitiānwúgōng, níshānchì)
Yunnan, China	Water soaked form is use for cough and moistening the lungs. Whole lichen is used as medicinal food [98, 147, 148]
Yunnan, China	Decoction used in abdominal pain of women [148]
Yunnan, China	The decoction is drunk in rheumatism and inflamed lymph nodes. In injuries from falls and infected hair follicles, crushed form is applied as a poultice [149]
BC, Canada	Probably all *Usnea* species and *Alectoria sarmentosa* had beneficial effect in wound dressing, but *U. longissima* was preferably use. Wrapped around the wound and left for sometimes [150]
Usnea nidifica	
Rarotonga	Medicinal usage on Mangiai, in bleeding thallus is chew and apply to the cuts and stings [151]

Table 15.10 (Continued)

Country	Doctrinal use
Usnea orientalis	
Eastern Nepal	Whole plant is ground and finally applied on cut and wound [152]
Usnea pectinata [syn. *Eumitria pectinata*]	
China	Relieve from pain, swelling, bloody feces and bleeding from external injuries [13]
Usnea plicata	
Libya	One of ingredient in scíba with *Ramalina calicaris*, *R. farinacea* and *Pseudevernia furfuracea*[153]
Scandinavia	Apply on the sores of feet after walking long distances [154]
Europe (early modern)	Act as an astringent for internal and external application and strengthening agent in abdominal cavity. Used in whooping cough, jaundice, and restraining abortion [78, 80, 124]
Usnea sikkimensis	
India	Medical application in respiratory troubles, asthma, and hemorrhages; powder is used to strengthen hair [155]
India	Applied as bandage on surface wounds, boils, and skin eruptions; to stop nose bleeding inserted into the nostril; kept in the footwear to prevent blisters. Effective herbal remedy for respiratory health [31, 156]
Usnea strigosa	
Papua New Guinea	For headaches, concoction was taken orally [82]
Usnea subfloridana	
Ireland	Used for the treatment of sore eyes. It was mixed in butter and tobacco, boiled, cooled, and finally applied on the eyes [35]
China	Use for swelling, reddened and painful eyes, bleeding from external injuries [13]
Usnea subsordida	
India	Medicinal properties were same as *U. baileyi*. One of the adulterants in charila [21]
Usnea trichodeoides [syn. *Dolichousnea trichodeoides*]	
China	Used for blurred vision, inflamed cornea, coughs, sores, pus discharge, pulmonary tuberculosis, headaches, swellings, bleeding from external injuries, and bloody feces. In menstrual disorders like vaginal discharge and uterine bleeding. For treatment, apply or drink decoction or powdered lichen [13]

15.6 Conclusion

Due to their diverse composition of fungal and algal cells, entophytes, and other components, lichens are considered an affluent source of secondary metabolites. A good deal is likely due to their secondary metabolites; some may be physiologically active and act as antibiotics. Currently, many peculiar compounds have been uncovered in lichens that were traditionally used to heal several conditions. Lichens have had long-standing medicinal uses, far back in tradition. It also links with folklore and has been used medicinally for centuries. People have used lichens for centuries to treat medical conditions such as coughs, jaundice, and rabies. During the middle ages, they were often found in medicinal herbals used by healers. This history makes them excellent candidates for future pharmacological research.

References

1 Launert, E. (1981). *The Hamlyn Guide to Edible and Medicinal Plants*. London: Hamlyn.

2 Pereira, J. (1843). *The Elements of Materia Medica and Therapeutics*, vol. 2. Philadelphia, PA: Lea and Blanchard.

3 Elkhateeb, W.A., Elnahas, M., and Daba, G. (2021). Lichentherapy: highlights on the pharmaceutical potentials of lichens. *Journal of Microbiology and Biotechnology* 6 (1): 1–10.

4 Elkhateeb, W.A. and Daba, G.M. (2019). Lichens, an alternative drugs for modern diseases. *International Journal of Researchof Pharma and Biosciences* 6: 5–9.

5 Crawford, S.D. (2019). Lichens used in traditional medicine. In: *Lichen Secondary Metabolites: Bioactive Properties and Pharmaceutical Potential* (ed. B. Rankovic), 31–97. Cham: Springer International Publishing.

6 Verma, N. and Behera, B.C. (2019). Future directions in the study of pharmaceutical potential of lichens. In: *Lichen Secondary Metabolites: Bioactive Properties and Pharmaceutical Potential* (ed. B. Rankovic), 237–260. Cham: Springer International Publishing.

7 Behera, B.C., Verma, N., Sonone, A. et al. (2005). Antioxidant and antibacterial activities of lichen *Usnea* ghattensis in vitro. *Biotechnology Letters* 27: 991–995.

8 Alam, M.A., Khatoon, R., Huda, S. et al. (2020). Biotechnological applications of lichens. In: *Lichen-Derived Products: Extraction and Applications* (ed. M. Yusuf), 203–219. Wiley-Scrivener.

9 Barnes, J., Hawksworth, D., Shaw, C. et al. (2000). Pharmacognosy in the 21st century. *The Pharmaceutical Journal* 264: 701–703.

10 Aravind, S., Sreelekha, T.T., Dileepkumar, B.S. et al. (2014). Characterization of three depside compounds from a Western Ghat lichen *Parmelia erumpens* Kurok with special reference to antimicrobial and anticancer activity. *RSC Advances* 4 (65): 34632–34643.

11 Karagoz, I.D., Ozaslan, M., Kılıç, İ.H. et al. (2015). Hepatoprotective effect of diffractaic acid on carbon tetrachloride-induced liver damage in rats. *Biotechnology & Biotechnological Equipment* 29 (5): 1011–1016.

12 Jeon, Y.-J., Kim, S., Kim, J.H. et al. (2019). The comprehensive roles of ATRANORIN, a secondary metabolite from the antarctic lichen *Stereocaulon caespitosum*, in HCC tumorigenesis. *Molecules* 24 (7): 1414.

13 Wang, L. and Qian, Z. (2013). *Illustrated Medicinal Lichens of China*, 1–176. Yunnan: Yunnan Science and Technology Press.

14 Shailajan, S., Joshi, M., and Tiwari, B. (2014). Hepatoprotective activity of *Parmelia perlata (Huds.) Ach.* against CCl4 induced liver toxicity in Albino Wistar rats. *Journal of Applied Pharmaceutical Science* 4 (2): 070–074.

15 Pieroni, A. (2000). Medicinal plants and food medicines in the folk traditions of the upper Lucca Province, Italy. *Journal of Ethnopharmacology* 70 (3): 235–273.

16 Verma, N., Behera, B., and Makhija, U. (2008). Antioxidant and hepatoprotective activity of a lichen *Usnea ghattensis* in vitro. *Applied Biochemistry and Biotechnology* 151: 167–181.

17 Shukla, I., Azmi, L., Shukla, S.K. et al. (2020). Contribution of lichens in rectification of alcohol-induced liver damage: preventive or curative. In: *Innovations in Food Technology: Current Perspectives and Future Goals* (ed. P. Mishra, R.R. Mishra, and C.O. Adetunji), 201–210. Singapore: Springer.

18 Mastan, A., Sreedevi, B., and Kumari, J.P. (2014). Evaluation of the in vitro antioxidant and antibacterial activities of secondary metabolites produced from lichens. *Asian Journal of Pharmaceutical and Clinical Research* 193–198.

19 Sharma, M. and Mohammad, A. (2020). Lichens and lichenology: historical and economic prospects. In: *Lichen-Derived Products: Extraction and Applications* (ed. M. Yusuf), 101–118. Wiley-Scrivener.

20 Devkota, S., Chaudhary, R.P., Werth, S. et al. (2017). Indigenous knowledge and use of lichens by the lichenophilic communities of the Nepal Himalaya. *Journal of Ethnobiology and Ethnomedicine* 13 (1): 1–10.

21 Shah, N. (1998). Lichens of economic importance from the hills of Uttar Pradesh, India. *Journal of herbs, spices & medicinal plants* 5 (3): 69–76.

22 Kirmizigül, S., Koz, Ö., Anil, H. et al. (2003). Isolation and structure elucidation of novel natural products from Turkish lichens. *Turkish journal of chemistry* 27 (4): 493–500.

23 Toma, N., Ghetea, L., Nitu, R. et al. (2001). Progress and perspectives in the biotechnology of lichens. *Romanian Biotechnological Letters* 6: 1–16.

24 Thell, A., Crespo, A., Divakar, P.K. et al. (2012). A review of the lichen family *Parmeliaceae*–history, phylogeny and current taxonomy. *Nordic Journal of Botany* 30 (6): 641–664.

25 Thell, A., Feuerer, T., Kärnefelt, I. et al. (2004). Monophyletic groups within the *Parmeliaceae* identified by ITS rDNA, β-tubulin and GAPDH sequences. *Mycological Progress* 3: 297–314.

26 Singh, H., Husain, T., Agnihotri, P. et al. (2012). Biodiversity conservation through traditional beliefs system: a case study from Kumaon Himalayas, India. *International Journal of Conservation Science* 3 (1): 33.

27 Pennington, C.W. (1963). *The Tarahumar of Mexico: Their Environment and Material Culture*. Salt Lake City: University of Utah Press.

28 Nadkarni, K. (1954). *Indian Materia Medica*, vol. 7 946–948. Popular Book Depot. Bombay, Bombay Popular.

29 Uprety, Y., Asselin, H., Boon, E.K. et al. (2010). Indigenous use and bio-efficacy of medicinal plants in the Rasuwa District, Central Nepal. *Journal of Ethnobiology and Ethnomedicine* 6: 1–10.

30 Stopp, K. (1963). Medicinal plants of the Mt. Hagen people (mbowamb) in New Guinea. *Economic Botany* 17 (1): 16–22.

31 O'Neill, A.R., Badola, H.K., Dhyani, P.P. et al. (2017). Integrating ethnobiological knowledge into biodiversity conservation in the Eastern Himalayas. *Journal of Ethnobiology and Ethnomedicine* 13: 1–14.

32 Poudel, P. (2008). *Medicinal Plants of Change VDC of Taplejung*. Eastern Nepal, Kathmandu.

33 Kumar, S.K., Banskota, A.H., and Manandhar, M.D. (1996). Isolation and identification of some chemical constituents of *Parmelia nepalensis*. *Planta Medica* 62 (1): 93–94.

34 MacIntyre, D.L. (1999). Role of Scottish native plants in natural dyeing and textiles. https://doi.org/10.7488/era/1631.

35 Applequist, W.L. (2004). Medicinal plants in folk tradition: An ethnobotany of Britain an Ireland by David E. Allen and Gabrielle Hatfield. *Systematic Botany* 29 (4): 1021–1021.

36 Wangchuk, P., Namgay, K., Gayleg, K. et al. (2016). Medicinal plants of Dagala region in Bhutan: their diversity, distribution, uses and economic potential. *Journal of Ethnobiology and Ethnomedicine* 12: 1–19.

37 Hawksworth, D. (2003). Hallucinogenic and toxic lichens. *International Lichenological Newsletter* 36 (2): 33–35.

38 Marles, R.J., Clavelle, C., Monteleone, L. et al. (2000). *Aboriginal Plant Use in Canada's Northwest Boreal Forest*. Vancouver, BC: UBC Press.

39 Turner, N.J. and Hebda, R.J. (2012). *Saanich Ethnobotany: Culturally Important Plants of the WSÁNEC People*. Victoria, BC: Royal British Columbia Museum.

40 Londoño-Castañeda, P.A., Buril, M.L.L., Rego-Cunha, I.P. et al. (2017). Lichens used in the traditional medicine by the Pankararu Indigenous Community, Pernambuco-Brazil. *Global Journal of Science Frontier Research: C Biological Science* 17 (4): 15–22.

41 Llano, G.A. (1948). Economic uses of lichens. *Economic Botany* 2 (1): 15–45.

42 Chandra, S. and Singh, A. (1971). A lichen crude drug (*chharila*) from India. *Indian Journal of Medical Research* 6: 209–215.

43 Kaushal, K. and Upreti, D. (2001). *Parmelia spp.*(lichens) in ancient medicinal plant lore of India. *Economic Botany* 55 (3): 458–459.

44 Karadi, K. (2010). A pharmacognostical and analytical study of Shaileya (Parmelia species) WSR to different market samples. Doctoral dissertation. Rajiv Gandhi University of Health Sciences (India).

45 Biswas, K. (1947). The lichen flora of India. *Journal of the Royal Asiatic Society of Bengal Science* 13: 75–113.

46 Bugni, T.S., Andjelic, C.D., Pole, A.R. et al. (2009). Biologically active components of a Papua New Guinea analgesic and anti-inflammatory lichen preparation. *Fitoterapia* 80 (5): 270–273.

47 Madulid, D.A., Gaerlan, F.J.M., Romero, E.M. et al. (1989). Ethnopharmacological study of the Ati tribe in Nagpana, Barotac Viejo, Iloilo. *Acta Manilana* 38 (1): 25–40.

48 Elmore, F.H. (1943). *Ethnobotany of the Navajo*. Monographs of the School of American Research 8. Albuquerque, NM: University of New Mexico Press.

49 Wyman, L.C. and Harris, S.K. (1951). *The Ethnobotany of the Kayenta Navajo* [University of New Mexico Publication in Biology, No. 51. Albuquerque, NM: University of New Mexico Press.

50 Whiting, A.F. (1939). *Ethnobotany of the Hopi. Bull-Mus North Ariz* 15: 120.

51 Watt, J.M. and Breyer-Brandwijk, M.G. (1962). *The Medicinal and Poisonous Plants of Southern and Eastern Africa being an Account of their Medicinal and other Uses, Chemical Composition, Pharmacological Effects and Toxicology in Man and Animal*, 2e. Edinburgh: E. & S. Livingstone.

52 Herrick, J.W. (1995). *Iroquois Medical Botany*. Syracuse, NY: Syracuse University Press.

53 Curtin, L.S.M. and Moore, M. (1997). *Healing Herbs of the Upper Rio Grande: Traditional Medicine of the Southwest*. Santa Fe, NM: Western Edge Press.

54 Lipp, F.J. (1995). *Ethnobotanical method and fact: a case study*. In: *Ethnobotany: Evolution of a Discipline* (ed. R.E. Schultes and S. Von Reis), 52–59. Portland: Dioscorides Press.

55 Van Damme, P. and van Den Eynden, V. (1992). Plant uses by the Topnaar of the Kuiseb Valley Namib desert. *Afrika focus* 8 (3–4): 223–252.

56 Fonnegra, F.G. (2007). *Plantas medicinales aprobadas en Colombia*. Colombia: Universidad de Antioquia.

57 Acharius, E. (1803). *Methodus Lichenum supplementum*. Stockholm: Ulrich.

58 Meli, M.A. Desideri, D., Cantaluppi, C., et al. (2018). Elemental and radiological characterization of commercial *Cetraria islandica (L.) Acharius* pharmaceutical and food supplementation products. *Science of the Total Environment* 613–614: 1566–1572.

59 Kari, P.R. (1987). *Tanaina Plantlore, Dena'ina K'et'una*. Alaska Region: National Park Service.

60 McKennan, R. (1959). *The Upper Tanana Indians*, vol. 55, 1–226. New Haven, CT: Yale University.

61 Bank, T. (1953). Botanical and ethnological studies in the Aleutian Islands: II. Health and medical lore of the Aleuts. *Papers of the Michigan Academy of Science, Arts, and Letters*, vol. 38, 415–432.

62 Eidlitz, K. (1969). *Food and Emergency Food in the Circumpolar Area*. Stockholm: Almqvist & Wiksell.

63 Nissen, K. (1921). *Lapponian Lichen Names*. oslo: Videnskapsselskapet.

64 Berihuete-Azorín, M. (2013). First archaeobotanical approach to plant use among Selknam hunter-gatherers (Tierra del Fuego, Argentina). *Archaeological and Anthropological Sciences* 5: 255–266.

65 Smith, H. (1932). Ethnobotany of the Ojibwe Indians. *Bulletin of thePublic Museum City of Milwaukee* 4: 327–525.

66 Fraser, M.-H. (2006).Ethnobotanical investigation of plants used for the treatment of type 2 diabetes by two Cree communities in Quebec: quantitative comparisons and antioxidant evaluation. Thesis. McGill University.

67 Richardson, D.H. (1975). *The Vanishing Lichens: Their History, Biology and Importance*. Newton Abbot: David & Charles.

68 Rout, J., Kar, A., and Upreti, D. (2005). Traditional remedy for kidney stones from a high altitude lichen: *Cladonia rangiferina* (L.) Wigg (reindeer moss) of Eastern Himalaya. *Ethnobotany* 17: 164–166.

69 Leighton, A.L. (1985). *Wild Plant Use by the Woods Cree (Nihithawak) of East-central Saskatchewan*. Ottawa, ON: University of Ottawa Press.

70 Black, P.L., Arnason, J.T., and Cuerrier, A. (2008). Medicinal plants used by the Inuit of Qikiqtaaluk (Baffin Island, Nunavut). *Botany* 86 (2): 157–163.

71 Moskalenko, S. (1986). Preliminary screening of far-eastern ethnomedicinal plants for antibacterial activity. *Journal of Ethnopharmacology* 15 (3): 231–259.

72 Hu Sy, K. and But, P. (1980). *An Enumeration of the Chinese Materia Medica*. Hong Kong: The Chinese University Press.

73 Garrett, J.T. (2003). *The Cherokee Herbal: Native Plant Medicine from the Four Directions*. New York: Simon and Schuster.

74 Pérez-Nicolás, M., Vibrans, H., Romero-Manzanares, A. et al. (2017). Patterns of knowledge and use of medicinal plants in Santiago Camotlán, Oaxaca, Mexico. *Economic botany* 71: 209–223.

75 Garibaldi, A. (1999). *Medicinal Flora of the Alaska Natives*. Anchorage: University of Alaska, Alaska Natural Heritage Program.

76 Turner, N.J. (2004). Expert report: Tsilhqot'in and Xeni Gwet'in plant use and occupancy. *William vs. Her Magesty the Queen, Xeni Gwet'in/Tsilhqot'in Land Rights Trial*. BC Supreme Court, Canada.

77 Turner, N.J., Bouchard, R., and Kennedy, D.I. (1980). *Ethnobotany of the Okanagan-Colville Indians of British Columbia and Washington*. Victoria, BC: BC Provincial Museum.

78 Lindley, J. (1838). *Flora Medica (Indian Reprint, 1985)*. New Delhi: Ajay Book Service.

79 Willemet, P.R. (1787). *Lichénographie économique, ou histoire des lichens utiles*. National Library of Scotland.

80 Luyken, J.A. (1809). *Tentamen historiae lichenum in genere cui accedunt primae lineae distributionis novae*. Henry Dieterich, Gottingen.

81 Pereira, J. (1850). *The Elements of Materia Medica and Therapeutics*, vol II, part II. London: Longman, Brown, Green, and Longmans.

82 Jorim, R.Y., Korape, S., Legu, W. et al. (2012). An ethnobotanical survey of medicinal plants used in the eastern highlands of Papua New Guinea. *Journal of Ethnobiology and Ethnomedicine* 8: 1–17.

83 Petruzzello, M. (2021). Beard lichen. *Encyclopedia Britannica* (31 March). https://www.britannica.com/science/beard-lichen (accessed 9 February 2024)

84 Guo, L., Shi, Q., Fang, J.-L. et al. (2008). Review of usnic acid and *Usnea barbata* toxicity. *Journal of Environmental Science and Health, Part C* 26 (4): 317–338.

85 Okuyama, E., Umeyama, K., Yamazaki, M. et al. (1995). Usnic acid and diffractaic acid as analgesic and antipyretic components of *Usnea diffracta*. *Planta Medica* 61 (02): 113–115.

86 Ranković, B. (2019). *Lichen Secondary Metabolites: Bioactive Properties and Pharmaceutical Potential*. Cham: Springer.

87 Upreti, D.K., Divakar, P.K., and Nayaka, S. (2005). Commercial and ethnic use of lichens in India. *Economic Botany* 59 (3): 269–273.

88 Rodriguez, J.M. and Estrabou, C. (2008). Usnea amblyoclada «Barba de piedra»(Ascomycetes liquenizados) en Argentina. *Boletín de la Sociedad Argentina de Botánica* 43 (3–4): 221–225.

89 Kiringe, J.W. (2006). A survey of traditional health remedies used by the Maasai of Southern Kaijiado District, Kenya. *Ethnobotany Research & Applications* 4: 61–74.

90 Muthee, J.K., Gakuya, D.W., Mbaria, J.M. et al. (2011). Ethnobotanical study of anthelmintic and other medicinal plants traditionally used in Loitoktok district of Kenya. *Journal of Ethnopharmacology* 135 (1): 15–21.

91 Sharnoff, S.D. (1997).Lichens and people. https://www.anbg.gov.au/lichen/lichens-people.html (accessed 8 March 2011).

92 Hooper, D. and Field, H. (1937). *Useful Plants and Drugs of Iran and Iraq*, 73–237. United States: Field Museum of Natural History, Chicago.

93 Rauf, A., Latif, A., Afaq, S.H. et al. (2011). In-vitro antibacterial screening of extracts of *Usnea longissima* lichen. *International Journal of Applied Biology and Pharmaceutical Technology* 2: 14–18.

94 Razzack, M.A.a.F.U. (1993). *The Concept of Birth Control in Unani Medicine*, vol. 38. Central Council for Research in Unani Medicine, Ministry of Health & Family Welfare, Government of India.

95 Yavuz, M. and Çobanoğlu, G. (2010). Ethnological uses and etymology of the word *Usnea* in Ebubekir Razi's "Liber Almansoris". *The British Lichen Society Bulletin* 106: 3–12.

96 Li, F., Zhuo, J., Liu, B. et al. (2015). Ethnobotanical study on wild plants used by Lhoba people in Milin County Tibet. *Journal of Ethnobiology and Ethnomedicine* 11: 1–11.

97 Jing, S. (1998). *The Divine Farmer's Material Medica: A Translation by Yang Shou-zhong* (ed. Y. Shou-Zhong). Boulder, CO: Blue Poppy Press, Inc.

98 Kong, Y. and Chen, D. (1996). Elucidation of Islamic drugs in Hui Hui Yao Fang: a linguistic and pharmaceutical approach. *Journal of Ethnopharmacology* 54 (2–3): 85–102.

99 Lee, S.J. (1966). *Korea Folk Medicine*. Seoul: Seoul National University Publishing Center Press.

100 Best, E. (1905). Maori medical lore. Notes on sickness and disease among the Maori People of New Zealand, and their treatment of the sick; together with some account of various beliefs, superstitions and rites pertaining to sickness, and the treatment thereof, as collected from the Tuhoe tribe. Part I. *The Journal of the Polynesian Society* 14 (1(53)): 1–23.

101 Kerry-Nicholls, J.H. (1886). The origin, physical characteristics, and manners and customs of the Maori race, from data derived during a recent exploration of the king country, New Zealand. *The Journal of the Anthropological Institute of Great Britain and Ireland* 15: 187–209.

102 Goldie, W. (1904). *Art. I: Maori Medical Lore*. Transactions and Proceedings of the New Zealand Institute, London.

103 Brooker, S.G. and Cooper, R.C. (1961). New Zealand medicinal plants. *Economic Botany* 15 (1): 1–10.

104 Lebail, B. and Lebail, J.-B.E.F. (1853). *Des lichens: considérés sous le point de vue économique, médical et physiologique (nutrition)*. Paris: Rignoux.

105 Lev, E. (2007). Drugs held and sold by pharmacists of the Jewish community of medieval (11–14th centuries) Cairo according to lists of *materia medica* found at the Taylor–Schechter Genizah collection, Cambridge. *Journal of Ethnopharmacology* 110 (2): 275–293.

106 Christanell, A., Vogl-Lukasser, B., Vogl, C.R. et al. (2010). The cultural significance of wild gathered plant species in Kartitsch (Eastern Tyrol, Austria) and the influence of socio-economic changes on local gathering practices. In: *Ethnobotany in the New Europe: People, Health, and Wild Plant Resources* (ed. M. Pardo-de-Santayana, A. Pieroni, and R.K. Puri), 51–75. New York: Berghahn Books.

107 González-Tejero, M., Martínez-Lirola, M.J., Casares-Porcel, M. et al. (1995). Three Lichens used in Popular Medicine in Eastern Andalucia (Spain). *Economic Botany* 49: 96–98.

108 Vartia, K.O. (1973). *Antibiotics in lichens*. In: *The Lichens, 3rd edition, Academics, New York* (ed. V. Ahmadjiian and M.E. Hale), 547–561. Elsevier.

109 Ahmadjian, V. and Nilsson, S. (1963). *Swedish Lichens*. USA: The Chancellor Press.

110 Turner, N.J. (1973). The ethnobotany of the Bella Coola Indians of British Columbia. *Economic Botany* 25 (1): 63–104.

111 HunnG (2005) Unpublished 1976–1980 ethnobotany field notes.

112 Abdel-Malek, S., Bastien, J.W., Mahler, W.F. et al. (1996). Drug leads from the Kallawaya herbalists of Bolivia. 1. Background, rationale, protocol and anti-HIV activity. *Journal of Ethnopharmacology* 50 (3): 157–166.

113 Estomba, D., Ladio, A., and Lozada, M. (2006). Medicinal wild plant knowledge and gathering patterns in a Mapuche community from North-western Patagonia. *Journal of Ethnopharmacology* 103 (1): 109–119.

114 Molares, S. and Ladio, A. (2014). Medicinal plants in the cultural landscape of a Mapuche-Tehuelche community in arid Argentine Patagonia: an eco-sensorial approach. *Journal of Ethnobiology and Ethnomedicine* 10: 1–14.

115 Scarpa, G.F. (2004). Medicinal plants used by the Criollos of Northwestern Argentine Chaco. *Journal of Ethnopharmacology* 91 (1): 115–135.

116 Kokwaro, J.O. (1976). *Medicinal Plants of East Africa*. East African Literature Bureau, Kampala.

117 Brooker, S.G. and Cooper, R.C. (1987). *New Zealand Medicinal Plants*. Aukland: Heinemann.

118 Darias, V., Bravo, L., Barquin, E. et al. (1986). Contribution to the ethnopharmacological study of the Canary Islands. *Journal of Ethnopharmacology* 15 (2): 169–193.

119 Afolayan, A.J., Grierson, D.S., Kambizi, L. et al. (2002). In vitro antifungal activity of some South African medicinal plants. *South African Journal of Botany* 68 (1): 72–76.

120 Madamombe, I. and Afolayan, A. (2003). Evaluation of antimicrobial activity of extracts from South African *Usnea barbata*. *Pharmaceutical Biology* 41 (3): 199–202.

121 Baniya, C.B. and Tamang, R. (2020). *Lichens of Nepal*, 55–61. Kathmandu: Plant Diversity of Nepal. Botanical Society of Nepal.

122 Foxworthy, F.W. (1921). *Minor Forest Products of the Malay Peninsula*. Methodist Publishing House.

123 Silalahi, M., Nisyawati, Walujo, E.B. et al. (2015). The local knowledge of medicinal plants trader and diversity of medicinal plants in the Kabanjahe traditional market, North SumatraIndonesia. *Journal of Ethnopharmacology* 175: 432–443.

124 Lightfoot, J. (1777). *Flora Scotica*, vol. *2*, 531–1151. London: B. White.

125 Lachiondo, F.B. and LóPEZ, A.C. (1947). *Contribución al estudio de los antibióticos procedentes de líquenes*. In: *Anales del Jardín Botánico de Madrid*. Real Jardín Botánico.

126 Cadogan, L. (1949). *Síntesis de la medicina racional y mística mby'aguaraní*. *América Indígena* 9 (1): 21–36.

127 Uphof, J.C.T. (1959). *Dictionary of Economic Plants*. J. Cramer, Weinheim.

128 Salick, J., Byg, A., Amend, A. et al. (2006). Tibetan medicine plurality. *Economic Botany* 60 (3): 227–253.

129 Bandoni, A., Mendiondo, M.E., Rondina, R.V.D. et al. (1972). Survey of Argentine medicinal plants. I. Folklore and phytochemical screening. *Lloydia* 35 (1): 69–80.

130 Correche, E., Kurdelas, R., and Gomez Lechon, M. (2008). Antibacterial, cytotoxic and antioxidant activities of extracts and secondary metabolites obtained from shrubs and lichens of Argentine. In: *Pharmacological applications of South America plants* (ed. M.P. Vinardell and M. Mitjans), 112–123. Kerala, India: Research Signpost.

131 García, G., Campos, R.H., Torres, R. et al. (1990). Antiherpetic activity of some Argentine medicinal plants. *Fitoterapia* 41 (6): 542–546.

132 Osorio, H. (1967). Contribution to the lichen flora of Uruguay III. Some additional new localities. *Comunicaciones Botánicas del Museo de Historia Natural de Montevideo* 4 (46): 1–10.

133 Jin, B., Liu, Y., Xie, J. et al. (2018). Ethnobotanical survey of plant species for herbal tea in a Yao autonomous county (Jianghua, China): results of a 2-year study of traditional medicinal markets on the Dragon Boat Festival. *Journal of Ethnobiology and Ethnomedicine* 14 (1): 1–21.

134 Luo, B., Liu, Y., Liu, B. et al. (2018). Yao herbal medicinal market during the dragon boat festival in Jianghua CountyChina. *Journal of Ethnobiology and Ethnomedicine* 14: 1–25.

135 Clark, B. (2014). *The Quintessence Tantras of Tibetan Medicine*. Shambhala Publications.

136 Ruiz Leal, A. (1972). *Flora popular mendocina: aportes al inventario de los recursos naturales renovables de la Provincia de Mendoza*. Mendoza: Universidad Nacional de Cuyo.

137 Houghton, P. and Manby, J. (1985). Medicinal plants of the Mapuche. *Journal of Ethnopharmacology* 13 (1): 89–103.

138 Ohmura, Y. (2003). What species of Japanese lichens are edible. *Lichen: News Bulletin of the Lichenological Society of Japan* 13: 6–9.

139 Marcano, V. (1991). *Propriedades biomedicas de las embriofitas asifonogamas andinas: II liquenes*, 1–13. Rev. Fac. Farm.(Merida).

140 Marcano, V., Rodriguez-Alcocer, V., and Mendez, A.M. (1999). Occurrence of usnic acid in *Usnea laevis* Nylander (Lichenized Ascomycetes) from the Venezuelan Andes. *Journal of Ethnopharmacology* 66 (3): 343–346.

141 Odabasoglu, F., Cakir, A., Suleyman, H. et al. (2006). Gastroprotective and antioxidant effects of usnic acid on indomethacin-induced gastric ulcer in rats. *Journal of Ethnopharmacology* 103 (1): 59–65.

142 Yazici, K. and Aslan, A. (2003). Lichens from the regions of Gumushane, Erzincan and Bayburt (Turkey). *Cryptogamie Mycologie* 24 (3): 287–300.

143 Sharma, G. (1997). Ethnomedicinal flora: Ayurvedic system of medicine. *Journal of Tennessee Academy of Science* 72 (3–4): 53–55.

144 Lama, Y.C., Ghimire, S.K., and Aumeeruddy-Thomas, Y. (2001). *Medicinal Plants of Dolpo: Amchis' Knowledge and Conservation*. Katmandu: WWF Nepal Program.

145 Bai, L., Bao, H.Y., and Bau, T. (2014). Isolation and identification of a new benzofuranone derivative from *Usnea longissima*. *Natural Product Research* 28 (8): 534–538.

146 Zhang, L., Chai, Z., Zhang, Y. et al. (2016). Ethnobotanical study of traditional edible plants used by the Naxi people during droughts. *Journal of Ethnobiology and Ethnomedicine* 12 (1): 1–16.

147 Zhang, L., Zhang, Y., Pei, S. et al. (2015). Ethnobotanical survey of medicinal dietary plants used by the Naxi People in Lijiang Area, Northwest YunnanChina. *Journal of Ethnobiology and Ethnomedicine* 11: 1–11.

148 Weckerle, C.S., Franz KHuber, YangYongping, et al. (2006). Plant knowledge of the Shuhi in the Hengduan Mountains, southwest China. *Economic botany* 60 (1): 3–23.

149 Ji, H., Shengji, P., and Chunlin, L. (2004). An ethnobotanical study of medicinal plants used by the Lisu people in Nujiang, northwest Yunnan, China. *Economic Botany* 58 (1): S253–S264.

150 Turner, N.J. et al. (1983). *Ethnobotany of the Nitinaht Indians of Vancouver Island*. Province of British Columbia.

151 McCormack, G. (2007). *Cook Islands Biodiversity Database, Version 2007.2*. Rarotonga: Cook Islands Natural Heritage Trust.

152 Shrestha, N., Shrestha, S., Koju, L. et al. (2016). Medicinal plant diversity and traditional healing practices in eastern Nepal. *Journal of Ethnopharmacology* 192: 292–301.

153 De Natale, A. and Pollio, A. (2012). A forgotten collection: the Libyan ethnobotanical exhibits (1912-14) by A. Trotter at the Museum O. Comes at the University Federico II in Naples, Italy. *Journal of Ethnobiology and Ethnomedicine* 8 (1): 1–19.

154 Linnaeus, C. (1737). *Flora Lapponica*. Amsterdam: Apud S. Schouten.

155 Biswas, K. and Chopra, R.N. (1956). *Common Medicinal Plants of Darjeeling and the Sikkim Himalayas*.

156 Pradhan, B.K. and Badola, H.K. (2008). Ethnomedicinal plant use by Lepcha tribe of Dzongu valley, bordering Khangchendzonga Biosphere Reserve, in north Sikkim India. *Journal of Ethnobiology and Ethnomedicine* 4 (1): 1–18.

157 Cramer, W.C.P. (1780). *Dissertatio inauguralis medica de lichene Islandico... quam... pro gradu doctoris... Mart. MDCCLXXX. publico et placido eruditorum examini subiiciet Guilielmus Christianus Philipp Cramer Onoldinus*. Erlangae, Germany: Typis Kunstmannianis.

158 Huguenin, M. and Regnault, J.-B.É.B.O. (1802). *Observations on Pulmonary Consumption, or, an Essay on the Lichen Islandicus, Considered Both as an Aliment and a Medicine in That Disorder*. London: Printed for the Author by J. Smeeton and sold by Johnson etc.

159 Matthiolus, P. (1586). *De plantis Epitome Utilissima*. Frankfurt: Johann Feyerabend.

160 Ray, J. (1686). *Historia Plantarum*, 1842. London: Clark.

161 Scopoli, J.-A. (1760). *Joan.-Ant. Scopoli Flora Carniolica Exhibens Plantas Carniolae Indigenas et Distributas in Classes Naturales cum Differentiis Specificis, Synonymis Recentiorum, Locis Natalibus, Nominibus Incolarum, etc*. Bavaria, Germany: Trattner.

162 Withering, W. (1801). *A Systematic Arrangement of British Plants; with an Easy Introduction to the Study of Botany*, 4e, vol. 3. London: Musci. H. Baldwin & Son.

163 Muntané, J. (1991). *Aportació al coneixement de l'etnobotànica de Cerdanya*. Barcelona: Facultad De Farmacia, Universidad De Barcelona.

164 Agelet, A. and Valles, J. (2003). Studies on pharmaceutical ethnobotany in the region of Pallars (Pyrenees, Catalonia, Iberian Peninsula). Part II. New or very rare uses of previously known medicinal plants. *Journal of Ethnopharmacology* 84 (2–3): 211–227.

165 Novaretti, R. and Lemordant, D. (1990). Plants in the traditional medicine of the Ubaye Valley. *Journal of Ethnopharmacology* 30 (1): 1–34.

166 Bellia, G. and Pieroni, A. (2015). Isolated, but transnational: the glocal nature of Waldensian ethnobotany, Western Alps, NW Italy. *Journal of Ethnobiology and Ethnomedicine* 11: 1–20.

167 Lokar, L.C. and Poldini, L. (1988). Herbal remedies in the traditional medicine of the Venezia Giulia region (North East Italy). *Journal of Ethnopharmacology* 22 (3): 231–279.

168 Mattalia, G., Quave, C.L., and Pieroni, A. (2013). Traditional uses of wild food and medicinal plants among Brigasc, Kyé, and Provençal communities on the Western Italian Alps. *Genetic Resources and Crop Evolution* 60: 587–603.

169 Pieroni, A. and Giusti, M.E. (2009). Alpine ethnobotany in Italy: traditional knowledge of gastronomic and medicinal plants among the Occitans of the upper Varaita valley, Piedmont. *Journal of Ethnobiology and Ethnomedicine* 5: 1–13.

170 Vitalini, S., Iriti, M., Puricelli, C. et al. (2013). Traditional knowledge on medicinal and food plants used in Val San Giacomo (Sondrio, Italy)—an alpine ethnobotanical study. *Journal of Ethnopharmacology* 145 (2): 517–529.

171 Grasser, S., Schunko, C., and Vogl, C.R. (2012). Gathering "tea"–from necessity to connectedness with nature. Local knowledge about wild plant gathering in the Biosphere Reserve Grosses Walsertal (Austria). *Journal of Ethnobiology and Ethnomedicine* 8: 1–24.

172 Vogl, C.R., Vogl-Lukasser, B., and Walkenhorst, M. (2016). Local knowledge held by farmers in Eastern Tyrol (Austria) about the use of plants to maintain and improve animal health and welfare. *Journal of Ethnobiology and Ethnomedicine* 12: 1–17.

173 Ferrier, J., Šačiragić, L., Chen, E.C.H. et al. (2014). Ways the Lukomir Highlanders of Bosnia and Herzegovina treat diabetes. In: *Ethnobotany and Biocultural Diversities in the Balkans: Perspectives on Sustainable Rural Development and Reconciliation* (ed. A. Pieroni and C.L. Quave), 13–27. New York: Springer.

174 Ferrier, J., Saciragic, L., Trakić, S. et al. (2015). An ethnobotany of the Lukomir highlanders of Bosnia & Herzegovina. *Journal of Ethnobiology and Ethnomedicine* 11: 1–17.

175 Šarić-Kundalić, B., Dobes, C., Klatte-Asselmeyer, V. et al. (2010). Ethnobotanical study on medicinal use of wild and cultivated plants in middle, south and west Bosnia and Herzegovina. *Journal of Ethnopharmacology* 131 (1): 33–55.

176 Redžić, S. (2007). Syntaxonomic diversity as an indicator of ecological diversity—case study Vranica Mts in the Central Bosnia. *Biologia* 62: 173–184.

177 Pieroni, A., Rexhepi, B., Nedelcheva, A. et al. (2013). One century later: the folk botanical knowledge of the last remaining Albanians of the upper Reka Valley, Mount Korab, Western Macedonia. *Journal of Ethnobiology and Ethnomedicine* 9 (1): 1–19.

178 Nedelcheva, A. and Draganov, S. (2014). Bulgarian medical ethnobotany: the power of plants in pragmatic and poetic frames. In: *Ethnobotany and Biocultural Diversities in the Balkans: Perspectives on Sustainable Rural Development and Reconciliation* (ed. A. Pieroni and C.L. Quave), 45–65. New York: Springer.

179 Nelly, A., Annick, D.-D., and Frederic, D. (2008). Plants used as remedies antirheumatic and antineuralgic in the traditional medicine of Lebanon. *Journal of Ethnopharmacology* 120 (3): 315–334.

180 Kujawska, M., Łuczaj, Ł., and Typek, J. (2015). Fischer's Lexicon of Slavic beliefs and customs: a previously unknown contribution to the ethnobotany of Ukraine and Poland. *Journal of Ethnobiology and Ethnomedicine* 11 (1): 1–15.

181 Pranskuniene, Z., Dauliute, R., Pranskunas, A. et al. (2018). Ethnopharmaceutical knowledge in Samogitia region of Lithuania: where old traditions overlap with modern medicine. *Journal of Ethnobiology and Ethnomedicine* 14 (1): 1–26.

182 Sak, K., Jürisoo, K., and Raal, A. (2014). Estonian folk traditional experiences on natural anticancer remedies: from past to the future. *Pharmaceutical Biology* 52 (7): 855–866.

183 Sõukand, R. and Kalle, R. (2013). Where does the border lie: locally grown plants used for making tea for recreation and/or healing, 1970s–1990s Estonia. *Journal of Ethnopharmacology* 150 (1): 162–174.

184 Whitney, C.W., Gebauer, J., and Anderson, M. (2012). A survey of wild collection and cultivation of indigenous species in Iceland. *Human Ecology* 40: 781–787.

185 Lamont, S.M. (1977). The Fisherman Lake Slave and their environment: a story of floral and faunal resources. *University of Saskatchewan.*

186 Bastien, J.W. (1983). Pharmacopeia of qollahuaya andeans. *Journal of Ethnopharmacology* 8 (1): 97–111.

187 Hebda, R., Turner, N.J., Birchwater, S. et al. (1997). *Ulkatcho Food and Medicine Plants.* Anahim Lake, BC: Ulkatcho Band.

16

Food Values of Lichen

Ashoke Kumar Das[1], Subrata Sarkar[1], Lily Devi[2], Chetana Hasnu[3], and Sourav Bhattacharjee[3]

[1] *Department of Botany, Abhayapuri College, Abhayapuri, Assam, India*
[2] *Department of Botany, Bijni College, Bijni, Assam, India*
[3] *Department of Botany, Pandit Deen Dayal Upadhyaya Adarsha Mahavidyalaya, Tulungia, Bongaigaon, Assam, India*

16.1 Introduction

One of the most fascinating forms of life on Earth is the lichen, a mysterious and fascinating organism that develops from a symbiotic interaction between algae (phycobiont) and fungi (mycobiont) through the intricate process of "lichenization" [1]. They display incredible versatility, thriving in almost every imaginable habitat, including harsh ones. Surprisingly, lichens account for around 7% of the vegetation on Earth. Their importance goes beyond ecological functions because they provide a variety of organisms with essential nutrition, from invertebrates to wild and domesticated animals to even humans [2]. More than 3.1 billion people worldwide, or 42%, were unable to afford a healthy diet in 2021, indicating a decline in the ability of people to obtain healthy meals globally. In 2022, the state of nutrition and food security remained dire. According to the incidence of moderate or severe food insecurity, the report shows that 2.4 billion people, or about 29.6% of the world's population, did not always have access to food. Around 900 million of them experienced extreme food insecurity. According to recent studies, there are approximately 735 million hungry people worldwide, up from 613 million in 2019 [3]. There is a growing emphasis on healthy and alternative food systems as the globe deals with tremendous issues like climate change and rising health concerns. Plant-based foods are gaining popularity as a replacement. By 2050, the population of the globe is projected to increase quickly to 9.7 billion people. Food production might not be sustainable, but healthy and environmentally sustainable products must be adopted [4].

Several lichen species have been accepted as essential components of traditional diets and medicinal practices throughout a variety of regions, including China, Japan, India, Nepal, Iceland, South and North America, Europe, and beyond [5–11]. Notably, 142 lichen species are used alone in traditional medicine in the Himalayan region and parts of

Chemistry, Biology and Pharmacology of Lichen, First Edition. Edited by Ashoke Kumar Das, Ajay Sharma, Deepika Kathuria, Mohammad Javed Ansari, and Garima Bhardwaj.

southeast China. Additionally, 36 species are used as a source of vegetable food, and 6 species are used to create energizing teas [8]. An extensive analysis of the nutritional composition of lichen reveals a strikingly balanced profile. These creatures have high crude fiber, ash, and carbohydrate contents but have relatively little fat. The essential amino acids are present in significant amounts, exceeding the optimum protein benchmarks despite the slight protein level. Additionally, lichens have high concentrations of important minerals including potassium, calcium, and iron, while having low sodium levels [9]. Lichens are positioned as a possible natural source for dietary enrichment for humans due to their distinctive nutritional makeup. They are good sources of plant-based diets for human wellbeing.

As we get deeper into this chapter, we undertake a thorough investigation of lichens, revealing their nutritional power, cultural significance, and ability to considerably add to the human diet. In doing so, we hope to bridge conventional nutritional science with old wisdom and provide insight into the ongoing interaction between lichens and people.

16.2 Historical Background

The history of lichens, those fascinating and mysterious organisms produced by the association between fungi and algae through a process called "lichenization," is one characterized by both neglect and rediscovery. Their progression through botanical history is evidence of the understated but significant functions they play in our ecosystems and cultures. Due to their apparent economic insignificance on a worldwide scale, lichens were historically confined to the margins of botanical research [12]. Lichens were regarded by Carl Linnaeus, as little more than "waste materials of vegetation." Lichens were virtually ignored by botanists until the beginning of the 18th century, in contrast to the lavish attention paid to more conventionally glamorous flora and wildlife [13]. They did not belong to a distinct taxonomic class and were frequently eclipsed by their more ostentatious competitors. Despite appearing to be obscure plants, lichens have played a vital role in human history. Ancient people exploited their healing abilities to treat a variety of diseases. These resilient organisms also served as useful natural dyes for clothing, giving materials vibrant hues. Despite appearing to be obscure plants, lichens have been significant throughout history. To treat a wide range of illnesses, ancient people made use of their natural healing powers. In addition to providing materials with brilliant colors, these hardy organisms also served as useful natural dyes for garments [14, 15]. Only in the early 20th century with the complexities of the interaction between fungi and algae within lichens start to completely emerge. Extensive study of the complex symbiotic relationships between ascomycetes fungi and the related algae occurred during this time. According to the current understanding of lichens, they developed as intricate, mutually beneficial physical alliances between a variety of fungi, algae, and/or cyanobacteria [16, 17]. Because of the possible uses of lichens in both industrial products and natural cures, interest in lichens has recently increased. This revival continues a long history in which lichens were valued for their therapeutic qualities and incorporated into cultural practices [2, 18–23]. In addition to their conventional functions, lichens have shown promise as possible friends in the fight against desertification and in the breakdown of bioplastics [24]. Despite this revival, there is still a lot of unrealized potential in investigating lichen-derived industrial uses and products, which encourages more study and exploration.

16.3 Lichen as Food for Human

One of the most important aspects of lichen is their application as food. It has been reported that lichens or lichen products have been used as food by humans and fodder for animals in various parts of the world since many centuries [5]. Records from 1700–1800 BC show that lichens or products derived from them have been used historically as food for people and as animal feed [25]. It has been shown that lichens are consumed and incorporated into a variety of nutritional and culinary traditions in places including China, Japan, North America, Siberia, Korea, and several European nations [8, 26–28]. In recent years, lichen has gained attention in the scientific community for its food value. Lichens have a high level of carbohydrates (53.2–79.08%), fiber (5.386–16.36%), and low fat (1.3–6.5%), which contributes to their nutritional value. Mineral elements abound in lichens, which are also strong sources of protein (5.95–16.2%). Data from the literature showed the presence of tryptophan, methionine, phenylalanine, isoleucine, leucine, threonine, and valine in the African lichen thalli *Rimelia reticulate* [29].

Throughout history, lichens have been used as a last-resort food source in countries including the Czech Republic, Norway, and Estonia during famines and food shortages [5, 9]. In Nepal's highland indigenous communities, such as the Limbu, Sherpa, Gurung, Tamang, and Chhetri, lichens have been an important source of food, beverages, and even medical treatments [30].

Lichens have a long and rich culinary and nutritional history in India, with a legacy that spans millennia. Since ancient times, people have collected and used these extraordinary organisms as both traditional food and folk medicine. The Atharvaveda, an ancient Indian literature, refers to lichens as "Shila Pushp" or "Rock Flower" and provides the oldest known evidence of its usage in folk medicine in India from around 1500 BC [31]. Lichens are mentioned in old Indian medical writings as well, such as the Charaka Samhita (300–200 BC), where they are known as "Shailaya" [32]. According to estimates, there are currently 305 genera and 74 families with 2319 species making up the Indian Lichen Flora, largely geographically distributed in India's tropical, subtropical, temperate, and alpine regions [6]. Two exceptional Parmeliaceae species, *Parmotrema perlatum*, (Huds.) M. Choisy and *Everniastrum cirrhatum*, (Fr.) Hale are especially lauded for their gastronomic and nutritional benefits among this diverse lichen flora. In India, *Parmotrema perlatum*, (Huds.) M.Choisy also known as "Pattharphool," is not only grown but also traded commercially. Innumerable vegetarian and non-vegetarian meals incorporate this adaptable lichen [33]. *Everniastrum cirrhatum*, (Fr.) Hale a foliose lichen that grows in sections of Southern India, North and Central India, and the Himalayan regions, is prized for having strong antioxidant capabilities. This lichen is sold commercially throughout India and used as a prized vegetable in regional cuisines [9].

Only a small portion of the world's diverse array of lichen species have undergone thorough phytochemical screening [34]. Notably, certain lichen species contain poisonous substances unfit for eating by people or animals. As a result, only a few number of lichen species are thought to be suitable for human consumption (Table 16.1). These foliose or fruticose types of edible lichens are typically utilized in traditional cuisines or as food additives [9, 50].

16.4 Lichen as Spices and Flavor Enhancer

Only a small number of lichen species, mostly among the macro-lichen types, such as the fruticose and foliose forms, have been identified for their culinary qualities despite the overwhelming diversity of lichen species found all over the world [50]. These special lichens have been praised for their ability to enhance flavors and give dishes a unique flavor that cannot be replicated [31]. A few examples of lichens used as flavor enhancers are depicted in Table 16.2. At least 4000 BCE marks the beginning of the use of spices for flavoring, seasoning, coloring, and food preservation on the Indian subcontinent and nearby Chinese civilizations [59]. These secondary metabolite-rich food additives are well known for their antibacterial qualities, providing defense against human illnesses and foodborne microorganisms [60]. Several of the diverse lichen species have been accepted as essential components of culinary traditions throughout the world. In India, the Middle East, and Niger, for example, gastronomes use *Parmotrema tinctorum Hale., Permotrema perlatum (Huds.) M. Choisy, Everniastrum cirrhatum (Fr.) Hale, Ramalina conduplicans Vain*, and *Rimelia reticulata (Taylor) Hale & A. Fletcher* as spices or flavor enhancers [26]. *Permotrema tinctorum Hale*, often known as Al-Sheba in Arabic cooking, is a spice that gives a variety of meals a distinctive flavor [61]. The coveted foliose lichen *Pseudevernia furfuracea (L.) Zopf*, also referred to as tree moss, is used commercially in spice mixes, the making of curries, and even as a food preservative [62]. Foliose lichen, *Everniastrum cirrhatum (Fr.) Hale*, plays the role of a flavor enhancer for meat and vegetables, enhancing the culinary traditions of Madhya Pradesh's Bhaiga, Bhil, Bhilala, Gond, Korka, and Muria populations [63]. Even in the distant past, the lichen *Pseudevernia furfuracea (L.) Zopf* was utilized in ancient Egypt to retain the aromatic properties of spices used in mummification [64]. It was used in ancient Egypt to preserve the aromatic qualities of spices used in mummification even in the distant past [64]. Indian cuisine makes extensive use of the foliose lichen *Parmotrema perlatum (Huds.) M. Choisy*, often known as the black stone flower, Dagad Phool, Kalpasi, canary moss, or Mangalore spices. It embellishes the East Indian Bottle Masala, gives vegetarian dishes a special flavor, and improves the flavors of meat and fish preparations [58, 65]. Black stone blossom is referred to as "Shaileyam" in Sanskrit, and it is found throughout the world, including India, where it gives a variety of foods its distinctive flavor [66].

Chemical food additives that are used to preserve, improve, and color, flavor, and texture of food pose harm to human health when consumed. Some have a history of causing cancer [67]. Due to their unique symbiotic structure, lichen creates an abundance of nutrients and distinctive bioactive compounds. In addition to lichen's nutritional benefits, greater investigation and development are needed for the study of edible lichens [9]. When taken, chemical food additives that are used to improve, preserve, color, flavor, and texture of food are harmful to human health. Lichen produces a large amount of nutrients and unusual

Table 16.1 Lichen species used as food with available common names and distribution.

Scientific name	Common name	Distribution	Consumed as	References
Alectoria asiatica (Du Rietz) Brodo & D. Hawksw.	Witch's hair lichen	China, Japan	Food and food adjunct	[8, 28]
Bryoria confusa (D.D. Awasthi) Brodo & D. Hawksw.	Horsehair lichen	Himalayas and South-Western China	Vegetable	[8, 35]
Bryoria fremontii (Tuck.) Brodo & D. Hawksw.	Horsehair lichen	North America	Cooked food	[8]
Cetraria islandica (L.) Ach.	Iceland moss	Iceland, Elstonia, Sweden, Italy and Norway	Salad, bread ingredient	[9, 36]
Cetraria laevigata Rass.	Striped iceland lichen	Himalayas and South-Western China	Vegetable	[8]
Cladia aggregata (Sw.) Nyl.		Himalayas and South-Western China	Vegetable	[8]
Cladonia stellaris (Opiz) Pouzar & Vezda	Star-tipped cup lichen	Himalayas and South-Western China	Vegetable	[8]
Cladonia rangiferina (L.) Nyl.	Reindeer lichen	North America, Siberia	Partially digested lichen from reindeer rumen	[8]
Dermatocarpon miniatum (L.) W. Mann		China, Japan	Food and food adjunct	[37]
Evernia prunastri (L.) Ach.	Oakmoss	Norway, Iceland, Elstonia, Czech Republic	Bread ingredient	[38]
Lasallia pustulata (L.) Mérat	Rock tripe	North America	Food and food adjunct	[8, 39]
Lecanora esculenta (Pall.) Eversm	Rim lichen	Israel, Middle East	Cooked food	[5]
Leptogium delavayi Hue.		Himalayas and South-Western China	Vegetable	[8]
Lobaria pulmonaria (L.) Hoffm.	Tree lungwort	China	Ingredient of bread	[37]
Lobaria retigera (Bory) Trevis.		Himalayas and South-Western China	Vegetable	[40]
Lobaria yunnanensis Yoshim.		China	Vegetable	[41]
Nephromopsis pallescens (Schaer.) Y.S. Park		Himalayas and South-Western China	Vegetable	[8]
Parmotrema perlatum (Huds.) M. Choisy	Black stone flower	Nepal	Food and food adjunct	[33]

(Continued)

Table 16.1 (Continued)

Scientific name	Common name	Distribution	Consumed as	References
Pseudevernia furfuracea (L.)	Tree moss	Spain	Ingredient of bread	[42]
Ramalina conduplicans Vain.		China, Nepal	Food and food adjunct	[43]
Rimelia reticulata (Taylor) Hale & A. Fletcher	Cracked ruffle lichens	Africa	Ingredient of a folk food called "Yari"	[44]
Sulcaria sulcata (Lev.) Bystrek.		Himalayas and South-Western China	Vegetable	[8, 45]
Umbilicaria esculenta (Miyoshi) Minks.	Rock tripe	Korea, Japan	Food and food adjunct	[46]
Umbilicaria hypococcinea (Jatta) Llano.	Rock tripe	Himalayas and South-Western China	Vegetable	[8, 47]
Umbilicaria yunnana (Nyl.) Hue.	Rock tripe	Himalayas and South-Western China	Vegetable	[8, 40]
Usnea barbata (L.) Weber ex F.H. Wigg.	Old man's beard	Balkans	Ingredient of bread	[42]
Usnea longissima Ach.	Old man's beard	China	Food and food adjunct	[48, 49]

Table 16.2 Lichen species used as spice and flavor enhancer.

Scientific name	Common name	References
Platismatia glauca (L.) W.L. Culb. & C.F. Culb.	Ragbag lichen	[51]
Ramalina commixta Asahina	Strap lichen or cartilage lichen	[8, 52]
Ramalina fastigiata (Pers.) Ach.	Strap lichen or cartilage lichen	[53]
Ramalina roesleri (Hochst. ex Schaer.) Hue.	Frayed ribbon lichen	[54]
Ramalina sinensis Jatta	Burning bush lichen	[55]
Cretaria islandica (L.) Acharius s.l.	True iceland lichen or iceland moss	[36]
Peltigera aphthosa (L.) Willd.	Common freckle pelt, green dog lichen, leafy lichen or felt lichen	[56]
Evernia divaricata (L.) Ach.	Mountain oakmoss lichen or ring lichen	[8]
Hypotrachyna nepalensis (Taylor) Divakar, A. Crespo, Sipman, Elix & Lumbsch	Jhyauu	[57]
Parmotrema perlatum (Huds.) M. Choisy	Black stone flower	[58]

bioactive chemicals as a result of their particular symbiotic structure. More research and development are required for the study of edible lichens in addition to their nutritional advantages. Even if the nutritional benefit of lichens is becoming more widely acknowledged, its potential as a source of natural and safe food additives still has to be explored and developed [9]. We may discover sustainable substitutes for synthetic additives by utilizing the bioactive substances found in lichens, improving the quality and security of our food supply.

The relationship between lichens and our culinary pursuits may hold the secret to healthier, more natural, and environmentally responsible food choices as we advance in the fields of food science and sustainability. The traditional practice of eating lichens can be reconciled with our current knowledge of nutrition and safety in this potential area for research and innovation.

16.5 Lichens as Beverage

The practice of using lichens as a source of food and drink (a few important examples are provided in Table 16.3) has a long history that dates back to the prehistoric past. Incredibly, between 1700 and 1800 BC, the oldest well-documented evidence of people using lichens for food and medicine first appears [25]. As a result of this discovery, humans and lichens have developed a complex connection that includes using lichens as food, animal feed, beverages, food additives, flavor enhancers, ingredients in confectionery, delicacies, spices, and sauces. This intriguing link between lichens and food culture has mostly been documented all over the world in places as different as China, Japan, North America, Siberia, Korea, and a number of European countries [8, 26, 27, 37].

Only around one-third of the world's lichen species have undergone rigorous phytochemical screening to identify their biochemical components, despite their astounding diversity [34]. This means that many lichen species are still unknown, and their potential uses have not yet been identified. Furthermore, it has been shown that some lichens contain poisonous compounds that make them unfit for eating by either humans or animals, highlighting the need for thorough research. Only a small percentage of the world's lichen species have been designated as being suitable for human consumption. Foliose or fruticose types of edible lichens are frequently used in traditional foods or as food additives [9, 50]. It is interesting to note that some indigenous groups living in Nepal's mountainous regions, including the Limbu, Sherpa, Gurung, Tamang, and Chhetri, have long understood the nutritional and therapeutic benefits of lichens [57]. They use lichens extensively in their culinary traditions and for their therapeutic benefits. The Czech Republic, Norway, and Estonia are just a few places where lichens have historically been used as an essential emergency food source during times of hunger and food scarcity [5, 9, 38]. This ability of lichens to endure harsh conditions and support human existence only emphasizes their fascinating position in our culinary and dietary history. The possibility of incorporating lichens into different culinary practices and nutritional contexts is an avenue that is ripe for future investigation as scientific investigation continues to unravel the varied characteristics and applications of lichens.

Table 16.3 Lichen species used as beverage with use, available common names and distribution.

Scientific name	Common name	Distribution	Consumed as	References
Cladonia fenestralis Nuno		Nepal, Sri Lanka, Himalayas, and South-Western China	Tea	[8]
Flavocetraria nivalis (L.) Karnfelt & Thell	Crinkled snow lichen	Peru, Poland	Tea	[9]
Lethariella cladonioides (Nyl.) Krog.	Red snow tea	China	Tea, Wines	[9]
Lethariella flexuosa (Nyl.) J.C. Wei.		Himalayas and South-Western China	Tea	[8]
Lethariella zahlbruckneri (Du Rietz) Krog.		Himalayas and South-Western China	Tea	[8]
Lobaria retigera (Bory.) Trev.	Smoker's lung lichen	China	Tea	[37]
Pseudevernia furfuracea (L.)	Tree moss	Spain	Tea	[42, 68]
Thamnolia vermicularis (Sw.) Schaer.	White-worm lichens	Himalayas and South-Western China	Tea	[8, 11]
Umbilicaria esculenta (Miyoshi) Minks.	Rock tripe	USA, Canada	Tea	[9]

16.6 Lichens as Feed

In addition to serving as a source of food for humans, lichens have long been essential to animal husbandry. Lichens have long been used as animal feed due to their advantageous phytochemical components and therapeutic potential. The Golden Snub-Nosed Monkey (*Rhinopithecus roxellana* Milne-Edwards), an endemic and threatened primate species in central and Southwestern China, serves as an example. *Usnea* sp., which makes up around 35% of their diet for these monkeys, is especially important to them [28]. Similar to this, *Cladonia rangiferina (L.) Weber ex F.H. Wigg*, often known as reindeer moss and a member of the Cladoniaceae family, provides essential nutrition for animals like Reindeer and Bullocks in Siberia, North America, Canada, and other arctic nations. Lichens are not just found in the diets of primates and other mammals; many arthropods and mollusk, including mites, caterpillars, and snails, also eat them [5]. In many areas of China, the Parmeliaceae family plant *Bryoria confusa (D.D. Awasthi) Brodo & D. Hawksw* has been identified as a source of food for monkeys [69]. Additionally, animals are fed on species, under genus like *Cetraria*, *Parmelia*, and *Lecanora*. *Cetraria islandica (L.) Ach* is frequently used as bulk feed for pigs, horses, and cattle in many parts of the world [70].

Table 16.4 Lichen-based food supplement as a commercial product.

Product name	Manufacturer	References
Go Organic Vitamin D3	Pallavan Engineering And Marketing Services, India	[71]
Nutrazee Vegan Vitamin D3	Nutrazee, India	[72]
Sai Herbs Usnea Lichen Extract Powder	KR Impex Enterprises, India	[73]
Naturelo Vitamin D	Naturelo, USA	[74]
Stone Flower Lichen Parmelia Perlata	Trustayur	[75]
INLIFE Plant-Based Vegan Vitamin D3 K2 Supplement	Inlife Pharma Private Limited, India	[76]
Food Supplement Vitamin D3 from Lichen	L'Erbolario Società Benefit S.r.l., Italy	[77]
Erbavita Vita D Icelandic Lichen Food supplement	FARMACIA LORETO, Italy	[78]

At present, several lichen-based food supplements have been prepared and made available through online marketing (Table 16.4).

16.7 Conclusion

Lichens are a living illustration of the intricate relationships that exist between the natural world and food. They have been prized for their outstanding nutritional and therapeutic qualities across numerous civilizations and centuries. Lichens provide a multitude of health benefits, ranging from antioxidative and anti-inflammatory qualities to antibacterial and anti-tumor potential. They are also rich in carbohydrates, proteins, minerals, vitamins, and unique bioactive chemicals. Further enhancing their culinary relevance are their distinctive taste and natural preservation properties. But there are risks and difficulties along the way as we work to include lichens in our diets. Given that some lichen species may contain poisons or heavy metals that could endanger human health, caution is absolutely necessary. Protecting lichen populations and the environments they live in from overharvesting and harmful collecting practices is crucial.

Lichens have a bright future as a sustainable food source, supported by continuous research into their health benefits and therapeutic uses. Innovative food products and dietary supplements that address important challenges of human health and food security may be made possible by sustainable techniques of growth and harvesting. However, this investigation needs to be conducted with constant prudence and a firm commitment to environmental stewardship. In conclusion, lichens appeal as an alluring and uncharted territory in the worlds of food and medicine. The entire potential of these amazing creatures will become apparent as we go forward through ethical behaviour and more research, improving our quality of life and protecting the delicate balance of the natural world.

References

1 Pichler, G., Muggia, L., Carniel, F.C. et al. (2023). How to build a lichen: from metabolite release to symbiotic interplay. *New Phytologist* 238 (4): 1362–1378. https://doi.org/10.1111/nph.18780.

2 Llano, G.A. (1948). Economic uses of lichens. *Economic Botany* 2 (1): 15–45. https://doi.org/10.1007/BF02907917.

3 122 million more people pushed into hunger since 2019 due to multiple crises, reveals UN report. (n.d.). https://www.who.int/news/item/12-07-2023-122-million-more-people-pushed-into-hunger-since-2019-due-to-multiple-crises--reveals-un-report

4 Plant-based Foods Landscape (2023). *Plant-based Foods Landscape—Navigating Challenges and Opportunities*. Ernst & Young LLP.

5 Wa, E., De, E.-G., and Gm, D. (2022). Lichens uses surprising uses of lichens that improve human life. *Journal of Biomedical Research & Environmental Sciences* 3 (2): 189–194. https://doi.org/10.37871/jbres1420.

6 Upreti, D.K., Bajpai, R., and Nayaka, S. (2015). Lichenology: current research in India. In: *Plant Biology and Biotechnology* (ed. B. Bahadur, M. Venkat Rajam, L. Sahijram, and K.V. Krishnamurthy), 263–280. Springer India. https://doi.org/10.1007/978-81-322-2286-6_10.

7 Moore, D., Robson, G.D., and Trinci, A.P.J. (2020). *21st Century Guidebook to Fungi*, 2ee. Cambridge University Press. https://doi.org/10.1017/9781108776387.

8 Yang, M.-X., Devkota, S., Wang, L.-S., and Scheidegger, C. (2021). Ethnolichenology—the use of lichens in the Himalayas and Southwestern parts of China. *Diversity* 13 (7): 330. https://doi.org/10.3390/d13070330.

9 Zhao, Y., Wang, M., and Xu, B. (2021). A comprehensive review on secondary metabolites and health-promoting effects of edible lichen. *Journal of Functional Foods* 80: 104283. https://doi.org/10.1016/j.jff.2020.104283.

10 Svanberg, I. and Ægisson, S. (2012). Edible wild plant use in the Faroe Islands and Iceland. *Acta Societatis Botanicorum Poloniae* 81 (4): 233–238. https://doi.org/10.5586/asbp.2012.035.

11 Wang, L., Narui, T., Harada, H. et al. (2001a). Ethnic uses of lichens in Yunnan, China. *The Bryologist* 104 (3): 345–349.

12 Plitt, C.C. (1919). A short history of lichenology. *The Bryologist* 22 (6): 77–85.

13 Sharma, M. and Mohammad, A. (2020). Lichens and lichenology: historical and economic prospects. In: *Lichen-Derived Products*, 1e (ed. M. Yusuf), 101–118. Wiley. https://doi.org/10.1002/9781119593249.ch4.

14 Rather, L.J., Jameel, S., Ganie, S.A., and Bhat, K.A. (2018). Lichen derived natural colorants: history, extraction, and applications. In: *Handbook of Renewable Materials for Coloration and Finishing*, 1e (ed. M. Yusuf), 103–114. Wiley. https://doi.org/10.1002/9781119407850.ch6.

15 Yusuf, M., Shabbir, M., and Mohammad, F. (2017). Natural colorants: historical, processing and sustainable prospects. *Natural Products and Bioprospecting* 7 (1): 123–145. https://doi.org/10.1007/s13659-017-0119-9.

16 Hawksworth, D.L., Kirk, P.M., Sutton, B.C., and Pegler, D.N. (1995). *Ainsworth & Bisby's Dictionary of the Fungi*, 8e. CAB International.

17 Nash, T.H. III (1996). *Lichen Biology*. Cambridge University Press.

18 Choudhary, M.I., Azizuddin, J., and S., and Atta-ur-Rahman. (2005). Bioactive phenolic compounds from a medicinal lichen, *Usnea longissima*. *Phytochemistry* 66 (19): 2346–2350. https://doi.org/10.1016/j.phytochem.2005.06.023.

19 Elix, J.A. and Stocker-Wörgötter, E. (2008). Biochemistry and secondary metabolites. In: *Lichen Biology*, 2e (ed. T.H. Nash), 104–133. Cambridge University Press. https://doi.org/10.1017/CBO9780511790478.008.

20 Fahselt, D. (1994). Secondary biochemistry of lichens. *Symbiosis* 16: 117–165.

21 Kumar Kc, S. and Müller, K. (1999). Lichen metabolites. 2. Antiproliferative and cytotoxic activity of gyrophoric, usnic, and diffractaic acid on human keratinocyte growth. *Journal of Natural Products* 62 (6): 821–823. https://doi.org/10.1021/np980378z.

22 Laundon, J.R. (1975). The vanishing lichens: their history, biology and importance, by D. H. S. Richardson. David & Charles, Newton Abbot, Devonshire, England: 231 pp., 41 plates, 27 figs, 30 tables, 14 × 22 × 2 cm, £5.25, 1975. *Environmental Conservation* 2 (3): 239–240. https://doi.org/10.1017/S0376892900001624.

23 Müller, K. (2001). Pharmaceutically relevant metabolites from lichens. *Applied Microbiology and Biotechnology* 56 (1–2): 9–16. https://doi.org/10.1007/s002530100684.

24 Oksanen, I. (2006). Ecological and biotechnological aspects of lichens. *Applied Microbiology and Biotechnology* 73 (4): 723–734. https://doi.org/10.1007/s00253-006-0611-3.

25 Chevallier, A. (1996). *The Encyclopedia of Medicinal Plants*. DK Pub.

26 Kanwar, A. and De, D. (2010). Lichen planus in children. *Indian Journal of Dermatology, Venereology, and Leprology* 76 (4): 366. https://doi.org/10.4103/0378-6323.66581.

27 Rambhia, K., Kharkar, V., Pradhan, V. et al. (2018). A study of prevalence of autoantibodies in patients with lichen planus from Mumbai, India. *Indian Journal of Dermatology, Venereology and Leprology* 84 (6): 667. https://doi.org/10.4103/ijdvl.IJDVL_182_17.

28 Huang, Z.-P., Bian, K., Liu, Y. et al. (2017). Male dispersal pattern in golden snub-nosed monkey (*Rhinopithecus roxellana*) in Qinling Mountains and its conservation implication. *Scientific Reports* 7 (1): 46217. https://doi.org/10.1038/srep46217.

29 Ślusarczyk, J., Adamska, E., and Czerwik-Marcinkowska, J. (2021). Fungi and algae as sources of medicinal and other biologically active compounds: a review. *Nutrients* 13 (9): 3178. https://doi.org/10.3390/nu13093178.

30 Devkota, S., Chaudhary, R.P., Werth, S., and Scheidegger, C. (2017). Indigenous knowledge and use of lichens by the lichenophilic communities of the Nepal Himalaya. *Journal of Ethnobiology and Ethnomedicine* 13 (1): 15. https://doi.org/10.1186/s13002-017-0142-2.

31 Upreti, D.K., Divakar, P.K., and Nayaka, S. (2005). Commercial and ethnic use of lichens in India. *Economic Botany* 59 (3): 269–273.

32 Kaushal, K. and Upreti, D.K. (2001). *Parmelia spp*. (Lichens) in ancient medicinal plant lore of India. *Economic Botany* 55 (3): 458–459. https://doi.org/10.1007/BF02866567.

33 Gandhi, A.D., Sathiyaraj, S., Suriyakala, G. et al. (2020). Lichens in genus *Parmelia*: an overview and their application. *Current Pharmaceutical Biotechnology* 21 (13): 1289–1297. https://doi.org/10.2174/1389201021666200406105212.

34 Zambare, V.P. and Christopher, L.P. (2012). Biopharmaceutical potential of lichens. *Pharmaceutical Biology* 50 (6): 778–798. https://doi.org/10.3109/13880209.2011.633089.

35 Søchting, U. (1999). *Lichens of Bhutan: Biodiversity and Use*. Botanical Institute, Department of Mycology: University of Copenhagen.

36 Meli, M.A., Desideri, D., Cantaluppi, C. et al. (2018). Elemental and radiological characterization of commercial *Cetraria islandica* (L.) *Acharius* pharmaceutical and food supplementation products. *Science of the Total Environment* 613–614: 1566–1572. https://doi.org/10.1016/j.scitotenv.2017.08.320.

37 Huang, X., Ma, J., Wei, L. et al. (2018). An antioxidant α-glucan from *Cladina rangiferina* (L.) Nyl. and its protective effect on alveolar epithelial cells from Pb^{2+}-induced oxidative damage. *International Journal of Biological Macromolecules* 112: 101–109. https://doi.org/10.1016/j.ijbiomac.2018.01.154.

38 Pawera, L., Łuczaj, Ł., Pieroni, A., and Polesny, Z. (2017). Traditional plant knowledge in the white Carpathians: ethnobotany of wild food plants and crop wild relatives in the Czech Republic. *Human Ecology* 45 (5): 655–671. https://doi.org/10.1007/s10745-017-9938-x.

39 Kosanić, M., Ranković, B., Stanojković, T. et al. (2016). *Lasallia pustulata* lichen as possible natural antigenotoxic, antioxidant, antimicrobial and anticancer agent. *Cytotechnology* 68 (4): 999–1008. https://doi.org/10.1007/s10616-015-9856-y.

40 Wang, L. (2012). *中国云南地衣 [Zhong guo yunnan di yi= Illustrated Lichen in Yunnan, China].*

41 Wang, L.S. and Qian, Z.G. (2013). *Illustrated Medicinal Lichens of China*. Yunnan ke ji chu ban she.

42 Simkova, K. and Polesny, Z. (2015). Ethnobotanical review of wild edible plants used in the Czech Republic. *Journal of Applied Botany and Food Quality* 88: 4967. https://doi.org/10.5073/JABFQ.2015.088.009.

43 Luo, H., Yamamoto, Y., Jeon, H.-S. et al. (2011). Production of anti-helicobacter pylori metabolite by the lichen-forming fungus *Nephromopsis pallescens. The Journal of Microbiology* 49 (1): 66–70. https://doi.org/10.1007/s12275-011-0289-9.

44 Glew, R.S., Vanderjagt, D.J., Chuang, L.-T. et al. (2005). Nutrient content of four edible wild plants from West Africa. *Plant Foods for Human Nutrition* 60 (4): 187–193. https://doi.org/10.1007/s11130-005-8616-0.

45 Wang, L.S. (1993). Introduction to five edible lichens. *Edible Fungi China* 12: 32–33.

46 Perez-Llano, G.A. (1944). Lichens their biological and economic significance. *Botanical Review* 10 (1): 1–65.

47 Wang, T., Bengtsson, G., Kärnefelt, I., and Björn, L.O. (2001b). Provitamins and vitamins D2 and D3 in *Cladina spp*. over a latitudinal gradient: possible correlation with UV levels. *Journal of Photochemistry and Photobiology B: Biology* 62 (1–2): 118–122. https://doi.org/10.1016/S1011-1344(01)00160-9.

48 Siddiqi, K.S., Rashid, M., Rahman, A. et al. (2018). Biogenic fabrication and characterization of silver nanoparticles using aqueous-ethanolic extract of lichen (*Usnea longissima*) and their antimicrobial activity. *Biomaterials Research* 22 (1): 23. https://doi.org/10.1186/s40824-018-0135-9.

49 Prateeksha, P., Paliya, B.S., Bajpai, R. et al. (2016). The genus *Usnea*: a potent phytomedicine with multifarious ethnobotany, phytochemistry and pharmacology. *RSC Advances* 6 (26): 21672–21696. https://doi.org/10.1039/C5RA24205C.

50 Calcott, M.J., Ackerley, D.F., Knight, A. et al. (2018). Secondary metabolism in the lichen symbiosis. *Chemical Society Reviews* 47 (5): 1730–1760. https://doi.org/10.1039/C7CS00431A.

51 Abdallah, E. (2019). Lichen thalli of *Platismatia glauca* possesses a remarkable antimicrobial activity. *Proceedings of 5th International Electronic Conference on Medicinal Chemistry*, p. 6314, Basel, Switerland (30 October 2019). https://doi.org/10.3390/ECMC2019-06314.

52 Moreira, A., Braz-Filho, R., Mussi-Dias, V., and Vieira, I. (2015). Chemistry and biological activity of *Ramalina* lichenized fungi. *Molecules* 20 (5): 8952–8987. https://doi.org/10.3390/molecules20058952.

53 Soundararajan, S., Shanmugam, P., Nagarajan, N. et al. (2019). In vitro study on screening antimicrobial and anti-oxidant potential of *Ramalina fastigiata*. *Journal of Drug Delivery and Therapeutics* 9 (1): 216–219. https://doi.org/10.22270/jddt.v9i1.2217.

54 Goel, M., Dureja, P., Rani, A. et al. (2011). Isolation, characterization and antifungal activity of major constituents of the Himalayan lichen *Parmelia reticulata Tayl*. *Journal of Agricultural and Food Chemistry* 59 (6): 2299–2307. https://doi.org/10.1021/jf1049613.

55 Karimi, H., Koopaie, M., Sohrabi, M., and Norouzi, H. (2022). Cytotoxic, anti-proliferative, and apoptotic evaluation of *Ramalina sinensis* (Lichenized fungus, *Ramalinaceae*) on oral squamous cell carcinoma cell line; in-vitro study [Preprint]. In Review. https://doi.org/10.21203/rs.3.rs-2246187/v1.

56 Chekanov, K., Feoktistov, A., and Lobakova, E. (2017). Spatial organization of the three-component lichen *Peltigera aphthosa* in functional terms. *Physiologia Plantarum* 160 (3): 328–338. https://doi.org/10.1111/ppl.12552.

57 Devkota, S. and Scheidegger, C. (2021). Hypotrachyna nepalensis (Taylor) Divakar, A. Crespo, Sipman, Elix & Lumbsch Parmeliaceae. In: *Ethnobotany of the Himalayas* (ed. R.M. Kunwar, H. Sher, and R.W. Bussmann), 1075–1079. Springer International Publishing https://doi.org/10.1007/978-3-030-57408-6_265.

58 Dwarakanath, P.R., Abinaya, K., Nagasathya, K. et al. (2022). Profiling secondary metabolites from lichen "Parmotrema perlatum (Huds.) M.Choisy" and antibacterial and antioxidant potentials. *Biomass Conversion and Biorefinery* https://doi.org/10.1007/s13399-022-03572-0.

59 Ramkumar, T.R. and Karuppusamy, S. (2021). Plant diversity and ethnobotanical knowledge of spices and condiments. In: *Bioprospecting of Plant Biodiversity for Industrial Molecules*, 1ee (ed. S.K. Upadhyay and S.P. Singh), 231–260. Wiley https://doi.org/10.1002/9781119718017.ch12.

60 Liu, Q., Meng, X., Li, Y. et al. (2017). Antibacterial and antifungal activities of spices. *International Journal of Molecular Sciences* 18 (6): 1283. https://doi.org/10.3390/ijms18061283.

61 Anupama, T.V., Sheela, K.B., and Gomez, S. (2017). Phytochemical screening and proximate composition of lichen *Parmotrema tinctorum* (Nyl.) Hale (*Parmeliaceae*) from Wayanad, Kerala. *International Journal of Chemical Studies* 5 (5): 1003–1007.

62 Kalra, R., Conlan, X.A., Areche, C. et al. (2021). Metabolite profiling of the Indian food spice lichen, *Pseudevernia furfuracea* combined With optimised extraction methodology to obtain bioactive phenolic compounds. *Frontiers in Pharmacology* 12: 629695. https://doi.org/10.3389/fphar.2021.629695.

63 Prashith, K.T., Ramamoorthy, R.H., LakshmanaShetty, S.D. et al. (2012). Antifungal and cytotoxic activity of *Everniastrum cirrhatum* (Fr.) Hale. *Chiang Mai Journal of Sciences* 39 (1): 76–83.

64 Odabasoglu, F., Aslan, A., Cakir, A. et al. (2005). Antioxidant activity, reducing power and total phenolic content of some lichen species. *Fitoterapia* 76 (2): 216–219. https://doi.org/10.1016/j.fitote.2004.05.012.

65 Verma, S. (2016). Medicinal plants with anti-inflammatory activity. *The Journal of Phytopharmacology* 5 (4): 157–159.

66 Neelima, K., Sony, G., and Sabitha, Y. (2022). Estimation of phytochemical analysis and anti-inflammatory activity of fresh extract of *Parmotrema Perlatum* and *Vitex Negundo. International Journal for Research in Applied Science and Engineering Technology* 10 (8): 1037–1052. https://doi.org/10.22214/ijraset.2022.46348.

67 Shazia, K.M., Asema, U.K., and Kasim, S.S. (2017). To study the harmful effects of food preservatives on human health. *Journal of Medicinal Chemistry and Drug Discovery* 2 (2): 610–616.

68 Emsen, B., Togar, B., Turkez, H., and Aslan, A. (2018). Effects of two lichen acids isolated from *Pseudevernia furfuracea* (L.) Zopf in cultured human lymphocytes. *Zeitschrift Für Naturforschung C* 73 (7–8): 303–312. https://doi.org/10.1515/znc-2017-0209.

69 Grueter, C.C., Li, D., Ren, B. et al. (2009). Fallback foods of temperate-living primates: a case study on snub-nosed monkeys. *American Journal of Physical Anthropology* 140 (4): 700–715. https://doi.org/10.1002/ajpa.21024.

70 Sánchez, M., Ureña-Vacas, I., González-Burgos, E. et al. (2022). The genus *Cetraria* s. Str.—a review of its botany, phytochemistry, traditional uses and pharmacology. *Molecules* 27 (15): 4990. https://doi.org/10.3390/molecules27154990.

71 Go Organic Vitamin D3 Plant Based from Lichen. (n.d.). https://www.amazon.in/Go-Organic-Vitamin-Plant-Lichen/dp/B0BNNM4R3B.

72 Nutrazee Vegan Vitamin D3. (n.d.). Nutrazee Vegan Vitamin D3

73 Sai Herbs Usnea Lichen Extract Powder. (n.d.). https://www.amazon.in/SAI-HERBS-Lichen-Extract-Powder/dp/B0C567CLXX?th=1.

74 Naturelo Vitamin D. (n.d.). https://www.amazon.com/NATURELO-Vitamin-Natural-Supplement-Capsules/dp/B078TXVTXT.

75 Stone Flower – Lichen – Patthar ka fool – Parmelia Perlata. (n.d.). https://shop.trustayur.com/product/stone-flower-lichen-patthar-ka-fool-parmelia-perlata/?attribute_pa_form=powder&attribute_pa_weight=250-grams.

76 Inlife Plant Based Vegan Vitamin D3 K2 Supplement. (n.d.). https://www.amazon.in/Vitamin-Supplement-Natural-Organic-Pressed/dp/B09S3HDGVS?th=1.

77 Food Supplement Vitamin D3 from Lichen. (n.d.). https://www.erbolario.com/en/erbamea/food-supplement-vitamin-d3-from-lichen-4087.html.

78 Erbavita Vita D Icelandic Lichen Food supplement. (n.d.). https://pharmacyloreto.com/erbavita-vita-d-lichen-islandico-food-supplement-100-oral-soluble-tablets.

17

Lichen as a Raw Material in Perfumery and Cosmetic Industries

Shahnaz Alom[1,4], Farak Ali[2,4], Bibhuti Busan Kakoti[4], Sandipan Choudhury[2], and Abdul Baquee Ahmed[3]

[1] *Department of Pharmacology, Girijananda Chowdhury Institute of Pharmaceutical Science-Tezpur, Girijananda Chowdhury University, Sonitpur, Assam, India*
[2] *Girijananda Chowdhury Institute of Pharmaceutical Science, Tezpur campus, Girijananda Chowdhury University, Assam 784501, India*
[3] *Department of Pharmaceutics, Girijananda Chowdhury Institute of Pharmaceutical Science-Tezpur, Girijananda Chowdhury University, Sonitpur, Assam, India*
[4] *Department of Pharmaceutical Sciences, Dibrugarh University, Dibrugarh, Assam, India*

17.1 Introduction

Human behavior is significantly influenced by odors. Unpleasant odors can negatively affect our mood and cause worry and discomfort, whereas pleasant scents might have a relaxing impact or make us feel better. Odors' significance in influencing human behavior has been documented scientifically. The Egyptians' use of plants, gums, and resins in religious rites thousands of years ago is the earliest known instance of humans employing perfumes [1]. Perfumes are becoming an essential component of our daily life and well-being. Based on its incense origins, the word "perfume" is derived from the Latin words "per," which means through, and "fumare," which means to smoke [2]. India has a more than 5000-year history in the production of perfume. A water distillation still and receiver found during excavations at Harappa and Mohanjodaro from the Indus Valley civilization provide evidence of the presence of a fragrance industry in our nation even during those times. Hina, Shamama, Rose, Kewra, and other flower attars are produced in traditional Indian perfumery [3].

Lichens are composite organisms made up of cyanobacteria (Nostoc) or algae (Trebouxia or Trentepohlia) that coexist in mutualistic communities with various fungi. On around 7% of the planet's surface, lichens predominate in terms of vegetation kinds; in addition, they are crucial primary producers in a variety of substrates and ecosystems, including some of the worst environmental conditions on earth, such as North and South Pole, desert, even glass surfaces, and others. Lichens are a well-known example of a symbiotic interaction between algae and fungi that is beneficial for the food

Chemistry, Biology and Pharmacology of Lichen, First Edition. Edited by Ashoke Kumar Das, Ajay Sharma, Deepika Kathuria, Mohammad Javed Ansari, and Garima Bhardwaj.

business as well as for the prevention or treatment of a variety of human ailments. As a protection mechanism against outside predators, lichens frequently create secondary metabolites [4]. Extracts of lichen and other plants are increasingly popular ingredients in personal care products, including baby creams, deodorants, face, and body lotions. According to one estimate, there are more than 90 deodorants, perfumes, or colognes that specifically include lichen-derived chemicals, as well as countless more items that use potentially allergic Compositae species [5].

Lichens have been used since ancient times in perfumery, food, incense, and spice material. Lichen species collected from the Himalayas are used in the preparation of perfumes called "Hina" and "Attar" in various perfumery industries. The members of the lichen family Parmeliaceae including species of the genera Cetrelia, Flavoparmelia, Parmelaria, Everniastrum, Myelochroa, Leptogium, Lobaria, Nephromopsis, Ramalina, are used in making perfume. The maximum contribution in lichen material used in perfumery industry is made by the species of lichen genera *Everniastrum* and *Parmotrema* [3]. Because of the fragrant compounds found in them, lichens are used to make a variety of cosmetic products, perfumery items, *dhoop*, *hawan samagri*, etc. [4, 6]. The following article attempts to identify and quantify the many species of lichens used in the creation of various products in the perfumery and cosmetic industries.

17.2 Historical Background of Lichens Used in Perfume Industry

Lichens are remarkable plants having a unique morphology of two microorganisms, one that is of alga and the other of fungus existing in a symbiotic relation. The full nature of the lichen relationship did not begin to come into light until Swiss botanist Simon Schwendener proposed his dual theory of lichens in 1867, even though lichens had long been recognized as living organisms. Schwendener's hypothesis states that lichens are a microbial concoction of fungi and algae or cyanobacteria. In order to build his theory, which at the time lacked experimental backing, Schwendener meticulously examined the lichen, algae, and fungal growth and anatomy under a light microscope. Many of the leading lichenologists, like James Crombie and Nylander, disagreed with Schwendener's assertion despite the widespread belief that all living things were autonomous at the time [7]. Other eminent biologists, such as Heinrich Anton de Bary, Albert Bernhard Frank, Beatrix Potter, Melchior Treub, and Hermann Hellriegel, took longer to reject Schwendener's theories, and the idea quickly spread to other areas of study, such as the investigation of microbial, plant, animal, and human pathogens. When the complex interactions between pathogenic bacteria and their hosts were eventually identified, Schwendener's theory gained popularity. Further experimental proof of the dual nature of lichens was obtained when Eugen Thomas published his findings on the first successful re-synthesis experiment in 1939 [8]. In 2010, a brand-new facet of the interaction between fungi and algae was discovered. According to studies by Toby Spribille [9], several species of lichen that were previously thought to be ascomycete-algae pairings were ascomycete-basidiomycete-algae trios.

Since ancient times, lichens have been employed in food, incense, fragrance, and spice products. The Parmeliaceae lichen family, which includes species of the genera Bulbothrix, Cetrelia, Everniastrum, Flavoparmelia, Myelochroa, Nephromopsis, Parmelaria, Parmotrema, and Usnea, along with species of Heterodermia, Leptogium, Ramalina, and Lobaria, is extensively used in the preparation of perfume [3]. *Evernia prunastri* (oakmoss) and *Pseudevernia furfuracea* (tree moss) have been extensively employed in the perfume and cosmetics industries since the discovery of lichenol from lichens due to their high aroma value and superior fixative characteristics [10]. Evernia and Pseudevernia are the lichens most frequently utilized in Western Europe, although Ramalina, Lobaria, Parmelia, and Usnea have also been used. *Parmelia nepalensis*, *Permelia nilgherrensis*, *Ramalina subcomplanata*, and *Usnea lucea* are mainly used in India. The lichens *E. prunastri* and *P. furfuracea* commonly known as "oakmoss" and "tree moss," respectively, are collected from forests of Mediterranean areas are nowadays mainly used in the manufacturing of perfumes. Evernia is a lichen with a strap-shaped branching having a white lower surface and a pale gray–green upper surface. Pseudevernia also have a similar shape but are black on the underside and gray on the above. They grow on trees and are attached to a single point on the trunk or branch, from where they can be easily detachable. The most highly developed lichens are termed as "fruticose." The simple leafy lichens are known as "foliose" and the crusty surface lichens as "crustose" [10].

Lichens from the Himalayas are extensively employed in the North Indian traditional perfumery industry of Uttar Pradesh's Kannauj area to create the scents "Hina" and "Attar," which fall into the category of herbal and spicy attar. Lichens are recognized by several trade names such as "Oakmoss," "Stone Flowers," "Pathar Fool," "Daggarphool," "Chadila," "Charrila," "Chadilo," "Iceland moss," and "Rock moss" [3]. Approximately 900–1000 tonnes of lichens are harvested from the North of India each year, according to Moxham [10]. *E. prunastri's* lichen acid extract is utilized as a scent ingredient. However, it is said that Evernia, Ramalina, and Pseudorina contain fragrant volatile oils. Because of the fragrant compounds found in the thallus, lichens are used to make a variety of cosmetic products, perfumery items, *dhoop, hawan samagri*, etc. [4, 6].

17.3 Commercially Viable Lichen Species in Perfumery and Cosmetic Industries

Some lichen species have a distinctive and delicate herbal scent. Heterodermia species with a certain economic value are used to make fragrances and medicines. *E. prunastri*, sometimes known as oak moss, is a lichen that has been harvested in south-central Europe and used to produce high-end cosmetic ingredients and perfumes. In France, oak moss was also used to create perfumes and a variety of smells. In order to add mossy aromas to cosmetics, body lotion, perfume, and soaps, *P. furfuracea* (tree moss) and *Ramalina* spp., two other major lichen species, were also used. Commonly referred to as tree mosses, *E. prunastri* var. prunastri (L.) and *P. furfuracea* (L.) are exceptionally prolific on the wood of pine and cedar trees. Currently, the French perfumery business uses the extracts of both lichen species [11, 12].

Apart from these lichen species, some other lichen species are also used in perfume and cosmetic industries, which are listed in Table 17.1.

Table 17.1 Various lichen species are used in the perfumery and cosmetic industries.

Serial number	Lichen species	Family	Geographical distribution	Important phytoconstituents	References
1	*Bulbothrix meizospora*	Parmeliaceae	India, Nepal, Pakistan, Africa, South America	Norstictic acid, protocetraric acid	[3]
2	*Bulbothrix setschwanensis*	Parmeliaceae	India, China, and Nepal	2,3-bis (2 methylpentanoyloxy) propyl 2-methylpentanoate, Ethyl 2-[(2R,3R,4aR,8aS)-3-hydroxy-2,3,4,4a, 6,7,8,8a-octahydropyrano [3,2-b]pyran-2-yl]acetate	[13]
3	*Cetrelia braunsiana*	Parmeliaceae	India, Bhutan, China, Japan, Nepal, New Zealand, Philippines, and Taiwan	atranorin, α-collatolic acid, alectoronic acid	[14]
4	*Cetrelia cetrarioides*	Parmeliaceae	India, Bhutan, China, Japan, Europe, and North America	Atranorin, perlatolic acid, and imbricaric acid	[14]
5	*Everniastrum cirrhatum*	Parmeliaceae	India, Bhutan, China, Japan, Europe, and North America	Atranorin, salazinic acid, and protolichesterinic acid	[15]
6	*Everniastrum nepalense*	Parmeliaceae	India, Bhutan, and Nepal	Usnic acid	[16]
7	*Flavoparmelia caperata*	Parmeliaceae	India, Bhutan, China, Japan, and Nepal	Usnic acid, protocetraric acid, and caperatic acid	[17]
8	*Heterodermia boryi*	Physciaceae	India (Jammu and Kashmir, Kerala, Maharashtra, Sikkim, Tamil Nadu, Uttarakhand)	Atranorin, zeorin, salazinic acid, and norstictic acid	[18]
9	*Heterodermia diademata*	Physciaceae	India, Bhutan, China, Taiwan, Nepal, Japan, Africa, Europe, and South America	Atranorin, zeorin, salazinic acid, and norstictic acid	[18]
10	*Leptogium askotense*	Collamataceae	India (Arunachal Pradesh, Manipur, Sikkim, West Bengal, and Uttarakhand)	Not available	[3]

Table 17.1 (Continued)

Serial number	Lichen species	Family	Geographical distribution	Important phytoconstituents	References
11	*Leptogium pedicellatum*	Collamataceae	India, Bhutan, China, and Japan	Two unknown compounds were isolated.	[19]
12	*Lobaria retigera*	Peltigeraceae	India, Bhutan, China, Australia, New Zealand, Japan, South Africa, Thailand	Not available	[3]
13	*Myelochroa denegans*	Parmeliaceae	India, Bhutan, China, Japan, Europe, and North America	Not available	[3]
14	*Myelochroa subaurulenta*	Parmeliaceae	India, Bhutan, China, Japan, Europe, and North America	Atranorin, zeorin, leucotytic acid, leucotylin, secalonic acid	[20]
15	*Nephromopsis laii*	Parmeliaceae	India, Bhutan, China, Japan, Europe, and North America	Usnic acid, α-collatolic acid, protolichesterinic acid, alectoronic acid	[21]
16	*Nephromopsis stracheyi*	Parmeliaceae	India, Bhutan, China, Japan, Europe, and North America	Usnic acid, atranorin, protolichesterinic acid, alectoronic acid	[21]
17	*Parmelaria subthomsonii*	Parmeliaceae	India (Sikkim, Uttarakhand, and West Bengal)	Orcinol, orsellinic acid	[22]
18	*Parmelaria thomsonii*	Parmeliaceae	India (Arunachal Pradesh, Kerala, Manipur, Nagaland, Sikkim, Uttarakhand, and West Bengal) and Nepal	Atranorin, physodalic acid, orcinol, orsellinic acid	[22]
19	*Parmotrema austrosinense*	Parmeliaceae	India, Australia, Bhutan, China, Indonesia, Japan, Nepal, Sri Lanka, North and South America	Atranorin, lecanoric acid, erythrin, methyl orsellinate, methyl-α-orcinolcarboxylate	[23]
20	*Parmotrema nilgherrense*	Parmeliaceae	India, Bhutan, China, Thailand, and East Africa	Not available	[3]
21	*Parmotrema reticulatum*	Parmeliaceae	India, Bhutan, and Sri Lanka	Catechin, tannic acid, purpurin, reserpine	[24]

(Continued)

Table 17.1 (Continued)

Serial number	Lichen species	Family	Geographical distribution	Important phytoconstituents	References
22	*Pyxine meissnerina*	Physciaceae	India, Japan, Nepal, Africa, and South East Asia	Unknown triterpenoids	[25]
23	*Ramalina conduplicans*	Ramalinaceae	India, Nepal, and Eastern Asia	Atranol, sekikaic acid, 2,4-dimethyldivaric acid, homosekikaic acid, divaricatic acid, decarboxydivaricatic acid, olivetolic acid	[26]
24	*Ramalina sinensis*	Ramalinaceae	India, Bhutan, China, Nepal, Japan, and Pakistan	Usnic acid, iso-usnic acid, usninic acid, salazinic acid	[27]
25	*Usnea longissima*	Parmeliaceae	India, Bhutan, China, Nepal, Japan, Europe, and North America	Usnic acid, evernic acid, barbetic acid, Methyl-β-orcinol carboxylate, methyl 2,5-dihydroxy-3-methylbenzoate, orsellinic acid, squamatic acid, diffractaic acid	[28, 29]
26	*Usnea orientalis*	Parmeliaceae	India, Bhutan, China, and Japan	Usnic acid, salazinic acid, atranorin, eumitrin, orsellinate	[30, 31]
27	*Usnea perplexans*	Parmeliaceae	India (Himachal Pradesh, Jammu and Kashmir, and Uttarakhand), China, Europe, and America	Usnic acid, salazinic acid, diffractaic acid, stictic acid	[29]
28	*Usnea pseudosinensis*	Parmeliaceae	India, Nepal, Sri Lanka	Usnic acid, barbatic acid, cryptostictic, galbinic acid, isousnic acid, diffractic acid, stictic acid, salazinic acid	[32]
29	*Usnea subfloridana*	Parmeliaceae	India (Himachal Pradesh, Sikkim, Jammu and Kashmir, and Uttarakhand), Bhutan, China, Europe	Usnic acid, thamnolic acid, evernic acid, 3-hydroxyphysodic acid, barbatic acid	[33]

17.4 Lichen as Perfume and its Chemistry

More than 1000 secondary metabolites have been documented so far, many of which are only present in lichens. Lichens produce a wide range of natural chemicals. The bulk of these distinctive secondary metabolites found in lichens are fungal in origin. Many secondary metabolites of lichens play crucial ecological roles, such as those of light-screening, chemical weathering, allelopathy, and herbivore defense [11]. Furthermore, due to the numerous biological properties of lichen chemicals, which include antiviral, antibiotic, anticancer, allergic, plant growth inhibitory, antiherbivore, and enzyme inhibitory activity, lichens are a viable source for pharmaceutically intriguing natural products [11, 34]. The production of perfumes is currently one of the most significant economic uses of lichens. For extracts added to perfumes to give them a sweet and mossy scent and to ensure permanence on the skin, combined lichen material is employed. Every year in France, Morocco, and Southeast Europe, hundreds to thousands of tons of *E. prunastri* and *P. furfuracea* are gathered for this reason [35].

Specific and subtle fragrances are the characteristics of some lichens. Lichen species like *Heteroderma* are commonly known for their perfumery and medicinal properties. *E. prunastri* (oakmoss), which is mainly found in south-central Europe mostly used in cosmetics and perfumery industries [18]. Phytochemicals obtained from lichen species like *P. furfuracea* (tree moss) and *Ramalina sp.* are mostly used to add the mossy fragrance in production of perfumes, cosmetics, soaps, and body lotions [7]. Perfume industries in France have been using the ingredients of *E. prunastri* for a long period of time. Apart from these species, *Evernia mesomorpha* Nyl. and *Cetrariastrum nepalensie*, which are commonly known as Chinese Oakmoss has been extensively used in perfume industries for its perfume like odors [36]. Some other lichen species, such as *R. subcomplanata*, *L. pulmonaria*, *P. nepalensis*, *P. nilgherrensis*, and *U. lucea*, were also used to a lesser extent for their perfume-like properties [10, 11].

17.4.1 *Evernia prunastri* (Oakmoss)

E. prunastri, commonly known as Oakmoss, is greatly considered to be a perfume lichen. As per the gas chromatography–mass spectroscopy study revealed that, oakmoss contains several volatile oils as well as secondary metabolites, which are responsible for imparting perfumery properties to this oakmoss. Previously, some researchers carried out isolation of several compounds from the methanolic extract of *E. prunastri* with the help of high-performance liquid chromatography (HPLC). They have found that, extract of *E. prunastri* contains compounds like evernic acid, orcinol (3-hydroxy-5-methylphenol), orsellic acid (2,4-dihydroxy-6-methyl benzoic acid), orcinol monomethyl ether (3-methoxy-5-hydroxytoluene), everninic acid (2-hydroxy-4-methoxy-6-methylbenzoic acid), evernyl (methyl 2,4-dihydroxy-3,6-dimethylbenzoate), methyl evernate, ethyl everninate, atranorin, usnic acid, barbatic acid, lecanorin, divaricatic acid, thamnolic acid, prunastric acid, evernin, etc. The structure of the above-mentioned compounds is shown in Figure 17.1.

Among these phytochemicals, evernyl is found in the maximum amount of 15%, which is responsible for the typical odor of oakmoss. Due to structural similarity,

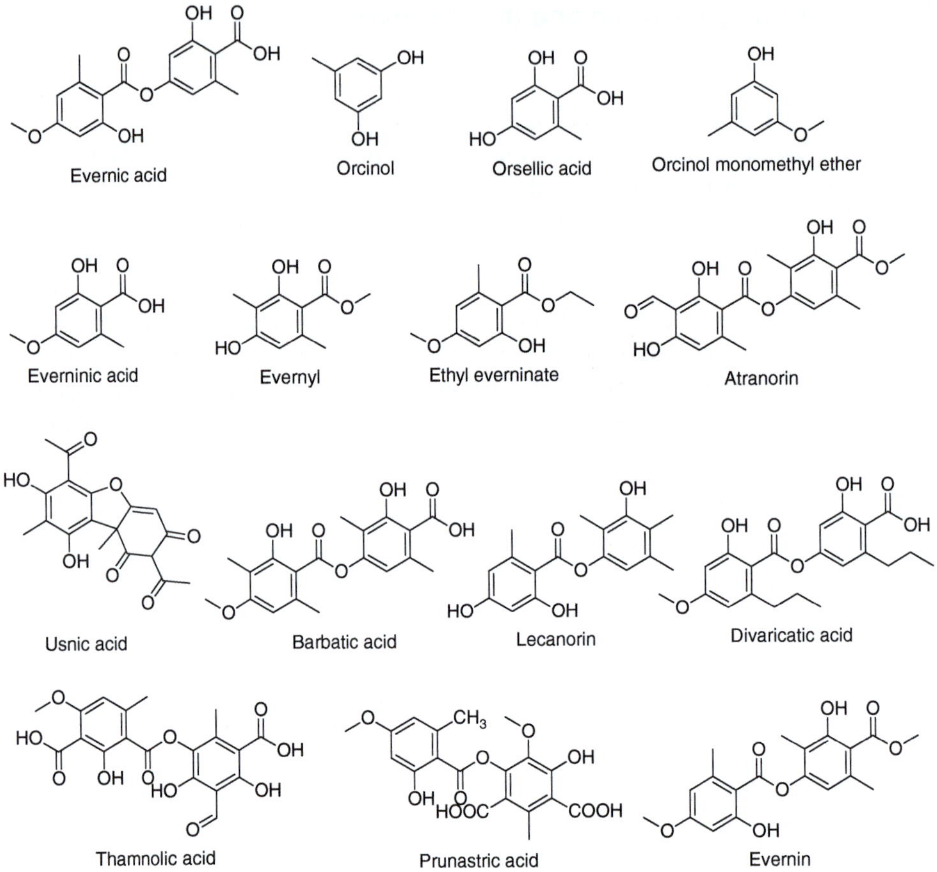

Figure 17.1 Important mono aromatic phytochemicals present in *Evernia prunastri.*

compounds like orsellic acid, evernic acid, and evernyl is also believe to be contributing factor for the odor. It has been observed that, concentration of evernic acid goes on decreasing after one hour due to thermal metabolism. Evernic acid undergoes hydrolysis to obtain orsellic acid and everninic acid, which are further decarboxylated to orcinol and orcinol monomethyl ether, respectively [11]. The metabolism of evernic acid is shown in Figure 17.2.

17.4.2 *Pseudovernia furfuraceae* (Tree moss)

P. furfuraceae (Tree moss) is another lichen species that is known for its fragrance, and it grows mainly on coniferous trees like cedar and pine. There are four races of tree moss based on the content of phytoconstituents, the first one having physodic acid, the second one with physodic acid and olivetoric acid, the third one with only olivetoric acid, and the fourth one with olivetoric acid, physodic acid, and oxyphysodic acid. Sometimes, resinoids obtained from lichen species like *Ramalina fraxinea* and *Usnea caucasia* in the presence of *E. prunastri* are called as tree moss resinoids [11].

Figure 17.2 Thermal degradation of evernic acid.

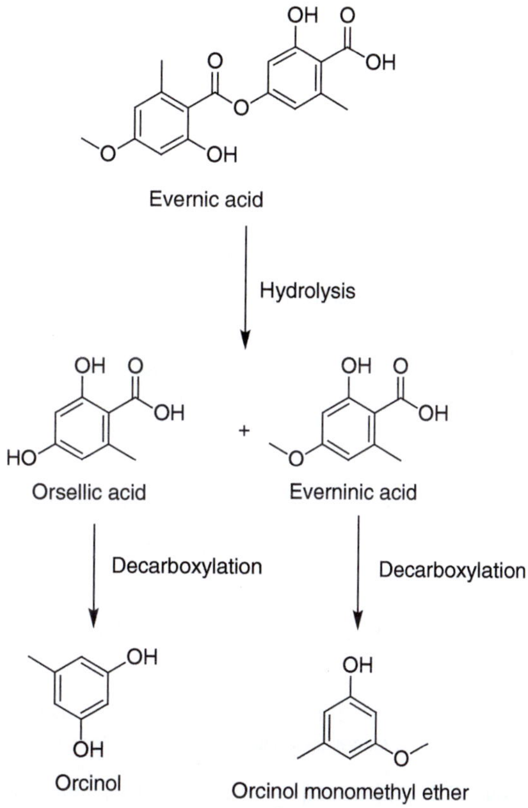

In-depth research has been done on the chemistry of tree moss (*P. furfuracea*), starting with the lichen itself and common industrial preparations. When conventional tree moss is gathered on pine trees in the Massif Central region of France, it typically contains 40–70% by weight of tree-derived materials (wood, bark, twigs, needles, etc.). In certain circumstances, this percentage may potentially reach 80%. A number of publications have been published on lichen compounds found in *P. furfuracea* in addition to these analytical studies, which were concentrated on the raw material utilized in the fragrance industry. Various monoaryl compounds, which are volatile in nature and might be responsible for the fragrance in tree moss, are shown in Figure 17.3.

Comparably, by using solid phase micro-extraction (SPME) sampling and comparable sorptive trapping, followed by GC-MS, volatile chemicals found in fresh lichens (*P. furfuracea* and *E. prunastri*) can be directly analyzed. In this approach, the volatile fraction from the crude lichen can easily be used to identify atranol and chloroatranol (as well as other mono-aryl compounds). This reveals that atranorin and chlorotranorin are hydrolyzed and decarboxylated in the "live" lichen without the need for any artificial treatments [12].

The intermediate benzoic acid derivatives can be thermally decarboxylated, and depsides are easily hydrolyzed. For instance, when the hydrolytic process is intensified by tepid, hot, or steam, the compound chloroatranorin almost completely degrades, yielding chloroatranol, chlorohaematommic acid, and methyl-orcinol carboxylate whose structures are

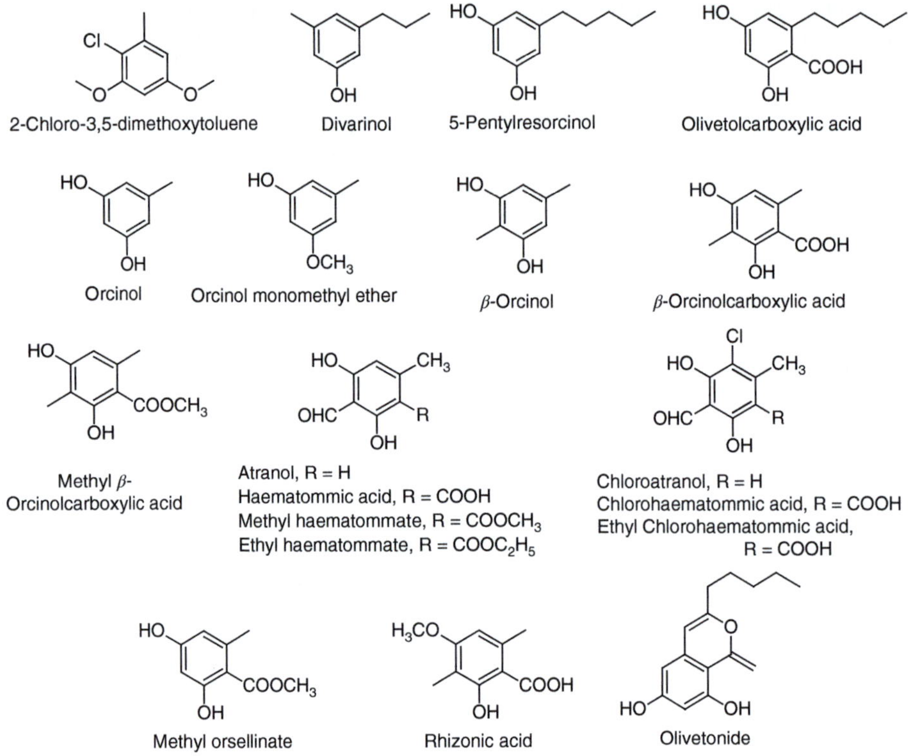

Figure 17.3 Monoaryl (phenolic compounds) obtained from tree moss.

shown in Figure 17.3. After carefully removing all other lichen species and any traces of wood and bark, 2.4 kg of *P. furfuracea* from Massif Central (France) was extracted to yield tiny levels of furfuric acid (2% of the depside-depsidone fraction). Synthesis and the one-step acid-catalyzed alkylation of methyl-orcinolcarboxylate or atranorin with physodalic acid served to confirm the structure of this particular depsidone. The primary component of commercial "tree moss" resinoid made from *P. nepalensis* in addition to the typical depsides and their derivatives, is protolichesterinic acid. As the extract of tree moss is rich in various volatile phenolic compounds with particular fragrances, for which these lichens containing fungal species are widely utilized as the main component in perfume industry.

Volatile orcinol compounds, which add to the product's woody richness, unmistakably impart phenolic undertones. The contribution of the host plant is evident in "pine tree moss" extracts, with tar-like and heavy pine needle/terpenic notes, unlike oak and cedar trees, where it is discrete but real. The plant material may occasionally be gathered at ground level along with genuine mosses or even fungi, which give the extracts their earthy, moldy flavor that is reminiscent of humus and forest undergrowth. Lichen (sometimes known as "moss") extracts' aroma still holds many secrets. Analytical chemists have found that when fractionating absolute oils in great detail and depth, regardless matter how complex the fractions are, they still have a "mossy" scent. This implies that trace elements have a significant impact on organoleptic function [12].

In general, perfumes are having the mixture of volatile oils, aromatic compounds, solvents or fixatives which is given to human, animals, food or living spaces for a pleasant smell and creating to fresh vibes. A scent is an action of an extremely light molecule that floats in the atmosphere. Typically, semi-volatile organic compounds with molecular weights that hardly exceed 260 AMU are the materials employed in perfumes. However, every lightweight molecule that floats in the air does not have perfumery properties such as carbon monoxide. The phenolic compounds are found to be volatile in nature as the hydroxyl bonds are easily cleaved at a low temperature. Since most of the important phytochemicals present in the lichen extracts are aromatic in nature, hence they are used in cosmetic and perfumery industries.

17.5 Conclusion

Lichens are composite organisms made up of cyanobacteria or algae that live in mutualistic relations with fungi. Lichens are a rich source of natural phytoconstituents, which have a wide range of applications, such as in the biomedical as well as perfumery and cosmetic industries. Various extracts of lichen species revealed that the extract of lichen contains various chemical compounds belonging to the category of phenols, flavonoids, terpenoids, volatile oils, etc., for which lichen is most used in the field of cosmetics and perfume production. As per our study, lichen species like *E. prunastri* (Oakmoss) and *P. furfuracea* (Tree moss) are the most commonly used lichen species in cosmetics and perfume production. These lichens are available in the Himalayan region as well as forest areas of Mediterranean regions. The extract of both oakmoss and tree moss possess significant flavoring properties because of having a high number of phenolic compounds like atranorin, evernic acid, orcinol, usnic acid, lecanorin, prunastric acid, thamnolic acid, rhizomic acid, methyl orsellinates, etc. Owing to the presence of these phytochemicals, the extracts of lichens are in high demand in the field of cosmetics and perfume production. Some other species of lichen were also discussed above whose phytochemicals investigations are not yet completely established and it leaves scope for upcoming researchers.

References

1 Chisvert, A., Salvador, A., Chaintreau, A., and Negri, R. (2007). Perfume in cosmetics. analytical methods. In: *Analysis of Cosmetic. Products* (ed. A. Salvador and A. Chisvert), 243–276. Elsevier.

2 Chisvert, A., López-Nogueroles, M., Miralles, P., and Salvador, A. (2018). Perfumes in cosmetics: regulatory aspects and analytical methods. In: *Analysis of Cosmetic Products*, 2e (ed. A. Salvador and A. Chisvert), 225–248. Boston, MA: Elsevier.

3 Singh, S., Upreti, D.K., Lehri, A., and Paliwal, A.K. (2015). Quantification of lichens commercially used in traditional perfumery industries of Uttar Pradesh, India. *Indian Journal of Plant Sciences* 4: 29–33.

4 Elkhateeb, W.A., El-Ghwas, D.E., and Daba, G.M. (2022). Lichens uses: surprising uses of lichens that improve human life. *Journal of Biomedical Research and Environmental Sciences* 3: 189–194.

5 Llano, G.A. (1948). Economic uses of lichens. *Economic Botany* 2: 15–45.

6 Manoharachary, C. and Deshaboina, N. (2021). Biodiversity, taxonomy and plant disease diagnostics of plant pathogenic fungi from India. *Indian Phytopathology* 74: 413–423.

7 Upreti, D.K., Bajpai, R., and Nayaka, S. (2015). Lichenology: current research in India. In: *Plant Biology and Biotechnology: Plant Diversity, Organization, Function and Improvement*, vol. 1 (ed. B. Bahadur, M.V. Rajam, L. Sahijram, and K.V. Krishnamurty), 263–280.

8 Lokhande, L. (2023). The immune microenvironment in mantle cell lymphoma targeted liquid and spatial proteomic analyses. Doctoral thesis. Lund University.

9 Spribille, T. (2010). How a guy from a Montana trailer park overturned 150 years of biology. *The Atlantic* https://www.theatlantic.com/science/archive/2016/07/how-a-guy-from-a-montana-trailer-park-upturned-150-years-of-biology/491702/ (accessed 6 February 2024).

10 Moxham, T.H. (1986). The commercial exploitation of lichens for the perfume industry. In: *Progress in Essential Oil Research, Proceedings of the International Symposium on Essential Oils, Holzminden/Neuhaus, Federal Republic of Germany* (18–21 September 1985) (ed. E.-J. Brunke), 491–504.

11 Joulain, D. and Tabacchi, R. (2009). Lichen extracts as raw materials in perfumery. Part 1: oakmoss. *Flavour and Fragrance Journal* 24: 49–61.

12 Joulain, D. and Tabacchi, R. (2009). Lichen extracts as raw materials in perfumery. Part 2: treemoss. *Flavour and Fragrance Journal* 24: 105–116.

13 Maurya, I.K., Singh, S., Tewari, R. et al. (2018). Antimicrobial activity of Bulbothrix setschwanensis (Zahlbr.) Hale lichen by cell wall disruption of *Staphylococcus aureus* and *Cryptococcus neoformans*. *Microbial Pathogenesis* 115: 12–18.

14 Mishra, G.K. (2015). The lichen genus Cetrelia (Parmeliaceae, Ascomycota) in India. *Phytotaxa* 236: 201.

15 Kekuda, T.R.P., Vinayaka, K.S., Swathi, D. et al. (2011). Mineral composition, total phenol content and antioxidant activity of a macrolichen everniastrum cirrhatum (Fr.) Hale (Parmeliaceae). *E-Journal of Chemistry* 8: 1886–1894.

16 Yonghang, S., Rai, S., and Sah, S.N. (2019). Antibacterial property of extract of Erveniastrum nepalense (Edible Lichen) collected from hilly regions of Eastern Nepal. *Tribhuvan University Journal of Microbiology* 6: 51–58.

17 Dieu, A., Mambu, L., Champavier, Y. et al. (2020). Antibacterial activity of the lichens Usnea Florida and Flavoparmelia caperata (Parmeliaceae). *Natural Product Research* 34: 3358–3362.

18 Kekuda, T.R.P., Vinayaka, K.S., and Sachin, M.B. (2018). Chemistry, ethnobotanical uses and biological activities of the lichen genus *Heterodermia Trevis.* (Physciaceae; Lecanorales; Ascomycota): a comprehensive review. *Journal of Applied Pharmaceutical Sciences* 8: 148–155.

19 Jayalal, U., Jang, S.H., Yu, N.H. et al. (2014). Notes on the lichen genus Leptogium (Collemataceae, Ascomycota) in South Korea. *Mycobiology* 42: 120–131.

20 Jayalal, U., Joshi, S., Oh, S.O. et al. (2012). A taxonomic study of the genus Myelochroa in South Korea. *Mycobiology* 40: 217–224.

21 Ureña-Vacas, I., González-Burgos, E., De Vita, S. et al. (2022). Phytochemical characterization and pharmacological properties of lichen extracts from cetrarioid clade by multivariate analysis and molecular docking. *Evidence-based Complementary and Alternative Medicine* 2022: 5218248.

22 Gómez-Serranillos, M.P., Fernández-Moriano, C., González-Burgos, E. et al. (2014). Parmeliaceae family: phytochemistry, pharmacological potential and phylogenetic features. *RSC Advances* 4: 59017–59047.

23 Rajaram, S.K., Innasimuthu, G.M., Alexander, R.A. et al. (2016). Antioxidant activity of lichen *Parmotrema austrosinense* and its antimicrobial effect on selected plant pathogens. *Cryptogam Biodiversity and Assessment* 1: https://doi.org/10.21756/cab.v1i2.7033.

24 Ghate, N.B., Chaudhuri, D., Sarkar, R. et al. (2013). An antioxidant extract of tropical lichen, Parmotrema reticulatum, induces cell cycle arrest and apoptosis in breast carcinoma cell line MCF-7. *PLoS One* 8: e82293.

25 Manas, P.C., Mandira, S., Sanjeeva, N., and Dilip, K.U. (2016). Distribution of Lichens on few ancient monuments of Sonitpur district, Assam, North East India. *International Journal of Biodiversity and Conservation* 8: 291–296.

26 Kumar, T.K., Siva, B., Anand, A. et al. (2022). Comprehensive Lichenometabolomic exploration of Ramalina conduplicans vain using UPLC-Q-ToF-MS/MS: an identification of free radical scavenging and anti-hyperglycemic constituents. *Molecules* 27: 6720. https://doi.org/10.3390/molecules27196720.

27 Moreira, A.S.N., Braz-Filho, R., Mussi-Dias, V., and Vieira, I.J.C. (2015). Chemistry and biological activity of ramalina lichenized fungi. *Molecules* 20: 8952–8987.

28 Ullah, M., Uddin, Z., Song, Y.H. et al. (2019). Bacterial neuraminidase inhibition by phenolic compounds from Usnea longissima. *South African Journal of Botany* 120: 326–330.

29 Shukla, P. (2014). Lichen genus Usnea (Parmeliaceae, Ascomycota) in Uttarakhand, India. *Current Research in Environmental and Applied Mycology* 4: 188–201.

30 Joshi, S. and Bastola, P. (2013). Phytochemical and Biological Studies on Usnea Orientalis and Rhododendron Campanulatum of Nepal. *Journal of Nepal Chemical Society* 30: 118–122.

31 Pathak, A., Upreti, D., and Dikshit, A. (2016). Antidermatophytic activity of the Fruticose Lichen Usnea orientalis. *Medicine* 3: 24.

32 Mesta, A.R., Rajeswari, N., and Kanivebagilu, V.S. (2019). Distribution of bioactive compounds in usneoid lichens from Western Ghats. *Plant Archives* 19: 2163–2168.

33 Sepahvand, A., Studzińska-Sroka, E., Ramak, P., and Karimian, V. (2021). Usnea sp.: antimicrobial potential, bioactive compounds, ethnopharmacological uses and other pharmacological properties; a review article. *Journal of Ethnopharmacology* 268: 113656.

34 Gonzalez, A.G., Barrera, J.B., Marante, F.J.T., and Castellano, A.G. (2002). Bioactivity of natural products. Ethnobotany. The chemistry and allelopathic effects of phenolic compounds from the lichen Evernia prunastri (L.) ach. *Natural Products in the New Millennium: Prospects and Industrial Application* 47: 195–210.

35 Lutnaes, B.F., Bruun, T., and Kjøsen, H. (2004). (22 S)-6- O -Acetyl-21 β H-hopane-3β,6β, 22,29-tetrol from oakmoss (Evernia prunastri). *Natural Product Research* 18: 379–385.

36 Sharma, M. and Mohammad, A. (2020). Lichens and lichenology. In: *Lichen-Derived Products* (ed. M. Yusuf), 101–118. Wiley. https://doi.org/10.1002/9781119593249.ch4

18

Lichen as Bio Indicators

Tridip Boruah, Himasri Devi, and Shilpa Sarkar

P.G. Department of Botany, Madhab Choudhury College, Barpeta, Assam, India

18.1 Introduction

In general terms, bioindicators are comprised of biological activities, communities, or species that are applied to assess the environmental quality and how the quality changes over time. Generally, there are three principal functions of bioindicators—(i) to detect biodiversity, (ii) to detect ecological activity, and (iii) to detect the physical and chemical changes in the environment [1]. In frequent utilization, the terms "bioindication" and "biomonitoring" are replaceable, but these terms have their distinct meaning in the scientific community. Bioindicators are organisms that are utilized for the recognition and qualitative resolution of anthropogenically created environmental constituents, while biomonitors are biological species that can be applied to the quantitative resolution of pollutants [2]. For example, in an area where *Lecanora conizaeoides* lichen is present, it indicates the poor quality of air; hence, it is used as a bioindicator. On the other hand, the loss of lichen chlorophyll content denotes the presence of air pollution. Hence, it is termed biomonitoring because it is quantitatively determined by air pollution. Different types of bioindicators are found in different habitats. Lichen, which is the symbiotic association of algae and fungi, is the most common and most applied bioindicator of air quality, toxic material, and heavy metal concentration. As lichen has no cuticles or roots and has a large surface area for the accumulation of nutrients directly from the airspace, hence they are mostly used as an air pollution detector. The most researched bioindicators of the air's cleanliness are lichens. Pignata et al. [3] explained lichen as a "permanent control system" for air contamination. Many experiments have been done regarding the uses of lichen as a biomonitor for air contamination during the previous 30 years. Several physiological frameworks are applied to assess environmental destruction to lichens, for example, breakdown and restoration of chlorophyll; ATP reduction, the difference in respiration process; production of ethylene and the conversion of

Chemistry, Biology and Pharmacology of Lichen, First Edition. Edited by Ashoke Kumar Das, Ajay Sharma, Deepika Kathuria, Mohammad Javed Ansari, and Garima Bhardwaj.

endogenous auxins and photosynthesis [4]. Generally, lichens may be applied as biomonitor or bioindicator in two dissimilar ways through visual observations:

1) By surveying all species are located in a particular region,
2) By calculating the morphological substitute in the lichen thallus, estimating the corporal parameters, or estimating the bioaccumulation of the contaminants when the lichens are transferred from an unpolluted region to a polluted region.

For the detection of the excellence of air in a region by lichen, the IAP (Index of atmospheric purity) method is mostly applied [5]. The IAP technique is based on the number (n), frequency (F), and tolerance capacity of the lichen located in that particular region. For measuring IAP value, there are present 20 different formulas. Lichens also have a vital part in accumulating heavy metals. Recent studies show the effective applications of lichen species in metal accumulation at different monitoring levels. Gadzała-Kopciuch et al. [6] suggested that Lichens have the capability to reflect persuade atmospheric contaminant levels in their tissues without notable unfavorable effects on their growth and development. *Evernia prunastri* is a good example of lichen, which has the capability to accumulate heavy metals including Lead [7]. Lichens are also applied for the detection of Cu dust emissions from mines. In Portugal, *Ramalina fastigiata* was applied as an indicator for the accumulation of mercury and methylmercury [8]. Conti [4] reported the geographical difference in the sulfur (S) isotope constitution of *Alectoria sarmentosa*, which is epiphytic lichen. The study reveals that lichens are also responsive to the sulfur salt that comes from the ocean. For the study of drizzle synthetic and phosphorus and nitrogen concentration, *Cladonia portentosa* is used as an excellent bioindicator, while *Hypogymnia physodes*, *C. portentosa*, are also applied as bioindicators of nitrogen deposition [9]. Lichens are also good bioaccumulators of radionuclides. *Parmelia sulcata* is a good example that is applied as a bioindicator for the presence of radionuclides. In *Xanthoria parietina*, the average organic lifetime of Cs 137 isotope is 58.6 months, which indicates that it is the most excellent biological indicator of radioactive consequence alongside mosses [10]. Since pollution is a major problem in today's time, hence there are the most needed pollution indicators by which we can recognize that the area is polluted. In this regard, lichen is the best and cost-free indicator that indicates different pollution levels. In this chapter, we are going to discuss the applications of different groups of Lichen as bioindicators in different fields including air pollution, heavy metal pollution, and indicators of toxic materials, and also discuss their monitoring method with giving special emphasis on emerging tools and techniques available for lichenologists.

18.2 Effective Biomonitoring of Lichen Species

Pollution is a significant problem in the present time. Depending on the origin of pollution, it can be divided into two chief categories. Point and nonpoint sources of pollution. Certini et al. [11] reported that from point sources, climatic depositions of contaminants, for example, acidic vapors, dirt, and alloys, are discharges that are tremendously changeable in time and space. Thus, it is not easy to gain complete information by applying customary monitoring processes in excess of an enormous region and it is a costly process. In this regard, natural monitoring can be a very effective process. It was suggested that Lichens are one of the best biomonitors that can be applied to examine the trace element content of constant contaminants and to guess the biological possessions of contaminants by calculating modification at the population level. Rusu et al. [12] reported that *H. physodes* (L.) Nyl., is one of the primarily applied lichens as biomonitor in Europe, which can accumulate trace elements from a point source pollution. It was also reported that the concentration of Pb content is highest globally in *H. physodes* from Zlatna. From Rusu's examination, four chief geographical outlines were eminent (Cu, Fe, Zn, and Pb), and they showed the characteristic point source pattern. The geographical pattern of Pb, Zn, Cu, and Fe verifies the accumulation of these metals by *H. physodes* in a very short duration. It was revealed that *H. physodes*, which is an epiphytic lichen, accumulate heavy metals from atmospheric pollution and are effectively used as biomonitors in Finland. According to Fantozzi et al. [13], in land ecosystems, by calculating the total Hg concentration in the body of the lichen, one can easily represent the geographical pattern of Hg pollution in the region of point sources or "hot spots." For example, at Abbadia San Salvatore, the epiphytic lichen *X. parietina* monitoring the concentration of Hg pollution in that region at a very low cost and a short period of time. Pino et al. [14] reported that lichens had been effectively applied to biomonitoring atmospheric Hg accumulation about additional point sources for example thermal influence plants or fortify plants. In comparison with the spread-out release sources, lichen biomonitoring is frequently helpful in areas with point sources. Active monitoring of lichen in the city atmosphere shows less enlargement of Hg attentiveness, apart from in specimens showing close-up to-point sources. Grangeon et al. [15] measured Hg accumulation rates with *X. parietina* in a region with solar influence plants and predicted for each power plant a stable representation of Mercury emission rate of 24 g/h during the year. In Australia, lichens were efficiently applied to detect the accumulation of heavy metals that are released from point sources of industrial origin; it was reported that a study was conceded within the Collie and Shire in south-western part of Australia, which concluded that the lichen population present in the local jarrah forest in Collie region might be pretentious by coal-fired emission [16]. In this study, two lichen species, namely *Cladonia rangiferina* and *Usnea inermis*, were reported, which indicate air pollution and a higher tolerance capacity in that area. *Hypogymnia subphysodes* is another lichen species that was also applied as an accumulator of heavy metal found in the same forest of Collie. This report suggested that the tolerance capacity of these lichens in Collie air-shed probably understands a contamination influence from the point source in the region. In Poland, another experiment was performed and found that epiphytic lichens accumulated more heavy metals than epigeic lichens, which were effectively applied as biomonitoring [17]. In developing countries like India metal biomonitoring by lichen is a very helpful process as it is a cost-free technique. A few scattered reports suggested that in the central region of India, *Phaeophyscia hispidula* and *Phaeophyscia praesorediosum*, which are foliose lichen enormously accumulate heavy metal from the industrial point sources and act as air pollution monitoring. Recently, Daimari et al. [18] reported that *Pyxine cocoes* have been effectively applied for biomonitoring of air pollution in Sonitpur district of Assam, which is a northeastern state of India. Kibble and Harrison [19] reported that energetic samplers are helpful for monitoring in extremely industrialized regions with identified point sources, while passive samplers are used to determine the nonpoint sources of potentially toxic elements (PTEs). Since epiphytic lichens have bulky exterior regions, a sluggish development rates with extended lifetime, and high-quality division with the simplicity of compilation, they can be applied as passive samplers for non-point sources. Rossbach and Lambrecht [20] reported that in Nova Scotia and Acadian Boreal forest of Canada *Usnea* lichen is used as biomonitoring of PTEs, which are emitted from nonpoint sources. Several scientific researchers suggested that nonpoint source of pollution is related to agrochemical activities. Antony et al. [21] reported that some lichen, such as *Letharia vulpine*, trapped agrochemical pollutants in their thallus. Some lichens change their color in agrochemical pollution. In *Cladonia uncialis* and *Bryoria furcellata*, the color of the thallus changes due to application of glyphosate and triclopyr, which are herbicides, and this indicates agrochemical pollution. Some reports suggested that in *Xanthoria* spp., these herbicides amplify stress-receptive chemical units, namely proline. Hence, this type of modification assists biomonitoring of the herbicides in a polluted area [22].

18.3 Methods of Lichen Biomonitoring

Over the last few decades, the monitoring of environmental problems by some globetrotter entities has been developed in an increasingly satisfying manner. According to Porter and Hajibabaei [23], the biomonitoring technique used by such organisms is more helpful than the customary analysis technique as the organisms indicate the rank of dynamism of living things, while the customary analysis method only shows the experiential values. Biological indicators can indicate the ecological tribulations either sensitive or accumulative way, but there are some organisms, such as lichen, which can operate in both biomonitoring ways [24]. Many scientists reported that during the last 50 years, lichens are used as biomonitor in a vast range and increase their application in different factors, sampling regions and methods of monitoring. Many studies were carried out to study biomonitoring of lichen mainly confined to artificial regions for example in manufacturing, discarding, and metropolitan regions. Several methods have been applied for the biomonitoring of lichen; among them, mapping, bioaccumulation, GIS-based diversity studies, and conductivity methods are frequently applied in lichen biomonitoring [25].

The mapping methodology of lichen is globally applied for monitoring the air quality or any environmental contamination. As lichen mapping is a cost-free technique and has the ability to afford a helpful assessment of ecosystem disturbance, hence it is widely applied. The first mapping method was approved in the 1970s, and the method is mostly applied in Europe, where lichen monitoring techniques are applied to expose lichen immigration of foliage quickly following the implementation of fresh air codification, as in London [26]. The Forest Health Monitoring programs of the United States also conducted this work with slight modifications. From these experiences, it is shown that if an ordinary sampling system is applied, then we can confirm the truthful valuation of modifications that appeared in lichen species profusion. Many studies revealed that a few species of lichen react unconstructively to the straight consequence of nitrogen dioxide (NO_2) and ozone (O_3) and also high to the attentiveness of sulfur dioxide (SO_2).

The lichen monitoring program also added some new methodologies including the Index of Atmospheric Purity (IAP) for calculating the diversity of lichen species in a polluted area [27]. In the IAP method, some algebraic values are added to the plant sociology uniqueness of lichens by a mathematical formula that was initiated in Canada in the late 1960s. According to Leus et al. [28], IAP is a commonly utilized method for encapsulating details about contaminant acceptance of species and their geographical differences of profusion and imitating a spectrum of lichen species affluence from high diversity to low diversity and recapitulating the consequence of the extensive period of ecological circumstances. The IAP value of a studied area can be measured by applying the subsequent method:

$$\text{IAP} = \sum \text{Fi}$$

where F is detected by the frequency of each variety measured as a figure of quadrilateral in the studied area.

According to this method, in a region, if there is a low IAP value in the area then the region has a low diversity of lichen species, whereas high species diversity will be detected if high IAP values are reported and if the area is disturbed by several pollutants then the area has low IAP values as lichens are very sensitive to air pollution. Since the IAP method was initiated in Canada, the application of this method is more popular in Europe than in Canada. In the eastern side of Canada, the IAP method was applied to map the impact of air contamination during the 1970s. Throughout the last few years, in Canada, several agencies like EMAN, and CEW have expanded their work in lichen monitoring procedures employing IAP with appropriate training technology. In an emergent economy like India, this IAP method also plays a very significant role in lichen monitoring programs as it is a cost-free technique. Several experiments have been carried out using the IAP method in different subcontinents of India including the northeastern region (especially in the Barak Valley of Assam state) of this country.

Trace element accumulation in lichen is a frequently applied biomonitoring method for evaluating trace element concentration in the surroundings. Generally, two different biomonitoring methods, namely active and passive methods, are applied for measuring more or less heavy metal attentiveness in a contaminated region [29]. In several instances, the active monitoring technique is only applied due to the extremely deprived foliage of the lichen community and hence a shortage of resources for element content purposes in the neighborhood of pollution resources [30]. The same study suggested that in contrast with Fruticose lichen, Foliose lichens are better accumulators. Garty [8] conducted an experiment with an active biomonitoring technique, especially using *H. physodes* lichen on different regions in Slovenia that are contaminated by heavy metals. In their experiment, lichens were relocated from the Rogla region, which is a contamination-free area to the zinc and lead ore refinery at Zerjav, which is a heavily contaminated region to observe the changes over a six-month period. From this experiment, it is revealed that the monthly accumulation of all four heavy metals was mathematically elevated in lichens exposed at Zerjav, which is a heavily polluted area. Sujetoviene and Sliumpaite [31] conducted an experiment in different metropolitan and manufacturing areas in Italy on the accumulation of heavy metals by *E. prunastri* lichen and concluded that Zn blending plant and an oil refinery are major contamination resources in that region as the lichen accumulated a huge quantity of trace elements from that region. For biomonitoring investigation by lichen in a different region of India for assessing trace element concentrations, *P. cocoes* are reported to be the more appropriate as the lichen has a more poisonous-forbearing capacity [32].

Geographic Information Systems (GIS) is one of the perfect methods available for monitoring the lichen diversity in a polluted area incorporating the expected values of the response changing as IAP values [33]. With the help of GIS operating system, the mapping method of lichen monitoring shows a better perceptible demonstration of lichen multiplicity. For mapping air contamination by lichen, there are incredibly diminutive assignments made available by the GIS method. In Sri Lanka, this method is applied for the mapping process of lichen monitoring, and the results show that corticolous lichens have improved qualifications to recognize air quality in the metropolitan ecological unit as compared with another assemblage of lichen. Popovic et al. [34] conducted an experiment on lichen monitoring using GIS implemented in Switzerland and concluded with plausible existence or nonexistence of lichen species in that area.

18.4 Lichen as Indicator to Air Pollution

The World Health Organization reported that each year, nearly seven million deaths around the world are caused due to air contamination. Gaseous pollutants such as carbon monoxide (CO), hydrocarbons (HC), hydrogen sulfide (H_2S), nitrogen oxide (NO), ozone (O_3), sulfur dioxide (SO_2), aldehyde, fluorides, and particulate air pollutants such as dust, coal, cement, liquid particles, heavy metals, and radioactive materials plays a detrimental role in maintaining the hazardous quality of air [35]. Lichens are very responsive to such air pollutants, and hence, they are extensively considered as functional indicators of air contamination. Some lichens are most susceptible to air pollution, while some are very tolerant to air pollution. Thus, studying the variations in richness and diversity of lichens can help in identifying the pollution source along with quantification of the relative intensity, duration, and distribution of the air pollutants [36]. Generally, researchers have found that epiphytic lichens are mainly affected by air pollution. Kularatne and De Freitas [37] reported that epiphytic lichens are one of the best-used biomonitors. During the last decades, it has been found that Sulfur dioxide (SO_2) has become the foremost air pollutant that impacts the diversity of epiphytic lichens in industrial and urban areas. The degradation of richness and distribution of lichen are identified around industrial and urban areas due to the extent of air pollution [38].

Basically, lichens mostly occur either as epiphytes or as separate trees on bare rocks or soil. Their morphological structure varies greatly from large shrubby forms known as fruticose, to large or small "leaf-like" forms known as foliose species that increase in size in the substrates and the species that are imbedded in the substrates are known as crustose [4]. When lichens are not specifically afflicted, they frequently serve as useful monitors of deposition trends. They have been extensively employed to track the steady overturning of metal ions at point sources and in urban regions. In Munich, England, and Paris, an independent observation has acknowledged that lichens were vanishing from urban regions at a surprisingly increasing rate [39]. Cities are covered with clouds and smoke due to the widespread use of coal for industrial and home heating purposes, and it is one of the main reasons behind the pollution of air that might cause the desertion of lichen. By the beginning of the 1900s, the European city effect was a broadly accepted incident, and In the city center of Stockholm, multiple studies found a "lichen desert zone" where nearly all lichen species had disappeared, a "struggle zone" wherein many species were destitute, but others managed to live, and the "normal zone" where all lichen species appeared to be unaffected or unchanged [40]. These urban zones are now regarded as representing all of the continents inhabited by humans. In recent years, a plethora of mapping studies using quantitative methodologies have been established for identifying species and community trends in connection to sources of air pollution. Over the 50 years of studies, many researchers have provided evidence that lichens were very easily affected by air contaminants such as SO_2, HF, and heavy metals. An experimental study in the mid-1970s accepted the sensitive nature of many lichens to SO_2 fumigations even at small concentrations existing in the field conditions [41]. Additional hypothetical studies documented the effects of zinc and fluoride on epiphytic lichens growing on metropolitan sites. Generally, there is a good interrelation between the field-established and decontaminate-established indication of lichen species responsive to sulfur dioxide (SO_2). For instance, *Parmelia caperata*, which is very sensitive to sulfur dioxide, was not recorded in cities popular for acid rains [42]. Correspondingly, *L. conizaeoides* has been shown to be one of the most tolerant lichen species. Various momentary fumigation studies revealed that Sulfur dioxide (SO_2) has adversely affected the lichen species and that contrasting susceptibility occurred among lichens. Lichen absorbs SO_2 throughout its whole surface, according to a recent explanation, but vascular plants may limit gas absorption by shutting their stomata. A researcher has noticed that in one of the presiding arctic lichens *C. rangiferina*, the rate of absorption of sulfur dioxide is the function of water content [43]. The greater amount of absorption of SO_2 would show greater physiological changes in them. Of all the discussed air pollutants explored, SO_2 is the one whose point-by-point robotic data on impacts on lichen was made available recently [44]. The components likely to be associated with SO_2 toxicity are (i) enzyme deactivation, (ii) enzyme incitement, (iii) communication with biomolecules with reactive interface, and (iv) arrangements of free radicals.

Fumigation experiments from a field study have shown that photosynthetic reduction occurs in, *Flavoparmelia caperata, Parmelia sulcata, and Hypogymnia enteromorpha* with ozone exposure [45]. Photosynthetic reductions were also found to occur due to exposure to Peroxyacetyl nitrate (PAN) in species such as *Parmelia sulcata*, *Collema nigrescens*, and *H. Enteromorpha*. In Los Angeles, *Ramalina menziesii*, is assumed to be an extremely sensitive species which even when fumigated with the maximum quantity, was unaffected by ozone [46]. Recently, acid rain has become one of the major environmental issues as both Sulfur dioxide and nitrogen oxide are the main constituents of acid rain. Electron microscope studies have documented that the lichens that are collected from the polluted sites have also changed their ultrastructure after controlled fumigations with sulfur dioxide, ozone, and peroxyacetyl nitrate [47]. Due to these air pollutants changes in the mitochondrial structures of photobionts and chloroplast occur, and as a result, early injury of the species seems to be happening.

Lichens' responses to exposure to air pollutants have been evaluated using a range of biochemical and physiological features. With exposure to some air pollutants, lichens vary in physiological and biochemical factors for instance changes in the activity of the enzyme phosphatase [48]. In *C. rangiferina*, due to air pollutants reduction in their enzyme activity was reported. Chlorophyll degradation and changes in protein and lipid biosynthesis occur due to exposure to air pollutants. Field studies provide evidence that air pollutants, acid rain, and fog also affect the nitrogen fixation of lichens. Chlorophyll is the main pigment that captures light energy used in photosynthesis; these were also influenced badly by air pollutants resulting in pigment degradation [49]. Air pollution altered the asexual reproduction in Lichens, and when exposed to gaseous Sulfur dioxide or sulfate, considerable reticence of phycobiont cell division of germinating soredia was also reported.

18.5 Lichen as Heavy Metal Indicator

Environmental contamination of heavy metals has developed into a great concern due to its adverse effects on agricultural ecosystems. Since lichen lacks a real root system, its mineral nutrition is mostly dependent on atmospheric inputs. Additionally, because they lack a shielding cuticle and stomata, their thalli are able to collect heavy metals at levels above the range required for metabolic function [50]. The most familiar heavy metals that can pollute the environment are Arsenic (As), Nickel (Ni), Copper (Cu), Iron (Fe), Mercury (Hg), Zinc (Zn), and Lead (Pb). They are defined as heavy metals due to their high density and high atomic weight. These heavy metals are found physically on the crust of the earth since its formation. Heavy metal contamination has mainly been due to human-dominated activity, especially due to smelting, mining, and leakage of metals from diverse sources such as waste dumps, landfills, automobiles, roadworks, etc. [51]. The usage of heavy metals in agriculture is a secondary source of pollution from heavy metals like herbicides and insecticides. Some natural phenomena such as volcanic activity, metal corrosion, geological weathering, and soil erosion, also increase pollution caused by heavy metals. Lichens are considered good bio-indicators as they can accumulate heavy metals and indicate heavy metal pollution. Lichens are also long-living, ectohydric creatures with little control over gas and water exchange, which makes them excellent bioindicators of pollution [52]. The use of lichens for biological monitoring of atmospheric heavy metals is widespread around the world, and several researchers have previously examined the development in this area. However, in recent years, the majority of studies have concentrated on the physiological and biochemical impacts of heavy metal buildup in lichens. This study reviews recent advances in our understanding of the interactions between lichens and heavy metals at various scales, from populations to cells and from ecology to molecular biology.

Numerous studies have shown how good lichens are in capturing particles both from the environment and from substrates. These particles can remain unmodified for a very long period and may be deposited onto the exterior of the lichen species or contained in the space between cells of the medulla. As a result, many lichen species may acquire and retain massive amounts of heavy metals that vastly exceed their physiological requirements. They can withstand these high concentrations by either securing the metals externally as oxalate crystals or by combining them with lichen acids [53]. Heavy metals cause alterations at the ecosystem level because lichens have a specific response to these substances. A study was conducted to monitor the survival strategies of epiphytic lichens growing nearby a copper smelter. The experiment was performed on the responses of *Tuckermanopsis sepincola* (sexually reproducing weak opponent) and *H. physodes* (asexually reproducing strong participant). It has been found that *T. sepincola* benefitted from the Copper-susceptibility of *H. physodes* [54]. When exposed to copper stress, *T. sepincola* grows more often on the birch stem, while *H. physodes* only produce esorediate or mild sorediate thalli. Bačkor and Loppi [52] have studied certain lichen communities on rocks and soil that are found in extensively metal-polluted places of the world, mostly as a result of metal mining. From their study, they have found that some lichen species that are linked with heavy metal-enriched substratum are very familiar species that endure metals and arise both in unpolluted and polluted sites. Some other species of lichen are constrained to heavy metals enriched substratum and have adversative dynamics that reflect the accessibility of appropriate sites. The lichens that require substrates with metal contents mostly belong to the genera *Porpidia, Lecidea, Aspicilia, Acarospora,* and *Rhizocarpon* [55].

In modern lichenology, one of the many extensively researched features is the way that lichens accumulate heavy metals because it is important for biomonitoring. There are many conditions that affect the accumulation of heavy metals by lichens and their individual components, and the processes are very complex. Some particular types of lichen accumulate a considerable amount of heavy metals; for instance, *Acarospora rugulosa* was found to accumulate up to 16% copper (Cu) on a dry weight basis [8]. *A. rugulosa* and *Lecidea lactea* from cupriferous pyritic rocks contained Copper (Cu) in amount as high as 5% concentrations. It was observed that apothecia of *Lecanora polytropa* accumulate Copper (Cu) up to 1.3%. However, there is some significant dissimilarity between the intracellular and total contents. Gathering of heavy metals by lichens is an ever-changing process. It has been reported by Tuominen and Jaakkola [56] from the short-term studies on the effects of overindulgence metal that lichens covered with water into the solution of metals and accumulated metals very fast; in most cases, it has occurred within a few hours. According to Rola [57], the concentrations of heavy metals in lichens would augment as a function of time, and it is a more complex situation. The concentration of a number of components fluctuates in transplanted lichens during the course of the investigation. The quantity of incident contaminants largely affects the concentrations in lichen thalli. Since lichens are metabolically more active in the rainy and colder seasons, these conditions are ideal for their development and mineral absorption. The rainfall is a rich contributor to the entirety element contents of lichens. These may give rise to a lofty degree of stress of biological origin, and as a result element uptake is altered. Bačkor and Loppi [52] found that Copper (Cu) pollution can affect the growth of lichens causing growth abnormalities, changes of color in the thalli of *T. sepincola*, distortion of apothecia and reduction in the fabrication of soredia in *H. physodes*. Heavy metals can also affect reproductive structures. Studies on lichen have demonstrated that enhanced asexual reproduction through conidia and decreased apothecia formation are involved in sexual reproduction [58]. Accordingly, it may be said that in urban settings, surface interception of air particles is more significant to accumulation than liveliness. In heavy metal accumulation, lichen bionts are differentially involved. Mycobionts make up more than 90% of lichen biomass, and they acquire more toxic substances from the atmosphere than they do. Sanità Di Toppi et al. [59] reported that the mycobiont hyphae that form particularly the upper portion of the cortex of the lichen thalli, were the major site of accumulation of cadmium (Cd). It has been reported that for a few metals like cadmium (Cd) and chromium (Cr) hyphae belonging to mycobionts are the major sites for accretion. In individual trees, the accumulation of heavy metals relies on the area of stem flow on a meticulous site and is more often than not higher next to ground level. In lichens, the first site of organic interaction of heavy metals is the plasma lemma. There are numerous physiological factors that are used to access the degree of heavy metal strain in lichens. It has been mentioned that the content of HPCD is considerably increased by nickel (Ni) treatment in two species of lichen, making them more susceptible to damage [38]. It has been reported that an excess amount of Manganese (Mn) in the photobiont of *Trebouxia jamesii* reduces the size of the cell and constricted thylakoids. The excess amount of heavy metals affects the ultrastructure of photobionts more than mycobionts. In the mycobiont hyphae, the concentric bodies are said to be the sites of intracellular Cd accumulation.

The advancement of technology has established photosynthetic pigment composition as among the most accurate and often used indicators of stress from heavy metals in lichens. It has been found by the experiment that in lichens and also in axenic photobiont cultures, reduced chlorophyll *a* concentrations are a susceptible marker for excess copper (Cu). However, an excess amount of decrease in mercury (Hg) is an indicators for both chlorophyll *a* and chlorophyll *b* content. In the presence of some particular types of heavy metals like copper (Cu) is also sensitive to the amalgamated content of carotenoids like pigments in the thallus of lichen. Karakoti et al. [60] have highlighted that chlorophyll degradation, which is also known as phaeophytinization quotient, is one of the most commonly utilized criteria in lichen stress physiology, reflects the ratio of *chlorophyll a* to *phaeophytin a*. In that short-term laboratory experiment, it has been found that due to excess copper (Cu) and cadmium (Cd) chlorophyll *a* decreases in an axenic lichen culture. Moreover, cysteine is an exclusive amino acid responsible for the attachment of heavy metals, and these amino acids have a thiol (–SH) group. It has been confirmed that the strain of photobiont of lichen *Trebouxia erici*, which is tolerant to when heavy metal levels in the growth medium are at nutritious levels, copper can produce more cysteine over the wild strain [61]. It has been found that at 10 μM Cu glutathione GSH increased in *Cladonia cristatella* lichen mycobiont cultures that were developed through aposymbiotic relationships. However, the same copper (Cu) concentration is very lethal and harmful for the culture of lichen photobiont *T. erici* in which GSH decreased. Despite the large number of articles discussing how lichens and heavy metals interact, there is still much to learn about their ecophysiological characteristics and cellular location. Due to the fact that only metals on the inner side of cells have a direct impact on physiological and metabolic functions connected to cancer, it is crucial to consider that total quantities have limited value for ecophysiological studies. It is necessary to do careful research on the processes governing the bioavailability, accumulation, toxicity, and detoxification of heavy metals in lichens as well as the limiting variables that affect lichen viability. Understanding the variability in the metal concentration in one species' thalli, which also responds to various environmental and experimental settings, can benefit from intraspecific genetic diversity studies. Again, the application of genotoxicity testing, along with research into the processes governing the bioavailability and biotoxicity of heavy metals, might increase the usefulness of lichens as biological air pollution monitors. A few examples of heavy metals have been provided in Table 18.1, which have been proven to have a negative impact on lichen thallus.

Table 18.1 Impact of heavy metals and toxic materials in the growth and development of Lichen thallus.

Serial number	Heavy metals and toxic materials	Name of the lichen species affected	Changes that occurs in the lichen thallus	References
1.	Zinc (Zn)	*Trebouxia gelatinosa*	Negative impact on growth	Bačkor and Fahselt [62]
2.	Cadmium (Cd)	*Trebouxia erici*	Inhibit solute transportation	Bačkor and Loppi [52]
3.	Copper (Cu) and mercury (Hg)	*Trebouxia irregularis*	Inhibit lipid metabolism	Guschina and Harwood [63]
4.	Copper (Cu) and nickel (Ni)	*Bryoria fuscescens*	Disturbance in membrane function	Tarhanen et al. [64]
5.	Mercury (Hg)	*Evernia prunastri*	Decreased 'chl a' and 'chl b'	Vannini et al. [65]
6.	Sulfur dioxide (SO_2)	*Lobaria pulmonaria*	Causes death	Paul and Hauck [66]
7.	Ozone (O_3)	*Hypogymnia enteromorpha*	Reduction in photosynthetic ability	Nash and Sigal [67]
8.	Ozone (O_3)	*Flavoparmelia caperata* and *Parmelia sulcata*	Negatively impact photosystem-I and II	Maslać et al. [68]
9.	Peroxyacetyl nitrate (PAN)	*Collema nigrescens*	Impact CO_2 exchange ratio	Scheidegger and Schroeter [69]
10.	Peroxyacetyl nitrate (PAN)	*H. enteromorpha*	Uncontrolled cell division	Eversman and Sigal [47]
11.	Peroxyacetyl nitrate (PAN)	*Parmelia enteromorpha*	Photosynthetic reduction	Ross and Nash [70]
13.	Copper (Cu)	*Ramalina menziesii*	pH reduction in Lichen thalli	Branquinho et al. [71]
14.	Copper (Cu)	*Lobaria pulmonaria*	Inhibit solute uptake mechanism	Kiliç et al. [72]
15.	High sulfate concentration	*Sticta stellaris*	Reduction in photosynthetic capacity	Albornoz et al. [73]
16.	Sulfate and nitrite	*Hypogymnia physodes*	Reduction in diaspore germination and early growth	Van Herk et al. [74]
17.	Sulfur dioxide (SO_2) in low concentration	*Evernia mesomorpha*	Reduction of net assimilation rate	Laxton et al. [75]
18.	Zinc (Zn)	*Cladonia uncialis*	Net photosynthetic reduction	Sveshnikova et al. [76]
19.	Zinc (Zn)	*Lasallia papulose*	Inhibition in uptake and retention of water	Tyler [77]

(*Continued*)

Table 18.1 (Continued)

Serial number	Heavy metals and toxic materials	Name of the lichen species affected	Changes that occurs in the lichen thallus	References
20.	Uranyl, vanady, cyanide, fluoride	*Cladonia rangiferina*	Imbalance in the nitrogen cycle	Trembley et al. [78]
21.	Low sulfur dioxide (SO_2) concentration	*Trebouxia* sp.	Reduction in proline production	Sehnal et al. [79]
22.	High sulfur dioxide (SO_2) concentration	*Peltigera camina*	Decrease in the number of carboxysome	Markert et al. [80]

18.6 Lichen as Indicator to Toxic Material

A large class of chemicals known as toxic substances is those that can harm both plants and animals, including people. Hazardous waste and electronic waste are examples of toxic materials. Individual metals and metal complexes, notably heavy metals that are toxic, have a negative impact on people's health. The heavy metal mercury is thought to be the most harmful to the environment. The mycobiont provides the photobiont with the necessary minerals, gives it a growth habitat, and shields it from harsh radiation and herbivory. The lichen's thallus is effective in absorbing minerals from the air, as well as through wet and dry atmospheric deposits, both of which may involve metal-rich airborne particles with a diameter of up to 100 mm [81]. Mineral distribution in lichen thalli is not uniform and is influenced by morphological characteristics. Apparently, chemical and physical parameters like particle abundance, metal concentration, chemical forms, pH, and temperature impact how hazardous metals behave. It was first noted that zinc and cadmium hindered the development of *Trebouxia gelatinosa* and *T. erici* photobionts demonstrating long ago that photobiont growth owing to metal overload [62]. Subsequently, copper, cadmium, and mercury, which are only irregular cultures, inhibited development and were relatively hazardous at 300 M concentrations in *Trebouxia* media: Hg is followed by Cd and then by Cu. Excess metals generated growth inhibition in the photobiont *T. erici* in the following order: Hg > Cd > Co >> Zn >> Cu.

According to Bačkor et al. [82], the eukaryotic Dictyochloropsis photobiont is less sensitive to manganese than the cyanobacterial Nostoc photobiont. However, eukaryotic photobiont sensitivity also varies. For instance, copper and lead have a greater impact on lipid metabolism in *Coccomyxa* than in *Trebouxia*. In the past ten years, excess copper and cadmium have been the focus of the majority of studies on metal toxicity in lichen photobionts; however, more recent research has looked at the effects of other metals, such as chromium, mercury, zinc, lead and manganese [83]. Both internal accumulations in vacuoles and adsorption of metals upon cell wall sites have been seen in lichen photobionts; however, none of these avoidance strategies completely shield against high quantities. The habitat circumstances, such as fertilization, soil aeration, pH, soil temperature, and the accessibility of the elements in the soil, are just a few of the variables that affect how much metals lichens absorb [57]. Comparing the capacity of *Cladonia cariosa* to accumulate metal concentrations with other lichen species investigated in the same geographic region, it was shown that in the most contaminated locations, *Cladonia cariosa* acquired relatively greater metal quantities in the thalli than a range of epiphytic lichen species [84]. High quantities of elements revealed in the many studies suggest that sturdy soil contamination considerably contributes to enhanced gathering and is the primary source of elements in epigeic lichens, despite *Cladonia* lichens being known as weak collectors of heavy metals. High extracellular Zn content in lichen thallus may result from two pathways. The complexation of heavy metals can be connected to the first process. For instance, *Diploschistes muscorum* accumulates Pb and Zn through increased oxalate production, which causes hazardous metals to precipitate as insoluble salts outside of the cells [17]. In the latter scenario, post-industrial waste-derived solid Zn-containing particles could be easily and effectively trapped in the medulla's intercellular gaps, increasing the proportion of extracellular fraction.

18.7 Conclusion

Lichens play a very important role in pollution monitoring in all the climatic regions of the world. Simply by surveying all species located in an area and calculating their morphological substitute in the lichen thallus, lichens can be applied as biomonitors in different polluted fields. Since lichen monitoring is a cost-effective and labor-free technology, this monitoring processing is very popular in developing countries. Several studies reported that lichen can be applied to estimate the rank of airborne rudiments proficiently, and it supplies the finest explanation if sufficient air monitoring tools cannot be attained. Lichen also indicated both point source and nonpoint source pollution. Thus, it is very helpful to identify the reason for that particular source of pollution. There are several biomonitoring methods available by applying lichen in a polluted area. Among them, GIS-based diversity studies, conductivity methods, bioaccumulation methods, and mapping methods are frequently applied. The photosynthetic pigment composition is a sensitive component extensively exploited during monitoring programs for measuring the heavy metal concentration in Lichens. *Aspicilia*, *Porpidia*, *Acarospora*, *Lecidea*, and *Rhizocarpon* are the main genus of Lichen that accumulate heavy metals in very large quantities and serve as major bioindicators for heavy metal pollution. Although several methods have been applied for lichen biomonitoring studies, there are still no standardized models for air pollution monitoring studies using lichen. Therefore, it is considered a major challenge for future researchers to develop a well-organized model for air pollution monitoring by lichens. There is a good opportunity for modifications of existing techniques of this lichen monitoring prospects, and future researchers should not only look for the upgradation of lichen diversity in the area but also look to address the problems arising from the methodological loopholes.

References

1 Holt, E.A. and Miller, S.W. (2011). *National Educuational Knowledge* 3: 8.

2 Oertel, N. and Salánki, J. (2003). *Modern Trends in Applied Aquatic Ecology*, vol. 1, 219–246.

3 Pignata, M.L., Plá, R.R., Jasan, R.C. et al. (2007). *International Journal of Environmental Health Research* 1: 29–46.

4 Conti, M.E. and Trans, W.I.T. (2008). *State-of-the-art in Science and Engineering* 30: 112–161.
5 Jeran, Z., Jaimovi, R., Batič, F., and Mavsar, R. (2002). *Environmental Pollution* 120: 107–113.
6 Gadzała-Kopciuch, R., Berecka, B., Bartoszewicz, J., and Buszewski, B. (2004). *Polish Journal of Environmental Studies* 13: 453–462.
7 Cansaran-Duman, D., Atakol, O., and Aras, S. (2011). *Journal of Environmental Sciences* 23: 1171–1178.
8 Garty, J. (2001). *Critical Reviews in Plant Sciences* 20: 309–371.
9 Rai, H. and Gupta, R.K. (2022). *Lichenized Fungi, a Primary Bioindicator/Biomonitor for Bio-Mitigation of Excessive Ambient Air Nitrogen Deposition Worldwide*, vol. 1, 267–301.
10 Shukla, V., Upreti, D.K., and Bajpai, R. (2014). Ecosystem monitoring. In: *Lichens to Biomonitor Environment* (ed. D.K. Upreti, R. Bajpai, and V. Shukla), 97–170. Springer.
11 Certini, G., Scalenghe, R., and Woods, W.I. (2013). *Earth-Science Reviews* 127: 1–15.
12 Rusu, A.M., Jones, G.C., Chimonides, P.D.J., and Purvis, O.W. (2006). *Environmental Pollution* 143: 81–88.
13 Fantozzi, L., Guerrieri, N., Manca, G. et al. (2021). *Toxics* 9: 172–188.
14 Pino, A., Alimonti, A., Conti, M.E., and Bocca, B. (2010). *Journal of Environmental Monitoring* 12: 1857–1863.
15 Grangeon, S., Guédron, S., Asta, J. et al. (2012). *Ecological Indicators* 13: 178–183.
16 Bahadur, B., Rajam, M.V., Sahijram, L., and Krishnamurthy, K.V. (2015). *Plant Biology and Biotechnology: Plant Diversity, Organization, Function and Improvement* 1: 1–827.
17 Osyczka, P. and Rola, K. (2019). *Ecotoxicology and Environmental Safety* 174: 26–34.
18 Daimari, R., Bhuyan, P., Hussain, S. et al. (2021). *Environmental Monitoring and Assessment* 193: 1–12.
19 Kibble, A. and Harrison, R. (2005). *Occupational Medicine* 55: 425–431.
20 Rossbach, M. and Lambrecht, S. (2006). *Croatica Chemica Acta* 79: 119–124.
21 Antony, A.C., George, S., Chandran, S., and Sebastian, A. (2020). *Agrochemicals Detection, Treatment and Remediation* 1: 241–261.
22 Dörter, M., Karadeniz, H., Saklangıç, U., and Yenisoy-Karakaş, S. (2020). *Ecological Indicators* 113: 106212.
23 Porter, T.M. and Hajibabaei, M. (2018). *Molecular Ecology* 27: 313–338.
24 Swaminathan, M.S. (2003). *Environmental Pollution* 126: 287–291.
25 Pająk, M., Halecki, W., and Gąsiorek, M. (2017). *Chemosphere* 168: 851–859.
26 Larsen, R.S., Bell, J.N.B., James, P.W. et al. (2007). *Environmental Pollution* 146: 332–340.
27 Gombert, S., Asta, J., and Seaward, M.R.D. (2004). *Science of the Total Environment* 324: 183–199.
28 Leus, K., Folens, K., Nicomel, N.R. et al. (2018). *Journal of Hazardous Materials* 353: 312–319.
29 Vingiani, S., De Nicola, F., Purvis, W.O. et al. (2015). *Water, Air, and Soil Pollution* 226. https://doi.org/10.1007/s11270-015-2504-5.
30 Loppi, S. (2019). *Diversity* 11: 36–49.
31 Sujetoviene, G. and Sliumpaite, I. (2013). *Atmospheric Pollution Research* 4: 222–228.
32 Bajpai, R. and Upreti, D.K. (2012). *Ecotoxicology and Environmental Safety* 83: 63–70.
33 Llop, E., Pinho, P., Matos, P. et al. (2012). *Ecological Indicators* 13: 215–221.
34 Popovic, D., Todorovic, D., Frontasyeva, M. et al. (2008). *Environmental Science and Pollution Research* 15: 509–520.
35 Augustine, C. (2012). *Journal of Environmental Science and Water Resources* 1: 46–51.

36 Després, V.R., Alex Huffman, J., Burrows, S.M. et al. (2012). *Tellus, Series B Chemical and Physical Meteorology* 64: 15598.

37 Kularatne, K.I.A. and De Freitas, C.R. (2013). *Environmental and Experimental Botany* 88: 24–32.

38 Conti, M.E. and Cecchetti, G. (2001). *Environmental Pollution* 114: 471–492.

39 Heblich, S., Trew, A., and Yanos, Z. (2021). *Journal of Political Economy* 129: 1508–1552.

40 Ristić, S., Stamenković, S., Stojković Piperac, M. et al. (2020). *Environ. Monit. Assessment* 192: 1–10.

41 Baselga, J., Brasseur, D., Brodniewicz-Proba, T. et al. (2011). *Lancet* 377: 719–720.

42 Pescott, O.L., Simkin, J.M., August, T.A. et al. (2015). *Biological Journal of the Linnean Society* 115: 611–635.

43 Tiwari, G. (2008). *Indian J. Air Pollut. Control* 8: 8–17.

44 Bartholmai, M. and Neumann, P. (2010). *International Conference on Electric Power Systems, High Voltages, Electric Machine, International Conference Remote Sensing – Proceedings* 9 (93): 149–152.

45 Sujetovienė, G. (2015). *Recent Advances in Lichenology: Modern Methods Approaches Biomonitoring Bioprospection* 1: 1–265.

46 Werth, S. and Sork, V.L. (2008). *American Journal of Botany* 95: 568–576.

47 Eversman, S. and Sigal, L. (2009). *Society* 87: 112–118.

48 Hogan, E.J., Minnullina, G., Smith, R.I., and Crittenden, P.D. (2010). *The New Phytologist* 186: 911–925.

49 Dawson, T.L. (2007). *Coloration Technology* 123: 129–142.

50 Weissman, L., Garty, J., and Hochman, A. (2005). *Applied and Environmental Microbiology* 71: 6508–6514.

51 Agrawal, A., Sahu, K.K., and Pandey, B.D. (2004). *Resources, Conservation and Recycling* 42: 99–120.

52 Bačkor, M. and Loppi, S. (2009). *Biologia Plantarum* 53: 214–222.

53 Hossain, M.A., Piyatida, P., da Silva, J.A.T., and Fujita, M. (2012). *Journal of Botany* 2012: 1–37.

54 Häffner, E.V.A., Lomský, B., and Hynek, V. (2001). *Water, Air, and Soil Pollution* 131: 185–201.

55 Çobanoğlu Özyiğitoğlu, G. (2020). *Environ. Concerns Sustainable Development* 61–95.

56 Tuominen, Y. and Jaakkola, T. (1973). *The Lichens*, 185–223. Academic press.

57 Rola, K. (2020). *Journal of Trace Elements in Medicine and Biology* 61: 126512.

58 Sanders, W.B. (2014). *American Journal of Botany* 101: 1836–1848.

59 Sanità Di Toppi, L., Musetti, R., Vattuone, Z. et al. (2005). *Microscopy Research and Technique* 66: 229–238.

60 Karakoti, N., Bajpai, R., Upreti, D.K. et al. (2014). *Environ. Earth Science* 71: 2177–2183.

61 Fahselt, D., Davidson, R.D., and Wu, C.T. (2003). *Archives of Environmental Contamination and Toxicology* 45: 159–167.

62 Bačkor, M. and Fahselt, D. (2008). *Symbiosis* 46: 1–10.

63 Guschina, I.A. and Harwood, J.L. (2006). *Phytochemistry* 67: 1731–1739.

64 Tarhanen, S., Metsärinne, S., Holopainen, T., and Oksanen, J. (1999). *Environmental Pollution* 104: 121–129.

65 Vannini, A., Jamal, M.B., Gramigni, M. et al. (2021). *Biology (Basel)* 10: 1–10.

66 Paul, A. and Hauck, M. (2006). *Flora Morphology, Distribution, Functional Ecology of Plants* 201: 451–460.
67 Nash, T. and Sigal, L. (1979). *American Bryological and Lichenological Society* 82: 280–285.
68 Maslać, A., Maslać, M., and Tkalec, M. (2016). *Acta Botanica Croatica* 75: 186–193.
69 Scheidegger, C. and Schroeter, B. (1995). *Environmental Pollution* 88: 345–354.
70 Ross, L.J. and Nash, T.H. (1983). *Environmental and Experimental Botany* 23: 71–77.
71 Branquinho, C., Catarino, F., Brown, D.H. et al. (1999). *Science of the Total Environment* 232: 67–77.
72 Kiliç, Z., Atakol, O., Aras, S. et al. (2014). *Journal of the Air and Waste Management Association (1995)* 64: 115–123.
73 Albornoz, L., Torres-Benítez, A., Moreno-Palacios, M. et al. (2022). *Metabolites* 12: 1–16.
74 Van Herk, C.M., Mathijssen-Spiekman, E.A.M., and De Zwart, D. (2003). *The Lichenologist* 35: 347–359.
75 Laxton, D.L., Watmough, S.A., Aherne, J., and Straker, J. (2010). *Journal of Limnology* 69: 171–180.
76 Sveshnikova, N., Yuan, T., Warren, J.M., and Piercey-Normore, M.D. (2019). *BMC Research Notes* 12: 1–6.
77 Tyler, G. (1989). *Water, Air, and Soil Pollution* 47: 321–333.
78 Trembley, M.L., Fahselt, D., and Madzia, S. (2009). *American Bryological and Lichenological Society* 100: 368–376.
79 Sehnal, L., Váczi, P., and Barták, M. (2014). *Czech Polar Reports* 4: 47–56.
80 Markert, B., Wunschmann, S., Baltrénaité, E. et al. (2014). *International Journal of Sustainable Water Environmental Systems* 6: 17–26.
81 Williamson, B.J., Purvis, O.W., Mikhailova, I.N. et al. (2008). *Environmental Monitoring and Assessment* 141: 227–236.
82 Bačkor, M., Peksa, O., Škaloud, P., and Bačkorová, M. (2010). *Ecotoxicology and Environmental Safety* 73: 603–612.
83 Rajakaruna, N. and Boyd, R.S. (2013). *Mycoecology on Serpentine Soil* 61: 1106–1116.
84 Sanders, W.B. and Masumoto, H. (2021). *The Lichenologist* 53: 347–393.

19

Lichen Based Nanoparticles

Swati Gajbhiye and Sanjay J. Dhoble

Department of Physics, R.T.M. Nagpur University, Nagpur, India

19.1 Introduction

Nanotechnology has undoubtedly revolutionized science, especially in the medical and pharmaceutical industries. This nanotechnology revolution has resulted in the creation of novel nanoscale products with distinctive physiochemical properties that are constrained in their precursor forms and that demonstrate improved functionality, sensitivity, efficiency, and specificity [1]. Undoubtedly, the development of new drugs and advances in drug design have had a significant impact on humane disease prevention and life quality [2]. The development of eco-friendly nanoparticles (NPs) is of utmost importance to ensure the protection of natural resources and avoid the use of hazardous chemicals during the synthesis processes. Nanoparticles have demonstrated immense potential in various sectors, including environmental applications, medicine, biotechnology, and pharmaceuticals. Therefore, prioritizing environmentally conscious approaches in their production becomes crucial for sustainable and responsible utilization of these tiny marvels. By using green chemistry, hazardous substances are no longer used or produced during chemical reactions. Among the reported methods that can produce nanoparticles with clearly defined sizes and morphologies, the biological method, such as green synthesis, is well known to be nontoxic, easy to use, quick, and reliable [3]. Green technology (GT) is a form of environmental healing that lessens the harm to the environment that human convenience products and technologies cause. According to speculation, GT will increase farm profitability while preventing environmental damage and protecting natural resources. Green technologies are sustainable and leave no trace when used in a variety of processes or applications [4]. Since chemistry, physics, material science, and biology worked together to develop these novel technologies, nanotechnology is not thought of as a single entity in the scientific community but rather as a collection of traditional science [5]. In contrast to conventional physical and chemical techniques, eco-friendly strategies for nanomaterial synthesis present unique benefits, allowing for the production of nanomaterials with precisely defined sizes and morphologies under optimal conditions. In these eco-friendly methods, secondary metabolites sourced from

Chemistry, Biology and Pharmacology of Lichen, First Edition. Edited by Ashoke Kumar Das, Ajay Sharma, Deepika Kathuria, Mohammad Javed Ansari, and Garima Bhardwaj.

plants, bacteria, fungi, and algae play a crucial role by serving as both reducing and capping agents. This unique function facilitates the amalgamation of metal ions, leading to the formation of nanoparticles with desired properties [3]. Despite some obstacles, this technology offers a bright future. Lichen is one such promising source; it produces a special secondary metabolite that aids in the production of nanomaterials. Lichen is now a growing source for green chemistry-based nanoparticle synthesis. The lichen-based nanoparticles have been shown in numerous studies to have potent antimicrobial effectiveness because they are known to be biologically compatible [6]. Lichens are known to be a complex symbiotic relationship between fungi and algae that produces special extracellular secondary metabolites [7]. These remarkable organisms produce a significant quantity of lichen acids, which are referred to as secondary metabolites. These acids are predominantly found in the cortex and give rise to unique formations of crystals covering the fungal hyphae. To date, researchers have identified nearly 1000 secondary metabolites, with the majority being exclusive to lichens. In addition to antibacterial, antiproliferative, antioxidant, antiviral, and anti-inflammatory properties, lichen secondary metabolites also exhibited allelopathic, antiherbivore, and photoprotective properties. Despite numerous studies focusing on the activity of lichen secondary metabolites, there is still a lack of understanding regarding the production of lichens and their ecological role. Lichens play a crucial role as a valuable repository of bioactive compounds, rendering them a significant source of bioactive chemicals. Among these chemicals, notable examples include evernic acid, lecanoric acid, umbilicaris acid, fumarprotocetraric acid, usnic acid, and zeaxanthin. These compounds hold great promise for diverse applications in ecological, biological, and pharmaceutical contexts. Their potential extends to antiproliferative, antiherbivore, antibacterial, antiviral, allelopathic, antioxidant, UV-protective, antipyretic, analgesic, and anti-inflammatory properties. Ongoing research continues to unveil their multifaceted potential [8]. The fabrication of silver nanoparticles demonstrates efficacy against a spectrum of both gram-positive and Gram-negative bacteria, which has been documented in a number of reports using lichens like *Parmotrema pseudotinctorum* and *Ramalinahossei*. Similar results were obtained for nanocomposites made of Fe_3O_4, SiO_2, ZnO, and TiO_2 by *Lecanora muralis* (LM), which were successful in inhibiting both bacterial and fungal pathogens. Apart from this, metals and their oxides such as iron, gold, silver, cadmium, titanium, copper, and zinc-derived nanoparticles can be synthesis from various lichen which include *Parmeliopsis ambigua*, *Punctelia subrudecta*, *Evernia mesomorpha*, *Xanthoparmelia plitti*, *Cetraria islandica*, *Acroscyphus* sp. and *Sticta* sp., *Parmelia perlata*, *Usnea longissimia*, *Parmotrema tinctorum*, *Parmelia sulcata*, *Protoparmeliopsis muralis*, *Pseudevernia furfuracea*, *Ramalina sinensis*, *Cladonia rangiferina*, *Lobaria pulmonaria*, *Usnea antarctica*, *Leptogium puberulum*, and *Xanthoria elegans* [6, 9].

Nevertheless, lichen extract is widely recognized for its exceptional efficacy serving dual roles in nanoparticle synthesis as both a reducing and capping agent. The swift expansion of lichen contributes to their abundance, while their environmental sustainability further enhances their appeal in this context. Lichen extract contains secondary metabolites that prevent nanoparticle aggregation and enhance the synthesis of stable nanoparticles, which gives it great potential as a therapeutic agent that functions as an antimicrobial, antidiabetic, and antioxidant [10]. This chapter will give a thorough introduction to lichen-based nanoparticles, their uses, biocompatibility, biosynthesis, and prospects for the future in light of recent research.

19.2 Lichen Based Nanoparticles and Their Application

In recent times, lichen-based nanoparticles have been attracting increasing attention for their diverse applications in various domains. The forthcoming section will delve into the distinctive properties of these nanoparticles and explore their promising potential across different fields. Comprising a unique amalgamation of fungi and algae, as lichen-based nanoparticles are bound together by a polysaccharide matrix [11]. Possessing remarkable stability and an elevated ratio of surface area to volume, these particles prove to be exceptionally versatile and suitable for a wide array of applications. Additionally, they have a high affinity for water, making them ideal for use in medical and pharmaceutical applications [12]. Furthermore, they are nontoxic and biodegradable, making them safe for use in the environment. Lichen-based nanoparticles have a variety of potential applications in medicine, energy, and environmental protection. In medicine, they can be used as drug delivery systems (DDSs), as they can easily penetrate cell membranes and deliver drugs directly to the target cells. Among the well-explored secondary metabolites of lichens, usnic acid stands out as a prominently researched compound, exhibits various biological characteristics that hold promise for potential medical applications. Several research studies have explored incorporating usnic acid into colloidal carriers for both nano and micro dimensions. Researchers have successfully incorporated usnic acid into liposomes, employing both traditional and surface-functionalized liposome formulations. The primary objective of this integration is to enhance the antibacterial properties of usnic acid (either alone or in conjunction with antibiotics). The latter has also been looked into to boost antioxidant activity. Moreover, the potential Integration of usnic acid-loaded liposomes within thin films based on polymers was explored as a promising approach for developing advanced wound dressings. DDSs based on polymers have been developed to administer usnic acid. The cytotoxicity of usnic acid was assessed through an *in vitro* study using NCI-H292 cells (mucoepidermoid lung cancer cells) with the 3-(4,5-dimethylthiazol-2-yl)-2, 5-diphenyltetrazolium bromide (MTT) assay. The results indicated similar IC_{50} values (10.0 and 11.5 g/mL) for both unconstrained and encapsulated usnic acid, demonstrating that the effects of usnic acid, unconstrained and encapsulated, are dependent on the dosage. However, as compared to the unconstrained medication, encapsulated usnic acid did not alter the morphological properties of NCI-H292 cells in repeated histological examinations, especially when employed at greater concentrations (5 g/mL). The *in vivo* anticancer effectiveness of usnic acid nanocapsules was subsequently assessed using male mice harboring Sarcoma-180 tumor. The mice were administered either usnic acid-loaded nanocapsules or an intraperitoneal administration of a 0.5% usnic acid suspension in Tween 80, this was carried out once daily for seven consecutive days, with a dosage of 15 mg/kg. The therapeutic regimen was initiated 24 hours following tumor inoculation. In contrast, the untreated animals exhibited a consistent escalation in tumor dimensions; those treated with unconstrained and encapsulated usnic acid showed a reduction in tumor mass.

In order to do this, it was demonstrated that encapsulating usnic acid increased its anticancer efficacy by 26.4% when compared to pure unconstrained usnic acid. Additional utilized histological investigation of tumors supported these findings [13]. The study of lichen-based secondary metabolites also encompassed the investigation of usnic acid's potential as an anticancer agent. In this research, usnic acid-loaded nanoparticles with controlled drug release were developed using the nanoprecipitation process, sonication, and the biodegradable polymer poly lactic-co-glycolic acid (PLGA). The study found that both usnic acid and its loaded nanoparticles significantly influenced the survival of mice bearing DAL tumors. The experimental groups received treatments with 5FU, usnic acid, and nanoparticles usnic acid-loaded, respectively, the average weight gain was 1.90 ± 0.09, 2.68 ± 1.32, and 2.16 ± 1.11 gm, compared to 8.16 ± 1.4 gm for tumor-bearing mice. In comparison to unloaded free usnic acid, usnic acid-loaded nanoparticles had greater anticancer efficacy and decreased packed cell volume. This study definitely demonstrated that usnic acid might be introduced into chemotherapy using nanoparticulate systems and that PLGA is a better choice to deliver usnic acid in a controlled manner to the desired spot. As a result, created usnic acid nanoparticles may be used to deliver anticancer medications for chemotherapy [14]. In conclusion, the potential of Usnic acid-loaded nanoparticles derived from lichens finds prominence across various fields, as illustrated in Figure 19.1 below.

In a recent study, an antimicrobial and antifouling study of silver nanoparticles was done, which was derived from lichen-associated bacteria. *Staphylococcus aureus*, *Bacillus subtilis*, and *Escherichia coli* were the three bacteria against which the antibacterial activity of lichen synthesis Ag nanoparticles showed a region of inhibition approximately 5 mm in diameter when investigated. For *E. coli*, the maximum zone of inhibition measured was around 31 mm. Also, with a region of inhibition 26 mm in diameter, the presence of Ag nanoparticles in the antifouling coating effectively inhibited the growth of bacterial colonies. This study showed that inexpensive Ag nanoparticles derived from lichen species could exhibit attributes of being both antibacterial and resistant to fouling [15]. Another research study investigated the production of silver nanoparticles from *P. furfuracea* and *L. pulmonaria* lichen and assessed their antibacterial and antioxidant properties. In this

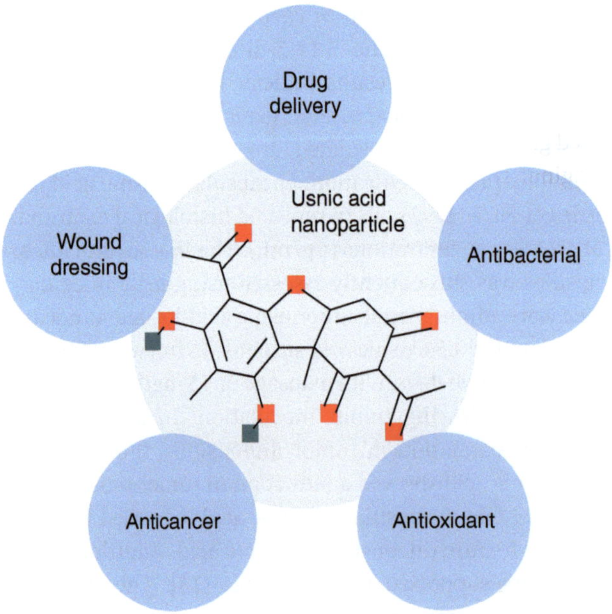

Figure 19.1 Applications of usnic acid loaded nanoparticles derived from lichens.

study, solid-state mechanochemical synthesis was used to create the silver nanoparticles. Both of the lichen species that were chosen for this investigation were successfully used as reducing agents to create AgNPs. Products containing AgNPs were examined for their antibacterial effectiveness against six distinct bacterial strains, and they were all found to be quite effective. Even though AgNPs were presented and were positively linked with the amount of total phenols and flavonoids, lichen extracts still had the lowest antioxidant efficacy. Natural antioxidants, phenols, and flavonoids interact with silver nitrate. Consequently, during the experimentation with lichen extracts and silver nitrate, a noteworthy observation was made regarding the reduction of total phenols and total flavonoid content, leading to a decrease in antioxidant activity. The study further demonstrated that employing phytochemicals for converting silver nitrate to silver nanoparticles resulted in an enhancement of antibacterial activity. However, this process also led to a reduction in antioxidant activity, as nanoparticle synthesis involved the participation of phytochemicals, which functioned as reducing agents. The conclusion was an intriguing one where antibacterial and antioxidant effects required distinct approaches [16].

In yet another study, the water-insoluble lichen metabolites that impede the green synthesis process were overcome using a mechanochemical synthesis technique. In this study, four lichen species (*C. islandica, X. elegans, U. antarctica, and L. puberulum*) and $AgNO_3$ were used as reducing agents and precursors for Ag(0) synthesis. Interestingly, a notable difference was observed between *U. antarctica* and *L. puberulum* compared to the other two lichen species. After six hours of milling, a significant amount of $AgNO_3$ remained unreacted in the former two species, whereas the reaction proceeded rapidly in the latter two species. However, it was observed that Ag(0) formation repeatedly occurred during remilling. A reversal of transformation was observed in products exhibiting a higher concentration of Ag(0), as evidenced by an increase in the content of $AgNO_3$ over time during storage. In every instance, transmission electron microscopy (TEM) investigation revealed a bimodal distribution of nano crystallite sizes. Yet, when silver nanoparticles were made using lichens with a better-reducing ability, the finer fraction was found to be more prevalent (*X. elegans* and *C. islandica*). Excellent antibacterial agents were presented in all products. The re-milled samples exhibited greater effectiveness against *S. aureus* and conversely, the products received in their original state demonstrated heightened efficacy against *E. coli*. This study is the finest illustration of how to choose the appropriate method for lichen-based nanoparticle manufacturing because it can alter the antibacterial properties against particular bacteria [17].

In a different research endeavor, scientists utilized an aqueous lichen extract from *Ramalina dumeticola* to produce silver nanoparticles. During the testing phase, these nanoparticles demonstrated notable antibacterial efficacy against bacteria of both Gram-positive and Gram-negative types. The AgNPs that were synthesized exhibited a cubic structure, featuring an average particle size of 13 nm. Consequently, these findings suggest that the application of these synthesized AgNPs holds significant potential for making important breakthroughs in various fields, such as medical device development, pharmaceuticals, and biomedical products [18]. The impact of both ursolic acid and ursolic acid-nanoparticles (UA-NPs) on melanoma cells (B16F10) was meticulously evaluated through *in vitro* and *in vivo* studies, as part of an investigation into the potential of synthetic ursolic acid-loaded PLGA nanoparticles as a drug delivery system. Ursolic acid-loaded PLGA nanoparticles were synthesized using the emulsion solvent evaporation technique. Subsequently, the zeta potential, polydispersity, and drug release efficiency of these nanoparticles were evaluated. Flow cytometric, confocal microscopy, and MTT assays were performed on B16F10 murine melanoma cell lines. The formulation's increased *in vitro* cellular absorption led to a higher level of cytotoxicity than the free drug. After individual incubation periods of six hours at 37 °C in sterile saline, freshly collected rat serum, and a histidine solution, the formulation underwent successful radiolabeling and demonstrated exceptional stability, with over 90% of the radiolabel remaining intact. According to biodistribution and scintigraphic investigations, the radiolabeled UA-NPs demonstrated delayed blood clearance and notable accumulation in the tumor site. UA-NPs' ability to target tumors has been demonstrated in both *in vitro* and *in vivo* investigations using the B16F10 melanoma cell line [19]. One of the studies of the synthesis of lichenan decorated selenium nanoparticles (L-SeNPs) was created because selenium nanoparticles (SeNPs) have poor water dispersibility, and severely limit the utilization of SeNPs therefore L-SeNPs were created. In this study, Selenium nanoparticles (L-SeNPs) fabricated with *U. longissima* lichenan were utilized. The results unveiled the existence of uniformly spherical, amorphous, orange-red nanoparticles with a zero-valent nature, showcasing an average diameter of 96 nm within the L-SeNPs. Notably, these L-SeNPs exhibited improved heating and storage stability, maintaining their integrity over a period exceeding one month in an aqueous solution at 25 °C. The observed stability was linked to the existence of C—O—Se bonds and hydrogen bonding interactions (O—H—Se) occurring between SeNPs and lichenan. Lichenan, a type of β-glucan featuring (1→3) and (1→4)-glycosidic linkages, was identified as a contributing factor to this phenomenon, had anticancer properties and might increase the levels of cytokines that reduce inflammation. Given its numerous biological functions, lichen might be the appropriate stabilizing agent for encasing SeNPs, increasing their stability and biological functions. The L-SeNPs have been given better antioxidant capabilities by the lichen decorating with SeNPs surface, and their capacity to scavenge free radicals was dose-dependent. Furthermore, the controlled-release capability of L-SeNPs for selenium was exceptional. The release kinetics of selenium from L-SeNPs in simulated intestinal fluids conformed well to the Korsmeyer–Peppas model, indicating a diffusion-controlled process following a Fickian mechanism; they followed the linear superimposition model in simulated gastric liquids. In this study, the lichen-decorated SeNPs (L-SeNPs) were first made via a chemical reduction technique using selenite and ascorbic acid. In an aqueous solution, *U. longissima* displayed a configuration characterized by a random coil shape, when employed for lichen extraction. The formation of C—O—Se bonds or hydrogen bonding interactions between the lichen and SeNPs resulted in the uniform dispersion of L-SeNPs. L-SeNPs have outstanding antioxidant effects, as shown by an *in vitro* antioxidant activity experiment. To investigate the prospective uses of selenium supplements, this study conducted an extensive examination of Se release patterns and mechanisms from L-SeNPs within a simulated gastrointestinal setting was conducted, employing four distinct kinetic models. The research findings underscore the remarkable contribution of lichen in enhancing the properties of nanoparticles [20].

19.3 Biocompatibility of Lichen Based Nanoparticle

The use of lichen-based nanoparticles in biocompatibility studies has been gaining attraction in recent years. Nanoparticles derived from lichens represent a category of nanomaterials suitable for investigating the impact of biocompatibility on living organisms. Lichen-based nanoparticles have emerged as a subject of considerable interest in recent times, primarily due to their promising potential for diverse biomedical applications. The assessment of biocompatibility, which refers to a material's capability to interact with living tissues without inducing adverse effects, assumes a crucial role in assessing the appropriateness of nanoparticles derived from lichens. For biomedical purposes, lichen-based nanoparticles are typically synthesized from lichen extracts or lichen-derived materials, and their biocompatibility has been evaluated in a variety of *in vitro* and *in vivo* studies. The research indicates that nanoparticles derived from lichens demonstrate minimal cytotoxic effects, meaning they do not cause significant harm to cells, and they do not induce significant inflammatory responses or adverse effects in living tissues. In this investigation, the cytotoxicity of free usnic acid and usnic acid encapsulated within microspheres was evaluated using the MTT technique against HEp-2 cells. Furthermore, a tumor-suppression experiment was carried out in mice, focusing on Sarcoma-180 tumors. The results showed a 21% increase in tumor inhibition when usnic acid was encapsulated into PLGA-microspheres compared to the treatment with free usnic acid. This successful encapsulation of usnic acid in microspheres not only improved its antitumor efficacy but also demonstrated excellent biocompatibility with cells [21]. One such study where the cytotoxicity effect of the silver nanoparticles prepared by lichen extract and mushroom extract was studied using A549 (human lung cancer) by MTT assay gave a promising result against cytotoxicity [22]. Mouse fibroblasts L929 cell culture was used for the *in vitro* biocompatibility investigations for the lichen-based study, and multiparametric analysis by flow cytometry was performed. This comprehensive assessment encompassed cell viability, cell cycle progression, apoptosis, and the levels of intracellular reactive oxygen species (ROS). This was the first investigation into the *in vitro* biocompatibility of a lichen microalga called Chlorophyta on mammalian cells. The findings showed that none of the lichen microalgal homogenates cause cytotoxicity or morphological changes. However, at the maximum concentration, they cause an increase in intracellular ROS and can reduce the phase of cell proliferation by preventing cells from entering the S phase, resulting in a decrease in the proportion of cells entering the G2/M phase. It has been discovered that xanthophylls, a type of chlorella carotenoids, prevent cancer cell proliferation. Further studies involving mycobionts without algae have revealed that a significant portion of lichen secondary metabolites is primarily synthesized by the fungal component. However, in certain lichen species, phycobionts originating from distinct green microalgae could potentially contribute to shaping the secondary metabolite composition of the lichen. Furthermore, only in collaboration with the right algal partners do lichenized fungus synthesize physiologically active compounds. These compounds exhibit bioactive qualities, making them potential sources of antibiotics, antioxidants, and anticancer medications. This work demonstrated for the first time that a lichen microalgal species is not poisonous and does not cause acute harm to mammalian cells in culture, despite the fact that at the greatest concentration used, they do cause a mild cytostatic effect and a modest increase in intracellular free radicals [21]. Likewise, a study showcased the biocompatibility of lichen extract within the context of Ag nanoparticle synthesis. Lichen extract (*C. islandica [L.] Ach*) served as both a reducing and stabilizing agent in the efficient and environmentally friendly method employed. The production of Ag nanoparticles for industrial application demonstrated economic feasibility using the statistically optimized medium. Utilizing lichen extract as a reducing agent offers not only environmental advantages but also significant benefits. The research results underscored how reaction time, temperature, and the ratio of $AgNO_3$ to lichen exerted influence on the Ag nanoparticles' synthesis process. These nanoparticles exhibited distinct surface plasmon absorption bands in the range of 405–438 nm. The application of response surface methodology (RSM) emerges as a dependable strategy for probing the manageable biosynthesis of Ag nanoparticles. The outcomes of the study exhibited the creation of Ag nanoparticles, each measuring a few nanometers, across all experimental iterations. Longer reaction times (20–220 minutes) led to an increased generation of Ag nanoparticles, as evidenced by higher absorbance values. Furthermore, a reduction in the $AgNO_3$ to lichen ratio yielded an elevated production of Ag nanoparticles, as evidenced by the augmentation of the surface plasmon resonance (SPR) band. This phenomenon was ascribed to the increased presence of biomolecules with reducing properties, which facilitated the reduction of Ag ions. Due to the absence of biomolecules, larger Ag nanoparticles are produced at higher ratios of $AgNO_3$ to lichen. Fourier Transform Infrared Spectroscopy (FTIR) analysis was performed to detect potential biomolecules accountable for the $Ag+$ ion reduction and the effective stabilization of Ag nanoparticles synthesized using lichen extract. Overall, FTIR results concur with studies in the literature that polysaccharides were in charge of Ag nanoparticle reduction and stabilization. The investigation unveiled the capability to finely regulate the mean size of Ag nanoparticles, spanning diameters from 5 to 29 nm, through manipulation of factors including the $AgNO_3$/lichen extract ratio, reaction duration, and temperature. Because lichen extract is biocompatible, the synthesis of Ag nanoparticles using this method might be advantageous for biological applications [23].

As with any new material, it is crucial to conduct thorough biocompatibility studies in accordance with regulatory guidelines before considering lichen-based nanoparticles for clinical use. This may involve *in vitro* cell culture studies, animal studies, and eventually, clinical trials in humans to assess their safety and efficacy. To guarantee the secure and efficient implementation of lichen-based nanoparticles in biomedical applications, it is crucial to engage with experts in biocompatibility and conduct comprehensive testing and research.

19.4 Biosynthesis of Lichen-Based Nanoparticles

An inventive technique for creating nanoparticles that have the potential to revolutionize the field of nanotechnology is the biosynthesis of lichen-based nanoparticles. Nanoparticles derived from lichen have exhibited potential within DDSs, wound healing, and anticancer therapy in preclinical studies, indicating their potential for biomedical applications. Nonetheless, it is crucial to acknowledge that biocompatibility can be impacted by various elements, including the particular species of lichen employed, the extraction and synthesis methods employed, and the intended application. Continued research and thorough evaluation are imperative to attain an all-encompassing comprehension of the biocompatibility and safety facets inherent in nanoparticles derived from lichen for diverse biomedical applications. Moreover, the growing popularity of these nanoparticles can be attributed to their cost-effectiveness, environmentally friendly nature, and excellent biocompatibility. The green chemistry-produced metal nanomaterials derived from lichens, in particular, have proven to be excellent substitutes for conventional antimicrobial drugs. The information on lichen-based nanomaterials and their synthesis are summarized in the current section [9]. Nanoparticles exhibit unique biological, physical, and chemical attributes attributed to their minute scale and expansive surface area. Secondary metabolites from lichens have the potential to act as capping and reducing agents. This fact fuels mounting fascination regarding the utilization of lichens in the fabrication of different metals (such as gold, silver, magnesium, and copper nanoparticles) along with metal oxide nanoparticles (MONPs) [24]. Most of the lichen is used for the biosynthesis of lichen-based Ag nanoparticles; controllable green synthesis is still a significant obstacle that needs to be overcome. A reliable approach for investigating the manageable biosynthesis of Ag nanoparticles is RSM. The researchers conducted the first-ever investigation on the eco-friendly fabrication of Ag nanoparticles using *C. islandica* extract, employing RSM. Their investigation delved into the influence of temperature, reaction duration, and the ratio of $AgNO_3$ to lichen (mL/mL), all aimed at achieving a manageable and environmentally friendly synthesis of Ag nanoparticles. Through the use of lichen extract (*C. islandica L.*) as both a reducing and stabilizing agent, this procedure demonstrated effective generation of Ag nanoparticles using green chemistry principles. The statistically fine-tuned medium facilitated a cost-effective biosynthesis of Ag nanoparticles, rendering them suitable for potential commercial utilization [23]. Similarly, AgNPs are produced through the biosynthesis process from lichen extracts like *Xanthoria parietina* and *Flavopunctelia faventior*, which have been shown to have antibacterial and cytotoxic properties. As the polarity index of water is approximately 10.2 interestingly, methanol has a polarity index of 5.1 and can dissolve polar compounds. Researchers in this study, extracted lichen using methanol. The interplay between the quantity of lichen extract and $AgNO_3$ in the mixture, the $AgNO_3$ concentration, temperature, and reaction duration collectively influenced the nanoparticle synthesis process. This study highlights that the time required for nanoparticle formation can be impacted by diverse elements like the solvent variety, application method, and environmental conditions. The successful synthesis of AgNPs using lichen extracts was evidenced by a transformation in the reaction mixture's color, shifting from a yellow–green hue to a deep brown shade. This alteration is ascribed to the conversion of silver ions into AgNPs, a mechanism linked to the SPR phenomenon. While numerous hypotheses regarding AgNP formation through green synthesis have been explored in biological systems, the precise mechanism underlying the process through biogenic agents remains elusive. One prevailing hypothesis suggests that intracellular synthesis occurs within the cells, as AgNPs are generated beneath the cellular wall surface and subsequently reduced in the presence of biomolecules or enzymes. However, further research is required to acquire a thorough comprehension of the exact mechanisms involved in the biogenic synthesis of AgNPs. Nicotinamide adenine dinucleotide (NADH) and NADH-dependent protein kinase have been proposed as being necessary for the silver ions [25]. Another study used an aqueous-ethanolic extract of *U. longissima* to make AgNPs and showed great results when the first impurity was gently washed off of the lichen with distilled water. Then, the material was subjected to drying at 600 °C and then finely ground into a powder. To dissolve the solid mass, 10 grams of this dry powder underwent reflux within a blend of ethanol and distilled water in a ratio of 50:50 for three hours. After the solution reached room temperature, it underwent centrifugation at 10,000 rpm. To initiate the transformation of silver ions into AgNPs, 1 mL of a 0.01 M solution of $AgNO_3$ was added to 10 mL of this extract at pH 7. The mixture was shielded from light and stirred vigorously under dark, using a magnetic stirrer under dark for 10–15 minutes. The change in color was seen repeatedly. Although the reaction took 72 hours to complete, it did show a color change. To separate the NPs, the reaction mixture was performed centrifugation. The liquid was poured off, and the clear liquid above a solid residue was spun around a central axis to separate any remaining nanoparticles. The resulting sample obtained using this method remained unchanged for several weeks, even though the yield of the experiment was very low (only 35%). All steps were executed under ambient conditions. The analysis of particle sizes revealed the presence of three different types of silver nanoparticles (AgNPs) exhibiting an average hydrodynamic diameter of 437.1 and 184.5 nm, which were widely distributed. The majority of the NPs were spherical, according to scanning electron microscope (SEM) images. Their size hardly varies at (9.4–11.3 nm). Silver nanoparticles (AgNPs) displayed antibacterial characteristics, and their continuous release could be achieved by employing an appropriate polymer coating. In this research, lichen served the dual role of both a capping and reducing agent for the AgNPs. Due to their stability over several weeks, these AgNPs can be stored for extended periods. However, they cannot be prevented from slowly oxidizing into silver ions [26]. In another independent study, researchers successfully produced silver nanoparticles by blending silver nitrate with an aqueous extract derived from the lichen *Parmotrema praesorediosum*, serving as a stabilizing agent. The silver nanoparticles obtained through this method exhibited an average particle dimension of 19 nm and structure in a cubic form. These nanoparticles displayed significant antibacterial activity, which underwent testing against a spectrum of eight microorganisms via the disk diffusion technique. Moreover, they demonstrated particular potential in combating Gram-negative bacteria [27]. Studies also showed that two lichen species, *P. praesorediosum*, and *R. dumeticola*, can be used in a simple biological process to make silver nanoparticles in a sustainable way. In this particular investigation, thorough characterization of silver nanoparticles was conducted utilizing two sophisticated methods: TEM and UV spectroscopy. The lichens *R. dumeticola* and *P. praesorediosum* played a crucial role as stabilizers in the synthesis process. Absorption peaks of the silver nanoparticles were detected at 407 nm for *R. dumeticola* and 423 nm for *P. praesorediosum*, respectively, within a short reaction time of 72 hours. TEM analysis provided valuable insights into the morphology of the silver nanoparticles. Those obtained from *R. dumeticola* exhibited a 20 nm average size, while those from *P. praesorediosum* with 42 nm average size. This study illuminates the distinctive potential of lichens in enabling the production of silver nanoparticles with varying properties. The findings hold promise for prospective uses, especially within the realm of antibacterial activities and other related areas of interest. Using environmentally friendly and cost-effective green chemistry methods, these two lichen species can produce silver nanoparticles [28]. Studies also showed that the lichen *Roccella phycopsis*, when used as an environmentally friendly method for producing silver nanoparticles (AgNPs), can produce excellent AgNPs. The validity of the study was validated through the observation of a prominent absorbance peak in the UV-visible spectroscopy (UV-Vis) spectrum at 425 nm. This observation provided further evidence and support for the existence of silver nanoparticles within the sample under investigation. The generation of AgNPs was discerned through a distinct alteration in color to brown in the lichen filtrate upon the addition of $AgNO_3$ solution. Spherical morphology was evident in the images captured by the SEM, with particle sizes ranging between 11 and 18 nm. The elemental composition was validated through X-ray Diffraction (XRD) analysis. Furthermore, the identification of the precise chemical compounds accountable for the reduction and stabilization of silver nanoparticles was facilitated through FTIR analysis. This comprehensive approach of characterizing the synthesized AgNPs provided valuable insights into their properties and potential applications. The study findings illustrated that the synthesized *R. phycopsis* silver nanoparticles exhibited stronger antibacterial and greater antioxidant efficacy was observed against Gram-positive bacteria when compared to Gram-negative bacteria. Gram-negative bacteria, overall, displayed elevated resistance was noted against the impact of these silver nanoparticles in both antibacterial and antioxidant assays. Additionally, the *R. phycopsis* AgNPs' had an effective IC_{50} value for inhibiting the acetylcholinesterase enzyme around 1.65 ± 0.07 mg/mL [29].

In the research titled "Lichens (*Parmotrema clavuliferum*) extracts as bio-mediators in silver nanoparticles formation and antibacterial potential," The research highlighted the transformation of a silver nitrate aqueous solution into silver nanoparticles (AgNPs) through the utilization of aqueous extracts from lichens (*P. clavuliferum*). Moreover, an assessment was conducted to ascertain the antibacterial capabilities of both AgNPs and the ethanolic extract of lichens against diverse bacterial strains, including *Pseudomonas aeruginosa* and *B. subtilis*, representing Gram-negative and Gram-positive bacteria, respectively. The well diffusion agar plate technique was employed for the detection and evaluation of the antibacterial efficacy of these substances against the mentioned bacterial strains. The outcomes of this investigation offer significant perspectives regarding the potential utility of lichens as bio-mediators in the creation of silver nanoparticles demonstrating compelling antibacterial attributes. The composite mixture exhibited a notable peak absorption at 440 nm as revealed by the experimental findings. Analysis of TEM images displayed the existence of spherical nanoparticle structures; furthermore, zeta potential measurements disclosed the nanoparticles' negatively charged nature and an average size of 106 nm. Moreover, FTIR was utilized to discern the functional groups within the lichen extract and how they interacted with the AgNPs. Regarding the antibacterial assessment, when various bacteria were exposed to the AgNPs mediated by lichens, a significant inhibitory zone ranging from 7.6 to 11.5 mm in length was observed. These results underscore the potential antimicrobial efficacy of the AgNPs synthesized using lichen extract as a bio-mediator, opening up possibilities for various applications in the field of antibacterial agents. Additionally, it was shown that the inhibitory zone for the ethanolic extract of lichens was between 3.6 and 19.8 mm. The research concluded that lichens (*P. clavuliferum*) serve as highly efficient bio-mediators in the synthesis of AgNPs, demonstrating excellent antibacterial properties; nevertheless, more experimental work is needed to clarify the possible function of bio-extract in the formation of AgNPs [30]. In the investigation of Ag-ZnO NCs, lichen, specifically *P. furfuracea* and *Usnea florida* extracts, played a crucial role in the biological synthesis process for potential application in the study of neurodegenerative diseases. The dynamic light scattering (DLS) measurement revealed a hydrodynamic diameter of 342 nm, while the UV absorbance was observed at 339 nm. Analysis of the SEM images indicated an average size of 35 nm. Subsequently, the Ag-ZnO NCs so obtained were characterized using UV spectroscopy, which demonstrated an absorption peak at 361 nm. Moreover, the hydrodynamic diameter of these NCs was determined to be 495 nm, while the SEM measurements indicated a diameter of 85 nm. Through Fourier Transform Infrared (FT-IR) analysis, this study unveiled the functional entities engaged in the synthesis of Ag-ZnO NCs derived from lichens. The elemental composition of Ag-ZnO NCs derived from both extracts and their crystal plane were identified using FT-IR and energy-dispersive X-ray spectroscopy (EDX) analysis, respectively.

The research demonstrated that Ag-ZnO NCs, at low concentrations, exhibited the potential to enhance neuronal survival. However, this neuroprotective effect was not observed at moderate or high concentrations. Investigating the neuroprotective properties of Ag-ZnO NCs against various neurotoxic agents proved to be valuable. Considering the diverse effects of NCs, it is postulated that their therapeutic or neurodegenerative impact depends on the concentration of exposure. Therefore, further investigation is imperative to gain a comprehensive grasp of the therapeutic capabilities of AgZnO NCs for neurodegenerative diseases as well as their potential neurotoxicity. By conducting more in-depth studies, it may be possible to prevent any neuronal damage that could arise from exposure to these NCs. In conclusion, the successful synthesis of AgZnO NCs using lichen extract demonstrated a therapeutic effect, particularly in promoting neuronal survival at lower concentrations [31]. Similarly, Ag@ZnO NP was also investigated for the antiproliferative effect on the human neuroblastoma SH-SY5Y cell line. Bimetallic nanoparticles possess a distinctive advantage in heightening both antimicrobial and photocatalytic efficacy, attributed to the synergistic interaction among their constituent elements. This unique characteristic sets them apart as promising materials with enhanced functional properties. In this specific study, Ag@ZnO nanoparticles (NPs) were effectively produced utilizing lichen extract (*Usnea sp.*). The NPs were then suspended in distilled water and subsequently diluted to the desired concentration, including 5, 10, 25, 50, 100, 200, and 400 μg/mL. A cell viability study using the MTT assay was employed to evaluate the influence of these different concentrations on cell viability. The findings indicated that in lower concentrations (5, 10, and 25 μg/mL), cell viability increased, indicating a potentially favorable effect on the cells. However, at higher concentrations (100, 200, and 400 μg/mL), cell viability decreased, suggesting a potential adverse effect on cell viability at these concentrations. This unique investigation sheds light on the concentration-dependent behavior of Ag@ZnO nanoparticles were produced through the utilization of lichen extract and provides valuable insights for their potential applications in cell-based studies, demonstrating that Ag@ZnO NP enhances cell proliferation at low doses while having antiproliferative effects on SH-SY5Y cells. The activation of cellular compensating mechanisms may determine the effects of the low dose. As a result, it can be concluded that lichen-based nanoparticles have beneficial effects at low concentrations and that their effects depend on their dosage [32].

Along with silver and zinc oxide nanoparticles, gold nanoparticles (AuNPs) are also studied extensively and are biosynthesized using lichen. In a particular research investigation concerning gold nanoparticles, the production process involved no external reducing or stabilizing agents. Alternatively, the scientists employed desiccated biomass from two unique high-altitude lichen species collected in the alpine region of the eastern Himalaya, situated in Arunachal Pradesh, a state in northeastern India. This innovative approach led to the successful synthesis of gold nanoparticles via a biogenic mechanism. UV-Vis, XRD, FT-IR, and TEM were employed for nanoparticle characterization. The gold nanoparticles as obtained displayed an SPR band at about 535 nm. XRD analysis validated the presence of a face-centered cubic structure within the gold nanomaterials. Notably, the nanoparticles derived from *Sticta sp.* showed a unique characteristic of being exclusively multiply twinned. On the other hand, the gold nanoparticles produced using *Acroscyphus sp.* exhibited a combination of multiply twinned quasi-spherical and prismatic shapes. Remarkably, these gold nanomaterials, when loaded with the biomatrix, demonstrated potent antioxidant activity [33]. In the process of producing gold nanoparticles through the utilization of the lichen *Parmelia sulcata* extract (PSE), bioactive lichen compounds played a crucial role in serving as both agents for the reduction and capping of the particles. The confirmation of face-centered cubic particle formation was established by observing XRD peaks at 111, 200, 220, and 311. TEM further confirmed that the AuNPs exhibited a spherical shape and possessed a distinct elemental composition. The zeta potential was measured to be −18, and the average particle size was approximately 54 nm. Additionally, the study demonstrated that these gold nanoparticles showed remarkable antioxidant properties and the ability to scavenge free radicals. The scavenging potential was evident through the IC_{50} values for 2,2-diphenyl-1-picrylhydrzyl (DPPH), which were found to be 1020 and 815 μg/mL, and for H_2O_2, which were 694 and 510 μg/mL, respectively. These results indicate the promising capacity of the synthesized AuNPs in effectively combating free radicals [34]. Another study in which lichen (*C. rangiferina*) extract was used to create magnesium nanoparticles in a sustainable manner. Soxhlet extraction with a nonpolar solvent was used in this study to extract lichen. Due to the capping of lichen bioactive components, this green synthesis method produced nanoparticles with a size range of about 23 nm that have a number of biomedical applications [35]. Silver, gold, and Ag–Au bimetallic nanoparticles (NPs) were synthesized through an environmentally friendly green chemistry method, utilizing the extract of the lichen *C. islandica (L.) Ach.* During the synthesis process, the Lichen extract acted as both a stabilizing and reducing agent. The resulting monometallic and bimetallic NPs characterization involved techniques such as TEM, UV–Vis, and FTIR spectroscopy. The study successfully demonstrated the synthesis of spherical and the synthesis resulted in stable monometallic and bimetallic nanoparticles characterized by small mean diameters and a precise control over particle size distribution. These NPs showed great potential as catalysts for various reactions due to their unique properties. Particularly, when employed in the reduction of 4-nitrophenols (4-NP) to 4-aminophenol (4-AP) with sodium borohydride ($NaBH_4$), the synthesized nanostructures exhibited excellent catalytic activity. Remarkably, the bimetallic NPs showed even higher catalytic activity compared to the monometallic silver and gold nanostructures. This research represents the initial successful exploration of 4-NP reduction using lichen extract-prepared Ag, Au, and Au–AgNP catalysts, opening up new possibilities for utilizing these environmentally friendly nanoparticles in catalytic applications [36]. One of the studies has explored the cytotoxicity of zinc oxide nanoparticles (ZnO NPs) on human neuroblastoma cells (SHSY-5Y) using an extract from the lichen *Ramalina fraxinea*. The eco-friendly fabrication of ZnO NPs were assessed for their impact on neurotoxicity and neuroprotection against hydrogen peroxide-induced cell damage in the SH-SY5Y cell line. The confirmation of ZnO nanoparticles was achieved using UV-Vis at 330 nm. SEM and field emission scanning electron microscopy (FE-SEM) images revealed an average ZnO NP diameter of around 21 nm. The hexagonal structure of the ZnO nanoparticles was revealed by XRD analysis. Moreover, the nanoparticles' structure was characterized by FT-IR analysis, which identified the presence of functional groups. The findings indicated that ZnO NPs had potential neurotoxic effects at higher concentrations, but they showed some benefits at lower doses. However, the ZnO NPs could not reverse cell death induced by hydrogen peroxide, and at higher doses, they even exacerbated the neurodegenerative effects of hydrogen peroxide. The study highlighted that ZnO nanoparticles, commonly utilized in a range of different products, the impact can vary, with potential benefits and drawbacks contingent on the dosage, potentially contributing to neuropsychiatric diseases, particularly neurodegenerative disorders. Importantly, this investigation marked the first exploration of the neurotoxicity of ZnO NPs derived from *R. fraxinea* extract on the SHSY-5Y cell line [37]. Previously, lichen organisms were not widely regarded as a secure origin of metal nanoparticles (MNPs) and MONPs. However, a recent study brought to light the presence of secondary metabolites in lichens, which were found to be responsible for the formation of MNPs and MONPs in various sizes and shapes, leading to this exciting discovery. In a similar vein, *P. muralis* lichen aqueous extract was utilized to environmentally friendly synthesize Ag and Cu MNPs, as well as TiO_2, ZnO, and Fe_3O_4 MONPs. These nanoparticles exhibited diverse shapes, including spherical, triangular, polyhedral, and cubic structures. The significant discovery was that these MNPs/MONPs possess exceptional antibacterial, antibiofilm, anti-quorum sensing, and antioxidant properties, making them effective combatants against multi-drug resistant (MDR) bacteria. The findings from agar disc diffusion plate technique and minimum inhibitory concentration (MIC)/minimum bactericidal concentration (MBC) assays demonstrated that the pathogen *P. aeruginosa* exhibited higher sensitivity to the synthesized nanoparticles compared to *E. coli* and *S. aureus*. This highlighted the strong antibacterial effectiveness of the nanoparticles. Importantly, the study introduced five distinct MNPs/MONPs, featuring innovative clean production techniques that prove to be efficient against pathogens, cost-effective, affordable, and environmentally friendly [38]. A separate investigation in which Silver nanoparticles (AgNPs) were produced utilizing the extract of *Heteroderimia leucomela* lichen, and the resulting nanoparticles were characterized through various methods. The lichen extract's phytocompounds acted as both capping and reducing agents during the synthesis process, which was confirmed through GC-MS analysis. *in vitro* tests were performed to assess the antioxidant and antibacterial characteristics of the produced AgNPs. The results from the synthesized AgNPs demonstrated that they were primarily in a spherical shape, and their average particle size of 20 ± 0.2 nm. The Ag nanoparticles showed remarkable capacity for scavenging free radicals, as evidenced by an IC_{50} value of 7.5 μg/mL against DPPH radicals. Additionally, the synthesized AgNPs demonstrated antibacterial activity against various test organisms. In terms of their antibacterial efficacy, the AgNPs exhibited vulnerability in the following order for candida strains: *Candida tropicalis, Candida albicans, Candida glabrata, Candida parapsilosis*, and *Candida krusei*. The AgNPs also exhibited antibacterial activity against *Escherichia coli, Bacillus subtilis, Vibrio cholerae, Streptococcus pneumonia*, and *Pseudomonas aeruginosa*. This study concluded that AgNPs synthesized from lichen extract obtained from high altitudes possessed potent antibacterial and antioxidant properties, making them highly relevant for applications in the nanomedicine industry. Moreover, the lichen extract proved to be an effective and green resource for producing stable, cost-efficient, and environmentally beneficial nanoparticles [39]. Lichens are not only utilized for synthesizing metal or metal oxide nanoparticles but also for producing nanocomposites. Lichens are characterized by the presence of crystals on fungal hyphae surfaces, alongside a diverse array of secondary bioactive metabolites contained within their thalli. These metabolites contribute to a range of pharmaceutical properties, such as antimicrobial, anti-inflammatory, antiviral, antioxidant, antigenotoxic, analgesic, anticancer, and antipyretic activities. Investigating the capacity of lichen extract as an antioxidant source to facilitate the biosynthesis of bioactive nanostructures holds significant importance. In this study, the researchers investigated the capability of LM lichen aqueous extract to synthesize Fe_3O_4/SiO_2 and $ZnO/TiO_2/SiO_2$ nanocomposites using a simple, cost-effective, environmentally friendly, and rapid approach. The lichen extract was analyzed through the utilization of the GC-Mass technique unveiled the existence of valuable bioactive phytochemicals. Moreover, the research on the antifungal and antibacterial properties of the green-synthesized nanostructures exhibited potent antibacterial effects [40]. Further, lichens possess the remarkable ability to produce diverse secondary metabolites, including polyphenols, which hold immense potential for nanoparticle biosynthesis. The process of lichen-based nanoparticle synthesis typically involves two essential steps. In the initial stage, lichen extracts are meticulously prepared, and the resulting extract is concentrated to serve as a rich source of polyphenols for nanoparticle synthesis. Subsequently, in the second step, the concentrated lichen extract effectively serves a dual role, acting as both a reducing agent and a capping agent in the process of nanoparticle synthesis. A precursor material is combined with the lichen extract and subjected to controlled conditions. The lichen extract's polyphenols serve as potent reducing agents, facilitating the conversion of precursors into nanoparticles. Additionally, they act as capping agents, ensuring the stability of the nanoparticles and preventing their agglomeration. The resultant nanoparticles derived from lichen exhibit a broad spectrum of potential uses across diverse domains, including catalytic applications, biomedicine, and environmental remediation. In biomedicine, these nanoparticles show promise in DDSs and targeted therapies, benefiting from their biocompatibility and controlled-release capabilities. Moreover, lichen-based nanoparticles exhibit remarkable catalytic activity, making them valuable candidates for diverse catalysis applications. Additionally, their efficient adsorption of pollutants indicates their potential for environmental remediation and cleanup efforts. One of the most significant advantages of lichen-based nanoparticles is their eco-friendliness. Unlike conventional chemical-based methods of nanoparticle synthesis, lichen-based approaches rely on natural compounds, reducing the dependence on toxic chemicals. Embracing lichen-based nanoparticles can, therefore, offer a more sustainable and environmentally friendly alternative to traditional methods. Overall, the abundance of secondary metabolites in lichens, particularly polyphenols, makes them nature's treasure troves for nanoparticle biosynthesis. The two-step process of lichen-based nanoparticle synthesis harnesses these bioactive compounds effectively, resulting in nanoparticles with diverse applications in biomedicine, catalysis, and environmental remediation. By embracing lichen-based nanoparticle synthesis, we can advance nanotechnology while prioritizing environmental sustainability. Table 19.1 illustrates the strains of lichen employed by the researchers for synthesizing a range of nanoparticles.

Table 19.1 Synthesis of diverse nanoparticles using lichen extracts from different strains.

Serial number	Lichen strains	Type of nanoparticle synthesis	Size of nanoparticles (nm)
1	*Cetraria islandica*	Silver, Ag–Au bimetallic nanoparticle	5–29
2	*Xanthoria parietina*	Silver nanoparticle	45
3	*Flavopunctelia flaventior*	Silver nanoparticle	69
4	*Parmotrema perlatum*	Silver nanoparticle	–
5	*Parmotrema praesorediosum*	Silver nanoparticle	19–42
6	*Cladonia rangiferina*	Silver and magnesium nanoparticle	23
7	*Usnea longissima*	Silver nanoparticle	9.4–11.3
8	*Ramalina fraxinea*	Silver, ZnO nanoparticle	20–21
9	*Roccella phycopsis*	Silver nanoparticle	11–18
10	*Parmotrema clavuliferum*	Silver nanoparticle	106
11	*Pseudevernia furfuracea*	Ag–ZnO nanocomposite	35
12	*Usnea florida*	Ag–ZnO nanocomposite	85
13	*Sticta*	Gold nanoparticle	15.75
14	*Acroscyphus*	Gold nanoparticle	11.36
15	*Parmelia sulcata*	Gold nanoparticle	54
16	*Protoparmeliopsis muralis / Lecanora muralis*	Ag, Cu metal nanoparticle/TiO_2, ZnO, Fe_3O_4 metal oxide nanoparticle/ Fe_3O_4/SiO_2 and $ZnO/TiO_2/SiO_2$ nanocomposites	55–90 50–85
17	*Heteroderimia leucomela*	Silver nanoparticle	20

19.5 Future Prospects and Conclusions

The future potential of lichen-based nanoparticles in pharmaceuticals, biomedicine, and biotechnology is undoubtedly promising, presenting a wide array of exciting applications in these fields. In the field of pharmaceuticals, lichen-based nanoparticles have emerged as a groundbreaking approach for targeted drug delivery, revolutionizing therapeutic efficacy while mitigating unwanted side effects. By encapsulating drugs and facilitating controlled release, these nanoparticles enable precise delivery to specific tissues or cells, maximizing the drug's impact and reducing systemic exposure. Their biocompatibility allows for various administration routes, including oral, topical, and intravenous, expanding their possibilities in drug delivery. Continued research may lead to innovative DDSs utilizing lichen-based nanoparticles, revolutionizing treatment options for various diseases. In biomedicine, lichen-based nanoparticles have demonstrated immense potential in wound healing, tissue engineering, and diagnostics. Their unique characteristics, such as small size, large surface area, and biocompatibility, make them ideal candidates for wound dressings and scaffolds for tissue regeneration. Moreover, the ability to synthesize nanoparticles with tailored properties opens up opportunities for customized biomedical applications. Lichen-based nanoparticles can encapsulate enzymes or bioactive molecules, safeguarding them from degradation and potentially enhancing the efficiency of biocatalysts. Nevertheless, extensive research is essential to fully comprehend the possibilities of lichen-based nanoparticles in pharmaceuticals, biomedicine, biotechnology, and cosmetics. Addressing challenges like optimizing biosynthesis methods, standardizing nanoparticle characterization, and evaluating long-term safety and environmental impact is crucial. Regulatory considerations and commercialization aspects must also be carefully examined for practical implementation.

In conclusion, lichen-based nanoparticles hold immense promise in pharmaceuticals, biomedicine, and biotechnology. The unique attributes of nanoparticles derived from lichen make them exceptionally attractive for a diverse array of applications. Continuing research and progress in this domain are anticipated to drive innovation across industries, offering new and environmentally friendly solutions.

References

1 Navya, P.N. and Daima, H.K. (2016). Rational engineering of physicochemical properties of nanomaterials for biomedical applications with nanotoxicological perspectives. *Nano Convergence* 3: 1.

2 Kumar Dash, D., Kant Panik, R., Kumar Sahu, A., and Tripathi, V. (2020). Role of nanobiotechnology in drug discovery, development and molecular diagnostic. In: *Applications of Nanobiotechnology* (ed. M. Stoytcheva and R. Zlatev). IntechOpen. https://doi.org/10.5772/intechopen.92796.

3 Balčiūnaitienė, A., Štreimikytė, P., Puzerytė, V. et al. (2022). Antimicrobial activities against opportunistic pathogenic bacteria using green synthesized silver nanoparticles in plant and lichen enzyme-assisted extracts. *Plants* 11 (14): 1833.

4 Aithal, S. and Aithal, P.S. (2021). Green and eco-friendly nanotechnology – concepts and industrial prospects. *International Journal of Management, Technology and Social Sciences (IJMTS)* 6 (1): 2581–6012. https://doi.org/10.5281/zenodo.4496913.

5 Saini, R., Saini, S., and Sharma, S. (2010). Nanotechnology: the future medicine. *Journal of Cutaneous and Aesthetic Surgery* 3 (1): 32–33.

6 Ahsan, T., He, L., Miao, Y. et al. (2021). Lichen-based nano-particles, an emerging antibacterial approach. *Journal of Materials Science and Chemical Engineering* 09 (8): 10–20.

7 Ranković, B. and Kosanić, M. (2019). Chapter 1: lichens as a potential source of bioactive secondary metabolites. In: *Lichen Secondary Metabolites Bioactive Properties and Pharmaceutical Potential*, 2e (ed. B. Ranković), 1–30. Springer.

8 Goga, M., Elečko, J., Marcinčinová, M. et al. (2018). Lichen metabolites: an overview of some secondary metabolites and their biological potential. In: *Co-Evolution of Secondary Metabolites* (ed. J.-M. Mérillon and K.G. Ramawat), 1–36. Cham: Springer.

9 Rattan, R., Shukla, S., Sharma, B., and Bhat, M. (2021). A mini-review on lichen-based nanoparticles and their applications as antimicrobial agents. *Frontiers in Microbiology* 12: 633090.

10 Hamida, R.S., Ali, M.A., Abdelmeguid, N.E. et al. (2021). Lichens—a potential source for nanoparticles fabrication: a review on nanoparticles biosynthesis and their prospective applications. *Journal of Fungi* 7: 291.

11 Spribille, T., Tagirdzhanova, G., Goyette, S. et al. (2020). 3D biofilms: in search of the polysaccharides holding together lichen symbioses. *FEMS Microbiology Letters* 367 (5): fnaa023.

12 Ureña-Vacas, I., González-Burgos, E., De Vita, S. et al. (2022). Phytochemical characterization and pharmacological properties of lichen extracts from cetrarioid clade by multivariate analysis and molecular docking. *Evidence-based Complementary and Alternative Medicine* 2022: 5218248.

13 Zugic, A., Tadic, V., and Savic, S. (2020). Nano- and microcarriers as drug delivery systems for usnic acid: review of literature. *Pharmaceutics* 12 (2): 156.

14 Sundara Pandian, M., Chitra Karthikeyini, S., and Nagarajan, M. (2017). Fabrication, characterization and pharmacological activity of usnic acid loaded nanoparticles. *International Journal of Pharmaceutical Sciences and Research* 8 (11): 4758–4766.

15 Subbaiyan, R., Ganesan, A., and Ramasubramanian, B. (2022). Self-potent anti-microbial and anti-fouling action of silver nanoparticles derived from lichen-associated bacteria. *Applied Nanoscience* 12 (8): 2397–2408.

16 Goga, M., Baláž, M., Daneu, N. et al. (2021). Biological activity of selected lichens and lichen-based Ag nanoparticles prepared by a green solid-state mechanochemical approach. *Materials Science and Engineering C* 119: 111640.

17 Baláž, M., Goga, M., Hegedüs, M. et al. (2020). Biomechanochemical solid-state synthesis of silver nanoparticles with antibacterial activity using lichens. *ACS Sustainable Chemistry and Engineering* 8 (37): 13945–13955.

18 Din, L.B., Mie, R., Samsudin, M.W. et al. (2015). Biomimetic synthesis of silver nanoparticles using the lichen *Ramalina dumeticola* and the antibacterial activity (Sintesis Biomimetik Nanozarah Perak Menggunakan Liken *Ramalina dumeticola* dan Aktiviti Antibakteria). *Malaysian Journal of Analytical Sciences* 19: 369–376.

19 Baishya, R., Nayak, D.K., Kumar, D. et al. (2016). Ursolic acid loaded PLGA nanoparticles: *in vitro* and *in vivo* evaluation to explore tumor targeting ability on B16F10 melanoma cell lines. *Pharmaceutical Research* 33 (11): 2691–2703.

20 Yang, Z., Hu, Y., Yue, P. et al. (2023). Structure, stability, antioxidant activity, and controlled-release of selenium nanoparticles decorated with lichenan from *Usnea longissima*. *Carbohydrate Polymers* 299: 120219.

21 Ribeiro-Costa, R.M., Alves, A.J., Santos, N.P. et al. (2004). *In vitro* and *in vivo* properties of usnic acid encapsulated into PLGA-microspheres. *Journal of Microencapsulation* 21 (4): 371–384.

22 Paul, S., Singh, A.R., and Sasikumar, C.S. (2015). Green synthesis of bio-silver nanoparticles by Parmelia perlata, Ganoderma lucidum and Phellinus igniarius & their fields of application. *Indian Journal of Research in Pharmacy and Biotechnology* 3 (2): 100.

23 Yildiz, N., Ateş, Ç., Yilmaz, M. et al. (2014). Investigation of lichen based green synthesis of silver nanoparticles with response surface methodology. *Green Processing and Synthesis* 3 (4): 259–270.

24 Zdenka, B., Matej, B., Aneta, S., and Michal, G. (2023). Synthesis of metal nanoparticles using lichens and their biological applications. In: *Nanobiotechnology for Plant Protection,*

Fungal Cell Factories for Sustainable Nanomaterials Productions and Agricultural Applications (ed. K.A. Abd-Elsalam), 163–203. Elsevier. https://doi.org/10.1016/B978-0-323-99922-9.00017-9.

25 Alqahtani, M.A., Al Othman, M.R., and Mohammed, A.E. (2020). Bio fabrication of silver nanoparticles with antibacterial and cytotoxic abilities using lichens. *Scientific Reports* 10 (1): 16781.

26 Siddiqi, K.S., Rashid, M., Rahman, A. et al. (2018). Biogenic fabrication and characterization of silver nanoparticles using aqueous-ethanolic extract of lichen (*Usnea longissima*) and their antimicrobial activity. *Biomaterials Research* 22 (1): 23.

27 Mie, R., Samsudin, M.W., Din, L.B. et al. (2013). Synthesis of silver nanoparticles with antibacterial activity using the lichen *Parmotrema praesorediosum*. *International Journal of Nanomedicine* 9 (1): 121–127.

28 Mie, R., Samsudin, M.W., Din, L.B., and Ahmad, A. (2012). Green synthesis of silver nanoparticles using two lichens species: *Parmotrema praesorediosum* and *Ramalina dumeticola*. *Applied Mechanics and Materials* 229–231: 256–259.

29 Ben Salah, M., Aouadhi, C., and Khadhri, A. (2021). Green *Roccella phycopsis* Ach. mediated silver nanoparticles: synthesis, characterization, phenolic content, antioxidant, antibacterial and anti-acetylcholinesterase capacities. *Bioprocess and Biosystems Engineering* 44 (11): 2257–2268.

30 Alqahtani, M.A.M., Mohammed, A.E., Daoud, S.I. et al. (2017). Lichens (*Parmotrema clavuliferum*) extracts: bio-mediator in silver nanoparticles formation and antibacterial potential. *Journal of Bionanoscience* 11 (5): 410–415.

31 Koca, F.D., Halici, M.G., Işik, Y., and Ünal, G. (2022). Green synthesis of Ag-ZnO nanocomposites by using *Usnea florida* and *Pseudevernia furfuracea* lichen extracts and evaluation of their neurotoxic effects. *Inorganic and Nano-Metal Chemistry* 1–8. https://doi.org/10.1080/24701556.2022.2078351.

32 Isik, Y., Unal, G., Koca, F. D. et al. (2020). Anticancer activity of lichen extract (*Usnea sp.*) based synthesized Ag@ZnO bimetallic nanocomposite. *The 3rd International Conference on Natural Products for Cancer Prevention and Therapy*, Kayseri, Turkey (18–20 December 2019), 51. Basel, Switzerland: MDPI.

33 Debnath, R., Purkayastha, D.D., Hazra, S. et al. (2016). Biogenic synthesis of antioxidant, shape selective gold nanomaterials mediated by high altitude lichens. *Materials Letters* 169: 58–61.

34 Gandhi, A.D., Murugan, K., Umamahesh, K. et al. (2019). Lichen *Parmelia sulcata* mediated synthesis of gold nanoparticles: an eco-friendly tool against *Anopheles stephensi* and *Aedes aegypti*. *Environmental Science and Pollution Research* 26 (23): 23886–23898.

35 Devasena, T., Ashok, V., Dey, N. et al. (2014). Phytosynthesis of magnesium nanoparticles using lichens [Internet]. *World Journal of Pharmaceutical Research* 3 (3): 4625–4632. https://www.wjpr.net/

36 Çlplak, Z., Gökalp, C., Getiren, B. et al. (2018). Catalytic performance of Ag, Au and Ag-Au nanoparticles synthesized by lichen extract. *Green Processing and Synthesis* 7 (5): 433–440.

37 Koca, F.D., Ünal, G., and Halici, M.G. (2019). Lichen based synthesis of zinc oxide nanoparticles and evaluation of its neurotoxic effects on human neuroblastoma cells. *Journal of Nano Research* 59: 15–24.

38 Alavi, M., Karimi, N., and Valadbeigi, T. (2019). Antibacterial, antibiofilm, antiquorum sensing, antimotility, and antioxidant activities of green fabricated Ag, Cu, TiO_2, ZnO, and Fe_3O_4 NPs via *Protoparmeliopsis muralis* lichen aqueous extract against multi-drug-resistant bacteria. *ACS Biomaterials Science & Engineering* 5 (9): 4228–4243.

39 Uma Maheshwari Nallal, V., Nisha Devi, K., and Razia, M. (2022). Biogenic fabrication and characterization of silver nanoparticles using high altitude lichen *Heteroderimia leucomela* extract and its potential applications. *Materials Today Proceedings* 50: 365–370.

40 Abdullah, S.M., Kolo, K., and Sajadi, S.M. (2020). Greener pathway toward the synthesis of lichen-based $ZnO@TiO_2@SiO_2$ and $Fe_3O_4@SiO_2$ nanocomposites and investigation of their biological activities. *Food Science & Nutrition* 8 (8): 4044–4054.

Index

Note: *Italic* refers to the keyword in figure; **Bold** refers to the keyword in table; *Passim* refers to that keyword is coming under of times in that particular chapter

Chemistry, Biology and Pharmacology of Lichen, First Edition. Edited by Ashoke Kumar Das, Ajay Sharma, Deepika Kathuria, Mohammad Javed Ansari, and Garima Bhardwaj.